Advances in Environmental Microbiology

Volume 7

Series Editor
Christon J. Hurst
Cincinnati, Ohio
USA

and

Universidad del Valle
Santiago de Cali, Valle
Colombia

Periodic spring located adjacent to the Miller-Leuser Log House in Anderson Township, Hamilton County, Ohio, United States of America. Photo by Christon J. Hurst, used with permission

More information about this series at http://www.springer.com/series/11961

Christon J. Hurst
Editor

The Structure and Function of Aquatic Microbial Communities

Springer

Editor
Christon J. Hurst
Cincinnati, Ohio, USA

Universidad del Valle
Santiago de Cali, Valle, Colombia

ISSN 2366-3324 ISSN 2366-3332 (electronic)
Advances in Environmental Microbiology
ISBN 978-3-030-16773-8 ISBN 978-3-030-16775-2 (eBook)
https://doi.org/10.1007/978-3-030-16775-2

Cover illustration: "Volvox reimagined" is a montage image created in 2015 and is being used with permission of the artist, Christon J. Hurst. Those images incorporated into this montage were: center image "Aspergillus flavus" (author: Hossein Mirhendi, image supplied by and used with author's permission); clockwise from upper right "Micrasterias" (author: Rogelio Moreno Gill, image supplied by and used with author's permission), "BrownGiantKelp3600ppx" (author: FASTILY, Creative Commons Attribution Share Alike 3.0 Unported license), "Stumpfungus" (author: Ecornerdropshop, public domain image), "Letharia vulpina JHollinger crop" (author: Jason Hollinger, Creative Commons Attribution-Share Alike 3.0 Unported license), "Phytoplankton—the foundation of the oceanic food chain", (author: NOAA MESA Project, public domain image), "Spirogyra cell" (author: Jasper Nance, Creative Commons Attribution-Share Alike 3.0 Unported license), "Morelasci" (author: Peter G. Werner, Creative Commons Attribution-Share Alike 3.0 Unported license).

This Springer imprint is published by the registered company Springer Nature Switzerland AG
The registered company address is: Gewerbestrasse 11, 6330 Cham, Switzerland

Can there be a combination of place and time which leaves a perfect memory, a place to which you do not return because you know that it will have changed? I believe that to be true. During the months of July, August and September of 1998 I was a Fulbright Senior Scholar in Cali, Colombia, staying in what was then named the Hotel Pacifico Royal. Because I was to be living there for three months, the hotel gave me a free upgrade to a two room suite complete with a small dining table. The university was in fiscal crisis, and so I agreed ahead of time to use my salary from the Fulbright Commission to pay for my own lodging and meals. The financial arrangement meant that my meals had to be inexpensive. My evening meal usually was only what I could heat in my rooms' microwave oven, and generally all that I could afford were the corn meal cakes called arepas plus some marmalade spread on top of them. Therefore, I initially had little need for a dining table in my hotel room.

The hotel employees were very kind to me and quickly became my friends. Breakfast fortunately was provided by the hotel and cooked by the breakfast buffet chef, Rocío. She smilingly reminded me that she could prepare whatever kind of omelet I might want. But, Rocío knew that for breakfast I only ate croissants with of course marmalade, and that I drank hot chocolate. On days when I seemed to arrive late for breakfast, Rocío would have saved for me a plateful of fresh croissants on a shelf under the buffet table and hot chocolate would be awaiting me. Once each two weeks I washed my laundry in the hotel room bathtub and then hung up the wet laundry all around my room for drying. The chambermaid, Blanca, told me that the hotel offered laundry service, but I said that I had no money to afford it. Blanca eventually told me that the hotel would dry all of my clothes for free! That was very kind, but I felt too guilty to ask the assistance. Two weeks later, when my suite was again filled with wet laundry, Blanca once more offered to have the clothes dried for me. I thought about it, and asked if they could dry just my jeans because those took three days to air dry in the summer humidity. Following that, each morning after I had done my laundry the wet jeans would disappear only to magically reappear a few hours later dried and carefully folded. The hotel men's soccer team adopted me as their unofficial mascot and made sure that I got to each of their practice sessions and games. I eventually learned that one time their team even had hired the car that transported me to the game and back. I remember José the doorman, who always was kind and never would accept a tip from me because I was his friend. Plus, of course, there was the concierge María

Fernanda whose kindness made everyone around her feel touched by her presence.

I eventually did find a good usage for that table. Each workday I walked through a shopping mall during trips between my hotel and the university. After a couple of weeks, I noticed that a stationery store in the shopping mall sold wet clay. My hobbies included pottery and ceramic sculpture, and with that I found a solution for occupying my spare time at the hotel. I covered the top of my dining table with plastic, paper, and on top of that I placed some heavy canvas purchased from a fabric store in the shopping mall. The clay contained sticks and pebbles that needed to be removed before the clay could be worked, and with a few simple tools plus a ceramic ashtray used as a rolling pin I made presents for the hotel staff. Almost all of those things that I made at the dining table can be seen in this picture.

Sculpture made in Cali Rm 902 Hotel Pacifico Royal 1998

I had no means of firing those objects, and so they were given away air dried. When my birthday came around, María Fernanda quietly collected enough money as donations from the hotel staff to buy a glass of wine for me from the hotel's bar. At her request, the hotel kitchen volunteered a slice a cake. I was out to a museum on that day, but everyone who could wait did stay until I returned to the hotel. Together, my friends at the hotel then presented to me the cake and wine, and they sang Happy Birthday to me. I again spent some time at the hotel in 2000, and although staying there was similar and very nice, something of what had seemed magical during that summer in 1998 was a bit changed. I have not returned to that hotel because, although I would imagine it currently is a very nice place, I want my memories to remain as perfect as it all seemed in 1998. To whom among those people would I dedicate my work on this book? Most certainly it should be María Fernanda, for that kind specialness in her soul which I am sure leaves everyone believing the world is a better place.

María Fernanda Gutiérrez Herrera in Cali 2000

Series Preface

The light of natural philosophy illuminates many subject areas including an understanding that microorganisms represent the foundation stone of our biosphere by having been the origin of life on Earth. Microbes therefore comprise the basis of our biological legacy. Comprehending the role of microbes in this world which together all species must share, studying not only the survival of microorganisms but as well their involvement in environmental processes, and defining their role in the ecology of other species, does represent for many of us the Mount Everest of science. Research in this area of biology dates to the original discovery of microorganisms by Antonie van Leeuwenhoek, when in 1675 and 1676 he used a microscope of his own creation to view what he termed "animalcula," or the "little animals" which lived and replicated in environmental samples of rainwater, well water, seawater, and water from snow melt. van Leeuwenhoek maintained those environmental samples in his house and observed that the types and relative concentrations of organisms present in his samples changed and fluctuated with respect to time. During the intervening centuries we have expanded our collective knowledge of these subjects which we now term to be environmental microbiology, but easily still recognize that many of the individual topics we have come to better understand and characterize initially were described by van Leeuwenhoek. van Leeuwenhoek was a draper by profession and fortunately for us his academic interests as a hobbyist went far beyond his professional challenges.

It is the goal of this series to present a broadly encompassing perspective regarding the principles of environmental microbiology and general microbial ecology. I am not sure whether Antonie van Leeuwenhoek could have foreseen where his discoveries have led, to the diversity of environmental microbiology subjects that we now study and the wealth of knowledge that we have accumulated. However, just as I always have enjoyed reading his account of environmental microorganisms, I feel that he would enjoy our efforts through this series to summarize what we have learned. I wonder, too, what the microbiologists of still future centuries would think of our efforts in comparison with those now unimaginable discoveries which they will have achieved. While we study the many wonders of microbiology, we also further our recognition that the microbes are our biological critics, and in the end they undoubtedly will have the final word regarding life on this planet.

Christon J. Hurst in Heidelberg

Indebted with gratitude, I wish to thank the numerous scientists whose collaborative efforts will be creating this series and those giants in microbiology upon whose shoulders we have stood, for we could not accomplish this goal without the advantage that those giants have afforded us. The confidence and very positive encouragement of the editorial staff at Springer DE has been appreciated tremendously and it is through their help that my colleagues and I are able to present this book series to you, our audience.

Cincinnati, OH Christon J. Hurst

Volume Preface

Microbial communities exist in all potential aquatic habitats including surface water, aquifers, and discarded materials on which precipitation collects. Some members of those microbial communities will naturally be floating on the surface, many others will live suspended at particular depths in the water column, a broad range of microorganisms make their existence in sediments, and still there are other communities that attach to solid matrices and contribute to the formation of biofilms.

The authors of this book describe how aquatic microbial communities are structured in ways that optimize the community's functioning and the authors further explain that community structuration often includes functional stratification. Structuration can be visibly obvious in biofilms including the presence of layers that typically extend from a community's surface of attachment outward into the aqueous surroundings. Vertical stratification often occurs within photic zones, and the photosynthetic pigments produced by microbes in photic biofilms may be evident as color banding, with the banding representing a community structured on the basis of optimal wavelength usage. In areas of low disturbance by metazoans, that vertical banding of photic communities can include seasonal laminations. The authors also explain that aquatic utilization of organic detritus as an energy source often begins with fungal colonization which starts a food chain. Further processing of sedimented organic material occurs by microbial communities that may be stratified on the basis of reduction potential, and their activity can include methanogenesis.

Preparing microbiologically safe drinking water requires eliminating hazardous components of the aquatic microbial community and often the processed water then needs to be safely distributed. *Vibrio cholerae* represents one example of pathogenic aquatic microorganisms for which consideration must be given whenever microbial ecology favors presence of that species in water that would be used for drinking. This book explains the techniques used for accomplishing microbial removal from water and also addresses the fact that drinking water distribution relies upon understanding and controlling aquatic environments which are designed to be isolated from their surroundings. The isolation is intended both to contain the product water and to reduce contamination caused by external impacts. Most often, municipal communities will have organized utility services which supply drinking water via enclosed piping distribution systems. Some human communities, particularly those which may be remote and isolated, are not serviced by piping networks and instead reply upon tanker trucks for the transportation of drinking water. Even with those options, many individuals choose to purchase bottled drinking water and very often emergency situations temporarily may necessitate the supplying of bottled water. It must be understood that drinking water provided by any of those means will not be sterile. The authors provide an understanding of those microbial communities that are suspended in drinking water and also explain the microbial biofilms that develop on the inner walls of water distribution systems. We need to understand those microbial communities and control the health risk which they may present, both because some of the microbes in drinking water are pathogenic and because drinking water biofilms can interfere with water disinfection practices.

Studying aquatic microorganisms often entails identifying them, and for that reason this book additionally provides knowledge of the techniques that have been developed for successful isolation and cultivation of bacteria.

I am tremendously grateful to Andrea Schlitzberger, Markus Späth, and Isabel Ullmann at Springer DE, for their help and constant encouragement which has enabled myself and the other authors to achieve publication of this collaborative project.

Cincinnati, OH Christon J. Hurst

Contents

Chapter 1
Understanding Aquatic Microbial Communities

Christon J. Hurst

Abstract Aquatic environments are divided both physically and functionally into ecosystems whose community organizations include competitions as well as cooperations. Bodies of water that are small and shallow are more likely to be energetically dependent upon allochthonous nutrient inputs from the land. And, at the same time, small and shallow bodies of water will have less buffering capacity against the potential abruptness of fluctuations in allochthonous inputs. Larger bodies of water will by nature of their size have more buffering capacity against allochthonous impacts, and larger bodies of water also will be more reliant upon their autochthonous energy resources. All of the surfaces within aquatic systems contain biofilms, and in a sense it often takes a biofilm to nurture a microbe. Being part of a biofilm has both its blessings and curses, its benefits as well as limitations. Our task of understanding the nature of aquatic microbial communities requires recognizing interrelationships between the good, the bad, and the ugly, with slimy and smelly being par for the course.

1.1 Introduction

The life that we know on this planet has a suggested starting point of perhaps 4.5 billion years ago. That beginning likely occurred sometime after the presumed collision of earth with either a very large asteroid or a small planet, and that object has since been named Theia. The collision eventually resulted in creation of our moon. It is understood that the heat arising from such an impact event almost certainly would have sterilized the earth, eliminating any previously existing life. Following that impact the earth's surface eventually would have cooled sufficiently for liquid water to appear. We presume that the life now recognized on this planet

C. J. Hurst (✉)
1814 Woodpine Lane, Cincinnati, OH 45255, USA

Universidad del Valle, Cali, Colombia

C. J. Hurst (ed.), *The Structure and Function of Aquatic Microbial Communities*,
Advances in Environmental Microbiology 7,
https://doi.org/10.1007/978-3-030-16775-2_1

began shortly after liquid water recollected, with microorganisms initially evolving and growing in that water. And so, in a sense, aquatic microbial life must be considered to represent the start of our biological heritage. Aquatic microbial life now exists from the very surface in all fresh, estuarine and oceanic waters, to the abyssal depths of the oceans. It has been suggested that we might even give a name to the common ancestor of all these life forms, with Glansdorff et al. (2008) having proposed that to be "Luca".

1.2 The Water Has Depth and Divides

In ponds and lakes, we describe vertical stratification of the water column from the surface to the bottom, according naming its three main layers the epilimnion, metalimnion and hypolimnion. The epilimnion is the uppermost layer and usually it will be the warmest because of sunlight. The epilimnion layer also will be oxic because water circulation within the epilimnion distributes atmospheric oxygen throughout that zone. The hypolimnion is the lowermost and generally coldest area of the water column, which also has a relatively low dissolved oxygen level because water circulation within that layer usually is limited to occurring between the bottom of the waterbody and the position of the thermocline. The metalimnion marks the thermocline, which is intermediate both with respect to its vertical position as well as its temperature and level of dissolved oxygen. Each of those zones has its microbial communities and their ecological activities. The thermocline can rise during the daylight hours and fall during the nighttime hours on a daily cycle during much of the year, depending upon the extent to which the surface water is heated by the sun. Very shallow water bodies may not have a thermocline at all. Moderately deep waterbodies, of perhaps a meter of so in depth, may have their thermocline sink to the bottom and disappear during the night and then the thermocline can subsequently reform and rise with the next days sunlight. Disappearance of the thermocline is important because it allows more oxygen to reach the otherwise relatively oxygen deprived bottom of the water column and sediments. I once received a teaching grant from El Instituto para el Control y la Conservación de la Cuenca Hidrográfica del Lago de Maracaibo (ICLAM), in Venezuela, where they interestingly designed the depth for a set of wastewater stabilization canals such that a thermocline would develop in those canals during the day, then the thermocline would sink and disappear each night. That daily thermocline pattern meant the sedimented solids could undergo alternating cycles of aerobic and anaerobic microbial activity. Moderately shallow lakes of perhaps a few meters may have their thermoclines disappear not daily but seasonally, completely dissipating during the winter and reforming during the spring. Quickly flowing bodies of water, such as streams, may experience sufficient mixing that thermoclines do not establish. The oceanic water column generally is divided into five layers called pelagic zones, and from top to bottom those are the epipelagic, mesopelagic, bathypelagic, abyssopelagic, and hadopelagic. Beneath the water column is the sediment, which has a low oxygen level at its surface and becomes anoxic not far beneath the sediment surface.

In a shallow lake with low turbidity, all of the water column may be photic zone meaning that sunlight penetrates even to the bottom such that photosynthesis can occur as a primary energy source throughout the system. In deeper lakes, and in the ocean, there will be a vast aphotic zone which underlies the photic zone. The amount of sunlight penetration into the aphotic zone is either too limited, or indeed nonexistent, such that other energy sources sometimes drive the deeper water lifecycles. However, with exception of locations such as deep sea hydrothermal vent plumes, most life in the column water and sediment is affected by what happens energetically higher in the water column, and most aquatic life utilizes food chains that are based upon photosynthesis (Cavan et al. 2018; Salter et al. 2012; Wolff et al. 2011).

There are aquatic microbial mechanisms, including those occurring among the communities in aquifers, which metabolize metals (Legg et al. 2012). Microbial mechanisms also help to cycle other natural elements including carbon (Painter et al. 2017; Poulton et al. 2017; Sanders et al. 2016), nitrogen (Painter et al. 2017), phosphorus (Davis et al. 2014; Painter et al. 2017; Poulton et al. 2017) and sulfur (Wasmund et al. 2017). Microbial life often converts those from elemental forms into organically useable forms, and sometimes back into elemental forms, done as enzymatic processes for deriving operating energy. Microbial processing of the minerals emitted from hydrothermal sources such as vent plumes have produced some amazing lifestyles (Dick et al. 2013). Aquatic microbial communities also have some capacity for degrading compounds such as spilled natural gas and oil (Redmond and Valentine 2012).

1.3 And Its Ecosystems Are Filled with Amazing Life Forms

Aquatic environments are divided into ecosystems. When we view from a large perspective some small body of water, such as a pond or stream which is narrow and shallow, we typically see that the water has an upper fluid surface exposed to the atmosphere plus there will be inanimate solid surfaces including the bottom and sides. Some of those solid surfaces are completely and permanently submerged, but other surfaces may not be constantly covered with water. We visibly may notice some living aquatic plants and macroalgae, plus some vertebrates and invertebrates, all of which also represent aquatic surfaces contained within that body of water. There additionally will be debris including detritus that once was living. When we view the ocean from a boat that is distant from shore, we mostly see only the surface of the water and floating debris, but we presume that the other types of solid surfaces are present within that ocean even though we cannot see them. When we touch any of those solid surfaces, even the debris, we will notice that none of it feels clean, and instead all of it feels slimy because in fact something is growing on it. Even some of the smallest aquatic organisms have other life forms growing on them (Huq et al. 1983).

Chapter 2 of this book, "Relationship Between Lifestyle and Structure of Bacterial Communities and Their Functionality in Aquatic Systems" by Luca Zoccarato and Hans-Peter Grossart, pp. 13–52, helps us to understand that when we view the same body of water from a more narrow perspective we can perceive it consists of numerous microhabitats filled with microscopic life. Broadly grouped, those microhabitats include: the diffusion controlled water phase; a colloidal phase of nanogels and microgels; particles which may be either exudates, or carcasses, or aggregates; and the living biosphere which has among its components such things as algae, zooplankton, and fish. Each of those microhabitats represents a different range of chemical and physical characteristics. Among those characteristics are the associated abiotic factors such as light, temperature, and oxygen. There also are differences in the concentration and composition of organic matter. Some of the organic material will be particulate in nature, and some will be dissolved, although there often is not an absolute distinction between those two broad categories of organic matter. Much of the organic matter still is living! A major goal always is to eat and not be eaten.

The variables of a water body's size and land proximity, or volume and distance from the center of the waterbody to the shoreline, can be grouped and described as a factor which contributes to an aquatic ecosystems buffering capacity, with the areas in smaller bodies of water experiencing less buffering against terrestrial inputs. The ecosystems within smaller bodies of water, and the near shore areas of larger bodies of water, will be more subject to short term disruptions effected by allochthonous materials that arrive from terrestrial environments and also will be more susceptible to the short term changes in terrestrial inputs caused by climate fluctuations associated with local and regional weather patterns (Davis et al. 2018) including precipitation. It is important to remember that precipitation drives both land surface runoff and groundwater discharge. Loading received from terrestrial inputs will change both the microbial community composition and the community functions. Those allochthonous materials include plant and animal debris which may drive the aquatic carbon cycle, along with both natural and anthropogenic chemicals that wash in from the land surfaces.

The flow of air also brings both nutrients and often contaminants to the water. Windborne solids can travel huge distances, although their impact may be most noticed near to the shore (Griffin and Kellogg 2004). Those inputs are factors which can vary daily, they will include larger scale seasonality patterns, and also encompass very long scale climate change patterns. There even are vast swathes of the earth that have cycled between aquatic environments and dry land in relationship to water and land usage patterns, with examples being the Mesopotamian Marshes in Iraq, Iran and Kuwait. Kenya's Lake Turkana Basin additionally has experienced very long scale climate change patterns (Bloszies et al. 2015; Goldstein et al. 2017).

Larger bodies of water and also the distantly offshore areas of oceans will have their ecologies and carbon cycles more driven by sunlight that directly reaches the water's surface. In some places oceanic deep sea thermal vents additionally will make their contribution as a local basic energy source. And thus, the carbon cycles in those aquatic areas will have an autochthonous base.

Oceanic currents bring changes including a supply of nutrients to aquatic areas that are far distant from the shore, acting in much the same way as do rivers carrying inputs from the land surface to the near shore areas. Some of those nutrient changes which are due to oceanic currents indeed occur very far distant from land while other changes due to currents can occur relatively near to the shore. Those oceanic flow patterns often carry contaminants in addition to bringing inputs of nutrients. Fluctuations associated with oceanic currents generally are less erratic than are fluctuations associated with terrestrial rivers, although the fluctuations associated with oceanic currents can be quite dramatic as in cycles of the El Niño Southern Oscillation. Oceanic currents which bring upwellings of cold, nutrient laden water (Hosegood et al. 2017) often result in plankton blooms that feed and increase fish populations. When those water flow patterns change they can disrupt planktonic growth, and have an opposite effect by causing fish populations to decline.

1.4 It Often Takes a Biofilm to Nurture a Microbe

Living organisms interact with their surrounding environment, being both affected by the environment and in return modifying it. Those actions involve the organisms collecting and taking in what they need from their surroundings and nearly simultaneously leaving behind what is, for them, unneeded refuse. The selection processes as to what is taken in versus left out are not random, they are in fact quite purposeful and for each species those processes both have driven evolution and been modified by evolution. And, generally, an organism never exists by itself. It is important to recognize the fact that life typically is a communal process and we must understand the key significance of microbial existence within a structured community.

When microorganisms exist in communities, the metabolic functions of individual organisms become interdependent and often competitive, and the community develops its communal characteristics. Interdependence within a community can mean selectively sharing metabolic products with other microorganisms and macroorganisms in ways that are mutually beneficial. One organism's requirements often are different from those of another, and what one microbe either intentionally overlooks or leaves as waste can represent the essential needs of another organism. That type of specialization leads to a community organization which efficiently allows one microbes trash to become another microbes treasure. The resulting community structuration may include metabolic zonation and also metabolic coupling. Competition often is only unilaterally beneficial and includes cheating, which can extend to the point of stealing needed resources from your competitors so as to energetically starve those competitors out of existence, or even just simply and outright eating your competitors.

There is an African proverb "It takes a village to raise a child", and by analogy it often takes a biofilm to nurture a microbe. Microbial communities very often develop on surfaces which form collection points. The resulting biofilms have both temporal and spatial heterogeneity as a consequence of the groups organizational development. That development is a process which strives to best utilize available nutrient and energy sources. Sometimes, the surface on which the biofilm exists has been chosen because the surface provides a basic energy source, such as the wall of a metal drinking water distribution pipe, a piece of organic detritus, or a living being that may become detritus due to actions of the accumulating biofilm. At other times, the surface may provide an anchorage point which facilitates communal exposure to an energy source such as sunlight. Energetically, there also may be organizational aspects which rely upon the availability of electron donors and acceptors, represented by reduction potential gradients within a biofilm. We humans are indeed a part of the biofilm which has formed on this rocky planet that we call our home. Our evolutionary path has been directed in part by the environment imposing requirements, and we undoubtedly have modified the environment by leaving a tale of changes in our wake. Through it all we must acknowledge and remember that microbes were our biological origin, biologically the microbes sustain us, and biologically the microbes will survive our lifetimes endpoint.

The characteristics and composition of microbial biofilms differ based upon where they form, the conditions under which they form, and the conditions under which they either continue to exist healthily or senesce. Chapter 3 of this book, "Biofilms: Besieged Cities or Thriving Ports?" by Otini Kroukamp, Elanna Bester and Gideon M. Wolfaardt, pp. 53–90 asks the question of whether biofilms which exist in flowing conditions are either more like thriving ports where there is organization and good to be found associated with everything that comes and goes, or if biofilms like besieged cities are potentially challenged by everything that comes near.

1.5 When Life Hits the Mat

Photosynthetic mats have a biofilm structure that is uniquely interwoven. Chapter 4 of this book, "Complex Structure but Simple Function in Microbial Mats from Antarctic Lakes" by Ian Hawes, Dawn Sumner and Anne Jungblut, pp. 91–120, describes and discusses the physical characteristics and cooperation found within the photosynthetic microbial mats which grow as benthic communities under the continuous ice cover of Antarctic lakes. Those authors explain that growth in such microbial communities shows a very high level of evolved structural organization. The community structure can include seasonal lamination because there is limited environmental disturbance and also because the habitats lack large metazoans that might disrupt or potentially even consume portions of the microbial community. Zonation in those communities is energetically driven and the authors examine the interacting dynamic forces which produce those mat structures.

1.6 The Fungi Will Get You If You Land in the Water

Fungi play a large role in aquatic recycling of organic carbon through their decomposition of detritus from plants and animals. That detritus represents a major source of nitrogen and phosphorus in addition to the obviousness of its carbon content. In chapter 5 of this book, "Fungal Decomposers in Freshwater Environments" by Vladislav Gulis, Rong Su and Kevin A. Kuehn, pp. 121–155, its authors explain that in aquatic environments the fungal biomass tends to have a dominate association with submerged leaf litter and wood. There, fungi valuably act by enzymatically breaking down the large plant polymers such as cellulose, hemicelluloses, lignins, and pectin. The kinetics of those enzymatic activities are affected by the chemical characteristics and oxygen levels of the surrounding water. Some of the fungal activity represents a direct recycling of nutrients to produce not only fungal elements which will be contained in the detritus but also fungal spores that will be released into the water. Much of the nutrients in that detritus, including the fungal elements themselves, will then be consumed by other aquatic microorganisms and macroorganisms. Photosynthetically active algae also play a role in the degradation of detritus, as do bacteria. Some algae are surface associated while others are suspended in the water column. The predominance of aquatic bacterial biomass is associated with fine particulate material.

1.7 Researching Microbiology Even When You Are Up to Your Waist in Alligators While Studying Respiration Without Oxygen, in a Swamp

Wetlands are a connecting point between aquatic and terrestrial life. They contain both aquatic and terrestrial characteristics and in a sense have their own combination of microbial activities. Seasonal wetlands can be aquatic environments during some time periods of the year, and terrestrial environments during other times, depending upon whether the water level is above versus below the ground surface. And, wetlands merge with one another and do as well merge with stream environments (Vanderhoof et al. 2016).

The Everglades is a wide and generally quite shallow river in Southern Florida of the United States, it now is a patterned peat land and serves as an example of a complex wetland with interdependent ecosystems. Peat is a product of partial plant material decomposition which occurs in wetlands (Hohner and Dreschel 2015) and represents long term carbon storage. Conversion of that plant material into coal would represent longer, geologic scale storage of its carbon. Chapter 6 of this book, "The Ecology of Methanogenic Archaea in a Nutrient Impacted Wetland" by Andrew Ogram, Hee-Sung Bae, and Ashvini Chauhan, pp. 157–172, explains that methane production is a key microbial characteristic of wetlands. The anoxic environment of peat systems results in slow decomposition of organic carbon. Therein, the methanogenesis that occurs is a form of anaerobic respiration which uses as its terminal electron acceptor any available atoms of carbon contained in either low molecular weight organic compounds or carbon dioxide, and the result is production of methane. Sulfate reduction also occurs in wetlands, where it is done by microorganisms that perform anaerobic respiration using sulfate as a terminal electron receptor and reduce that to hydrogen sulfide. Although it often is thought that methanogenesis generally occurs only when sulfates are depleted, both processes can occur in the same zonal area (Sela-Adler et al. 2017).

In a sense, methane production is an indicator of the microbial health of wetlands. Methane production is an anaerobic activity associated with archaea and that activity also takes place in the intestines of ruminants (Patra et al. 2017) and many other mammals. Methane production also occurs in anaerobic wastewater treatment processes (Qiao et al. 2015; Świątczak et al. 2017). Methanogenesis additionally occurs in landfills (Staley et al. 2012). The microbes which generate methane by those process are termed methanogens.

1.8 And Microbes Are even in the Water that We Would Want to Drink

Some of the microorganisms present in water cause disease in humans (Hurst 2018) including the notorious *Vibrio cholerae* (Ali et al. 2015; Azman et al. 2013) which is a commensal to crustaceans including those which are considered zooplankton, notably copepods (de Magny et al. 2011). *Vibrio cholerae* causes the disease cholera as explained in chapter 7 of this book, "Briefly Summarizing our Understanding of *Vibrio cholerae* and the Disease Cholera" by Hurst, pp. 173–184. Onc of our public health goals is removing hazardous microorganisms from drinking water, and for that purpose we use methods which vary in complexity and effectiveness. Water treatment processes are designed differently with regard to the volume of water that can be treated simultaneously (Huq et al. 2010; Hurst 2018) and the microbial community changes as a result of drinking water treatment processes (Liao et al. 2015). The procedures used for processing water to either remove hazardous microorganisms or destroy their infectiousness and thereby render the water safe for ingestion will differ depending upon the volume of water being treated. Sometimes the general treatment methodology is similar at large and small size scales but performed in different ways. The subject of supplying microbially safe drinking water is presented in chapter 8 of this book, "Options for Providing Microbiologically Safe Drinking Water" by Hurst, pp. 185–260.

Water treatment for large communities often is followed by efforts to safely distribute the treated water using community drinking water distribution networks that involving piping. Those drinking water distribution systems are enclosed man made aquatic environments colonized by a wide range of both microorganisms and small macroorganisms that arrive from several sources, as explained in chapter 9 of this book, "Microbiome of Drinking Water Distribution Systems" by Laurence Mathieu, Tony Paris and Jean-Claude Block, pp. 261–311. Organisms invariably will enter into the distribution system as constituents of the source water that is being distributed, even if that source water was processed by physical and chemical treatment processes. Those organisms will be joined by other biological contaminants which represent accidental arrivals that have entered the distribution system in association with infiltrations related to piping engineering failures. Such failures include inadequate sealing of the pipe connections, as well as breakage of piping and pipe connections. Accidental cross contamination of the drinking water and septage collection systems represents yet another source of microorganisms. Consequently, the plumbing of drinking water networks are aquatic ecosystems which contain archaea, bacteria, fungi, viruses, protozoa, and small invertebrate metazoa, many of which certainly can cause disease in humans who ingest the water from distribution systems. That accumulation of biomass and its ecosystem structure must be understood and controlled, because the biomass in drinking water interferes with chemical disinfectant processes upon which we rely for delivering safe product water to consumers.

Normally we think of removing phosphorus as something done for treating domestic wastewater to help protect the natural ecosystem from eutrophication (Fig. 1.1). However, there also is some consideration that removing phosphorus during drinking water treatment beneficially may reduce the microbial growth that occurs within water distribution systems (Wang et al. 2014).

1.9 They Are more than Just Nameless Faces: There Are Ways to Culture and Identify even the Seemingly Ungrowable

Aquatic microbial communities contain a vast range of organisms, each with its own activities and necessary requirements. Often, all but some viruses may be culturable if the appropriate conditions can be met. The subject of "Isolation and Cultivation of Bacteria" is addressed in chapter 10 of this book by Martin W. Hahn, Ulrike Koll and Johanna Schmidt, pp. 313–351. The information which they summarize helps us to focus upon and understand an important objective. Its successful achievement often requires complicated technical processes because those microorganisms which represent the study goal have evolved to naturally exist and thrive in their own environmental milieu, under conditions that perhaps require very specific physical and chemical parameter ranges.

Selective physical isolation procedures, including particle size class separations, often can be developed as a helpful early step to aid with identifying the microbes present in an environmental sample. The next task is a requirement to replicate the conditions under which the microorganisms would have been growing in their natural environment. If existing cultivation media formulations prove inadequate, then it may be necessary to design new cultivation media and include some provision of appropriate energy sources such as light of particular wavelength ranges. Flow cytometry is one of the techniques that can be used to assess cultivation efficiency. Gene sequencing is the most accurate testing approach for determining if the resultingly isolated and cultivated organisms either are newly discovered or

Fig. 1.1 The image shows eutrophication at a waste water outlet in the Potomac River, Washington, D.C. It is titled "Potomac green water" by Alexandr Trubetskoy and used under the Creative Commons Attribution-Share Alike 3.0 license

previously have been described, and for learning if perhaps those organisms have previously even been named by other researchers.

Compliance with Ethical Standards

Conflict of Interest Christon J. Hurst declares that he has no conflict of interest.

Ethical Approval This article does not contain any studies with human participants or animals.

References

Ali M, Nelson AR, Lopez AL et al (2015) Updated global burden of cholera in endemic countries. PLoS Negl Trop Dis 9(6):e0003832. https://doi.org/10.1371/journal.pntd.0003832

Azman AS, Rudolph KE, Cummings DA et al (2013) The incubation period of cholera: a systematic review. J Infect 66(5):432–438. https://doi.org/10.1016/j.jinf.2012.11.013

Bloszies C, Forman SL, Wright DK (2015) Water level history for Lake Turkana, Kenya in the past 15,000 years and a variable transition from the African humid period to holocene aridity. Global Planet Change 132:64–76. https://doi.org/10.1016/j.gloplacha.2015.06.006

Cavan EL, Giering SLC, Wolff GA et al (2018) Alternative particle formation pathways in the eastern tropical north Pacific's biological carbon pump. J Geophys Res Biogeosci 123:2198–2211. https://doi.org/10.1029/2018JG004392

Davis CE, Mahaffey C, Wolff GA et al (2014) A storm in a shelf sea: variation in phosphorus distribution and organic matter stoichiometry. Geophys Res Lett 41:8452–8459. https://doi.org/10.1002/2014GL061949

Davis CE, Blackbird S, Wolff G et al (2018) Seasonal organic matter dynamics in a temperate shelf sea. Prog Oceanogr. https://doi.org/10.1016/j.pocean.2018.02.021

de Magny GC, Mozumder PK, Grim CJ et al (2011) Role of zooplankton diversity in *Vibrio cholerae* population dynamics and in the incidence of cholera in the Bangladesh Sundarbans. Appl Environ Microbiol 77:6125–6132

Dick GJ, Anantharaman K, Baker BJ et al (2013) The microbiology of deep-sea hydrothermal vent plumes: ecological and biogeographic linkages to seafloor and water column habitats. Front Microbiol 4:124. https://doi.org/10.3389/fmicb.2013.00124

Glansdorff N, Xu Y, Labedan B (2008) The last universal common ancestor: emergence, constitution and genetic legacy of an elusive forerunner. Biol Direct 3:29. https://doi.org/10.1186/1745-6150-3-29

Goldstein S, Hildebrand E, Storozum M et al (2017) New archaeological investigations at the Lothagam harpoon site at Lake Turkana. Antiquity 91(360):e5. https://doi.org/10.15184/aqy.2017.215

Griffin DW, Kellogg CA (2004) Dust storms and their impact on ocean and human health: dust in earth's atmosphere. EcoHealth 1(3):284–295. https://doi.org/10.1007/s10393-004-0120-8

Hohner SM, Dreschel TW (2015) Everglades peats: using historical and recent data to estimate predrainage and current volumes, masses and carbon contents. Mires Peat 16:Article 01, 1–15

Hosegood PJ, Nightingale PD, Rees AP et al (2017) Nutrient pumping by submesoscale circulations in the Mauritanian upwelling system. Prog Oceanogr 159(12):223–236. https://doi.org/10.1016/j.pocean.2017.10.004

Huq A, Small EB, West PA et al (1983) Ecological relationships between *Vibrio cholerae* and planktonic crustacean copepods. Appl Environ Microbiol 45:275–283

Huq A, Yunus M, Sohel SS et al (2010) Simple sari cloth filtration of water is sustainable and continues to protect villagers from cholera in Matlab, Bangladesh. mBio 1(1):e00034–e00010. https://doi.org/10.1128/mBio.00034-10

Hurst CJ (2018) Understanding and estimating the risk of waterborne infectious disease associated with drinking water. In: Hurst CJ (ed) The connections between ecology and infectious disease. Advances in environmental microbiology, vol 5. Springer, Cham, pp 59–114. https://doi.org/10.1007/978-3-319-92373-4_3

Legg TM, Zheng Y, Simone B et al (2012) Carbon, metals, and grain size correlate with bacterial community structure in sediments of a high arsenic aquifer. Front Microbiol 2012(3):82. https://doi.org/10.3389/fmicb.2012.00082

Liao X, Chen C, Wang Z et al (2015) Bacterial community change through drinking water treatment processes. Int J Environ Sci Technol 12:1867–1874. https://doi.org/10.1007/s13762-014-0540-0

Painter SC, Hartman SE, Kivimäe C et al (2017) The elemental stoichiometry (C, Si, N, P) of the Hebrides shelf and its role in carbon export. Prog Oceanogr 159:154–177. https://doi.org/10.1016/j.pocean.2017.10.001

Patra A, Park T, Kim M et al (2017) Rumen methanogens and mitigation of methane emission by anti-methanogenic compounds and substances. J Anim Sci Biotechnol 8(1):13. https://doi.org/10.1186/s40104-017-0145-9

Poulton AJ, Davis CE, Daniels CJ et al (2017) Seasonal phosphorus and carbon dynamics in a temperate shelf sea (Celtic Sea) Prog Oceanogr. https://doi.org/10.1016/j.pocean.2017.11.001

Qiao S, Tian T, Qi B et al (2015) Methanogenesis from wastewater stimulated by addition of elemental manganese. Sci Rep 5:12732. https://doi.org/10.1038/srep12732

Redmond MC, Valentine DL (2012) Natural gas and temperature structured a microbial community response to the Deepwater Horizon oil spill. Proc Natl Acad Sci USA 109:20292–20297. https://doi.org/10.1073/pnas.1108756108/-/DCSupplemental

Salter I, Kemp AES, Moore CM et al (2012) Diatom resting spore ecology drives enhanced carbon export from a naturally iron-fertilized bloom in the Southern Ocean. Global Biogeochem Cycles 26:GB1014. https://doi.org/10.1029/2010GB003977

Sanders RJ, Henson SA, Martin AP et al (2016) Controls over ocean mesopelagic interior carbon storage (COMICS): fieldwork, synthesis, and modeling efforts. Front Mar Sci 3:136. https://doi.org/10.3389/fmars.2016.00136

Sela-Adler M, Ronen Z, Herut B et al (2017) Co-existence of methanogenesis and sulfate reduction with common substrates in sulfate-rich estuarine sediments. Front Microbiol 8:766. https://doi.org/10.3389/fmicb.2017.00766

Staley BF, de los Reyes FL III, Barlaz MA (2012) Comparison of Bacteria and Archaea communities in municipal solid waste, individual refuse components, and leachate. FEMS Microbiol Ecol 79(2012):465–473. https://doi.org/10.1111/j.1574-6941.2011.01239.x

Świątczak P, Cydzik-Kwiatkowska A, Rusanowska P (2017) Microbiota of anaerobic digesters in a full-scale wastewater treatment plant. Arch Environ Prot 43(3):53–60. https://doi.org/10.1515/aep-2017-0033

Vanderhoof MK, Alexander LC, Todd MJ (2016) Temporal and spatial patterns of wetland extent influence variability of surface water connectivity in the prairie pothole region, United States. Landsc Ecol 31:805–824. https://doi.org/10.1007/s10980-015-0290-5

Wang JL, Song J, Lu J et al (2014) Comparison of three aluminum coagulants for phosphorus removal. J Water Resour Prot 6:902–908. https://doi.org/10.4236/jwarp.2014.610085

Wasmund K, Mußmann M, Loy A (2017) The life sulfuric: microbial ecology of sulfur cycling in marine sediments. Environ Microbiol Rep 9(4):323–344. https://doi.org/10.1111/1758-2229.12538

Wolff GA, Billett DSM, Bett BJ et al (2011) The effects of natural iron fertilisation on deep-sea ecology: the Crozet plateau, Southern Indian Ocean. PLoS One 6(6):e20697. https://doi.org/10.1371/journal.pone.0020697

Chapter 2
Relationship Between Lifestyle and Structure of Bacterial Communities and Their Functionality in Aquatic Systems

Luca Zoccarato and Hans Peter Grossart

Abstract Aquatic bacteria dwell in an unexpectedly heterogeneous world defined by multiple specific microhabitats of differing complexity, size, and temporal-spatial dynamics. The diversity in microhabitats is proposed as a potential explanation for the "plankton paradox" wherein high levels of species and functional richness are maintained under limited resources. Microhabitats can be classified into broad groups including the diffusion-controlled water phase (DifP), colloidal phase (nano- and micro-gels; ColP), particles (exudates, carcasses, and aggregates; Par), and the living biosphere (algae, zooplankton, and fish; Bio). For each microhabitat, this chapter examined the various physiochemical properties and principal dynamics that define these environments and then linked these with major lifestyle strategies adopted by the bacteria inhabiting these systems. Within this context emphasis was placed on the revision of the concept of free-living bacteria.

Expectedly it was found that each microhabitat, in fact, selects for distinct functional guilds. The DifP-associated community is regulated by the dynamics of dissolved organic matter quality and quantity (e.g., enzymatic dissolution of polymeric materials). The communities associated with the ColP component were driven by the rapid ephemeral nature of gel structures assembled from physical polymer interactions. The Par microhabitats sustain dense bacterial communities with remarkable metabolic activity that can result in the complete or partial dissolution of the particle. Interestingly, Par-associated bacteria are in a state of constant attachment and detachment in response to physiochemical evolution of the micro-niches. Moreover, Bio-associated epi- and endobionts present additional levels of complexity due to the influence of the host-bacteria interfaces, which experience continued change in response to environmental and biological stressors. Understanding these interactions at local scale is critical for the comprehension of the impact of anthropogenic activities on microhabitat complexity and is necessary for enabling us to understand how microhabitat features scale up to major aquatic biomes.

L. Zoccarato (✉) · H. P. Grossart (✉)
Department of Limnology of Stratified Lakes, Leibniz Institute of Freshwater Ecology and Inland Fisheries, Stechlin, Germany
e-mail: zoccarato@igb-berlin.de; hgrossart@igb-berlin.de

C. J. Hurst (ed.), *The Structure and Function of Aquatic Microbial Communities*, Advances in Environmental Microbiology 7,
https://doi.org/10.1007/978-3-030-16775-2_2

Table 2.1 Overview on the proposed microhabitat definitions

Microhabitats		Size range	Major dynamics
Diffusion-controlled water phase	DifP	~0.1 μm to ~6 μm	• Temporal-spatial heterogeneity in autochthonous and allochthonous dissolved compounds
Colloidal phase	ColP		• Highly dynamic conditions at short time scales (nano- and micro-gel formation) • As matrices providing ephemeral nutrient hotspots leading to abrupt fluctuations of the bacterial community composition • Increased nutrient and polymer turnover compared to DifP
Freshwater and marine particles	Par	>0.8 μm to several cm and m	• Particle and bacterial community composition mutually affect each other • High temporal-spatial dynamics in particle composition leading to different degradation processes and time scales • Vertical bacterial dispersion due to particle sinking • Specific food web interactions
Living biosphere	Bio		• Host physiology changes in response to biological and environmental factors • Unspecific and specific interactions between bacteria and their host (symbioses, parasitism, commensalism, etc.) • Changes in bacterial adaptations and behavior shifts the association with the host, e.g., from commensalism to parasitism (may also kill the host)

2.1 Conceptual Idea

Bacteria—called "the unseen majority" less than 20 years ago (Whitman et al. 1998)—nowadays are known to represent the richest life domain in terms of species diversity and metabolic functionality. Dwelling in almost all ecosystems of the planet, bacteria show a nonrandom spatial community distribution; however, at this coarse observation scale, most factors causing the observed patchiness are still unclear. In this chapter, we try to tackle this issue by embracing the bacterial perspective in order to shed light on their microscale behavior taking into account the multitude of different temporal and spatial scales as well as different levels of environmental and organizational complexity.

The close surroundings of a bacterial cell in aquatic environments consist of several specific microhabitats extending from μm to several mm and cm (Table 2.1): the diffusion-controlled water phase, the colloidal phase (nano- and micro-gels), the marine and freshwater particles (exudates, carcasses, aggregates), and the living biosphere (algae, zooplankton, fish, etc.) (Grossart 2010). Each of these microhabitats is characterized by a different range of chemical and physical gradients and, therefore, harbors a certain level of heterogeneity and complexity providing multiple conditions for microbial micro-niche formation. These micro-niches form the basis for microbial diversity and functionality and thus may explain the so-called plankton paradox (Hutchinson 1961). Hutchinson was the first researcher wondering about "how it is possible for a number of species to coexist in a relatively isotropic or unstructured environment" since they were "all competing for the same sorts of material"? Nowadays, modern investigative technologies provide evidence of the remarked heterogeneity of chemical and physical properties associated with each of the microhabitats given in Table 2.1 and how aquatic microbial communities are adapted to exploit this diversity (e.g., Salcher et al. 2013) and how they interact (Garcia et al. 2015). The view of aquatic environments as "relatively isotropic or unstructured environments" has drastically changed during the past decades by recognizing different levels of complexity when moving from the diffusion-controlled water phase to the colloidal and particulate phase and lastly to the living biosphere.

Major abiotic factors influencing microbial microhabitats are light/UV penetration, temperature, oxygen, pH, pollutants, organic matter (OM) concentration and composition, chemical gradients, etc. (e.g., Church 2008) but also numerous biotic factors such as host physiology, symbiosis, commensalism (auxotrophy), parasitism, predation, competition (including antibiotics and toxins), and viral lysis (e.g., Burghardt and Oswald 2015). Moreover, abiotic and biotic factors dynamically interact and hence can mutually affect each other (Kirchman 2012). As a result, this microscale diversity affects bacterial community composition by selecting for those organisms which can rapidly adapt to the existing microhabitat conditions, and in turn the bacterial activity of the resulting microbial communities will modify and shape the microbial microhabitat. Interacting with the environment (exoenzymatic activity, uptake, and secretion), bacteria modify the chemical composition of their microhabitats by such processes as consuming and transforming available substrates, as well as secreting new compounds (metabolic by-products, antibiotics, toxins). In addition, the highly variable and heterogenic composition of micro-niches may result in different patterns of microbial presence and microbial activities which form as characteristic features of aquatic environments, e.g., eutrophic coastal waters with extremely high particle concentrations vs. the oligotrophic, open ocean with a much more uniform microbial microenvironment. This temporal and spatial variability of microhabitats at different temporal and spatial scales can further increase diversity, functionality, and complexity of aquatic microbial communities evolving in time and space.

Since microbial microhabitats and micro-niches are not constant in time and space, we propose certain features of the major, above defined microbial microhabitats:

The diffusion-controlled water phase (DifP) (Fig. 2.1a) represents microhabitats (for truly free-living microorganisms) that are controlled by the temporal-spatial dynamics of dissolved molecules, which, e.g., are generated by organismic exudation and enzymatic dissolution of polymeric materials such as autochthonous and

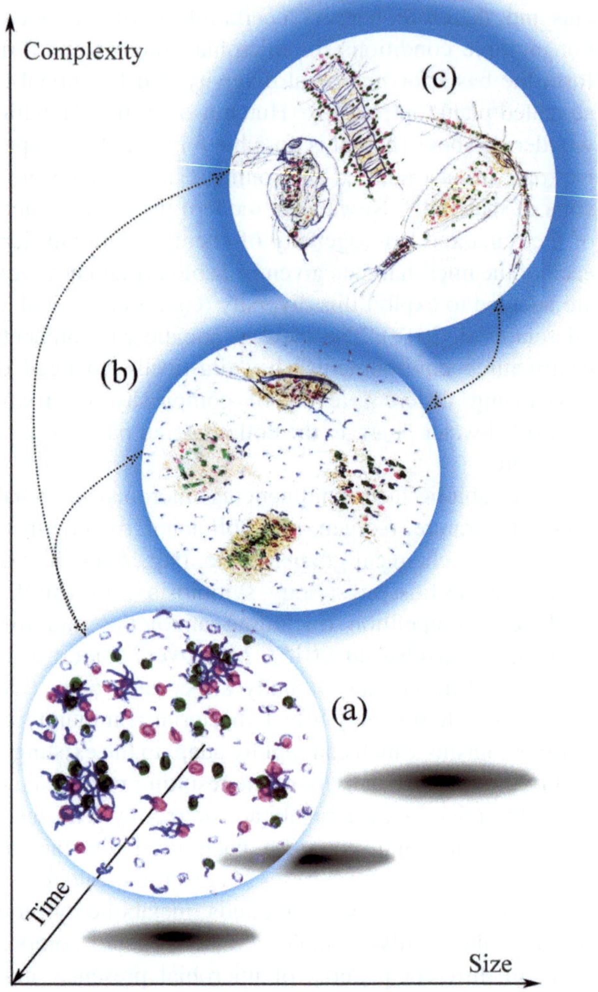

Fig. 2.1 Schematic representation of the principal aquatic bacterial microhabitats such as the diffusion-controlled water phase, colloidal phase (a), freshwater and marine particles (b), and living biosphere (c). The microhabitats are spatially organized according to the increase in size, the micro-niche's number and complexity, and the time scale of the dynamics which pervade the microhabitat. The blue halos indicate the gene richness of the bacterial genomes present in the microhabitats. The dotted lines show the possible attachment and detachment of the chemotactic and generalist bacteria that switch from free-living to particle-attached lifestyle or vice versa

allochthonous particulate OM. Bacteria adapted to these microenvironments are directly dependent upon the physical-chemical changes of the surrounding water phase, with any chemotactic bacteria being able to follow the spatial and temporal distribution of the dissolved compounds and the non-motile bacteria which remain randomly distributed consuming those compounds diffusing toward them (Stocker 2012).

The colloidal phase (ColP) (Fig. 2.1a) microhabitats encompass highly dynamic conditions at short time scales and, therefore, harbor the most ephemeral and dynamic micro-niches. The fast, spontaneous assembly and subsequent dispersion of the gel structures deeply affect the chemical and physical conditions of the microhabitats (Verdugo 2012) leading to variable kinetics of the uptake processes, different availabilities (e.g., labile vs. refractory) of the bacterial substrates and nutrients (Simon et al. 2002; Stocker et al. 2008), and abrupt changes in environmental conditions, and additionally the rapid exploitation of nutrient patches within a gel structure can induce spontaneous fluctuation of the bacterial population that was occupying that micro-niche.

Freshwater and marine particle (Par) (Fig. 2.1b) microhabitats can be completely or partially consumed by the associated bacteria, and the remaining refractory fraction can eventually sink controlling the efficiency of the oceanic carbon pump (i.e., the abiotic and biological processes of atmospheric carbon dioxide sequestration in the deep ocean and sediment) (Volk and Hoffert 1985; Jiao et al. 2010); different sinking rates can also influence the colonization process of the particles (Turner 2002) and favor bacterial dispersion. The bacterial community evolves over time in response to modifications of the physiochemical properties; e.g., a sinking particle enriched in organic substrates is colonized faster by motile bacteria as compared to a particle with low substrate concentrations since many bacteria can detect the OM plume surrounding OM-rich particles. Moreover, it has been suggested that the mean residence time of motile bacteria on a particle is about 3 h pointing to a continuous exchange between free-living and particle-attached bacteria. The dynamics of the colonization process is a feedback of the changing chemical environment in the Par microhabitats (Kiørboe and Thygesen 2001; Kiørboe et al. 2002). Shifts in community composition are correlated with alterations of the metabolic capability of the bacterial community (e.g., Fontanez et al. 2015), which in turn differently affects and modifies the composition and properties of the particle microhabitat. Par microhabitats also represent foraging hotspots for bacterivores and have relevant implications for the trophic food webs since the high prey density leads to a higher predation efficiency (e.g., limiting energy consumption for prey retrieval); thus the attached bacterial populations are top-down controlled by grazing activities which decrease the interspecies competition and increase the biodiversity of the microhabitats (Kiørboe et al. 2004; Corno et al. 2013).

The living biosphere (Bio) (Fig. 2.1c) microhabitats share part of the temporal-spatial dynamics of Par microhabitats. In addition, these microhabitats are influenced by changes in the host-bacteria interface such as modifications of the host physiology in response to environmental and biological stressors (Grossart 1999; Grossart et al. 2005; Amin et al. 2012; Glasl et al. 2016; Seymour et al. 2017). The diatom *Thalassiosira weissflogii*, for example, produces transparent exopolymer particles (TEP), which are responsible for aggregation dynamics, albeit that production occurs only in the presence of specific associated bacteria. This production process is controlled by the photosynthetic activity of the diatom, has relevant consequences on algal sinking, and therefore influences the efficiency of the carbon export (Gärdes et al. 2011). Bacteria similarly might change their behavior in response to environmental stressors with remarkable implications for the host (Vezzulli et al. 2012), and such is the case for the *Flavobacterium Kordia algicida*, which lives associated with several diatom species. When the population of *K. algicida* reaches a certain threshold, an algicidal protease is excreted through a quorum sensing-triggered mechanism, which renders the commensalistic relationship with the algal host to a parasitic one (Paul and Pohnert 2011). *Croceibacter atlanticus* was shown to prevent the algal host to divide and leading it to increase in size (i.e., colonizable surface) (van Tol et al. 2016).

To date, despite the outstanding heterogeneity of these microhabitats, some of them are largely overlooked in ecosystem-based investigations (in particular, bacteria associated either with particles or organisms) leading to inaccurate estimations of the measured microbial processes on a larger scale. Nowadays, new technologies provide novel and unexpected tools to investigate the genomic diversity and metabolic potential of aquatic bacteria offering the opportunity to tackle with unprecedented precision the task of understanding each single type of microhabitat. Techniques such as next-generation sequencing (also called high-throughput sequencing) and the related "omics" approaches help to uncover the composition of the prokaryotic communities as well as their manifold functionalities and physiological responses (Shendure and Ji 2008; Moran et al. 2013). Furthermore, advances in sequencing and chemistry methodologies enable single amplified genome (SAG)-based studies linking individual cell traits with potential functions in the ecosystem. The SAG studies, in combination with new mass spectrometric (MS) techniques such as MS-MS, and desorption electrospray ionization combined with MS (DESI-MS), allow the metabolome characterization of single organisms and thus facilitate deep investigations of both cell-to-cell and cell-to-particle interactions. These approaches help to shed light on microbial dynamics in the Par and Bio microhabitats.

2.2 How Much Free-Living Is Free-Living?

The differentiation between free-living (FL) vs. particle-attached (PA) bacteria is based on a conceptual distinction that often lacks a real ecological meaning. Bacteria are considered FL if they can pass through a filter of a certain pore size, whereas PA bacteria will be retained on the same filter. This distinction has been earlier introduced for separating organic matter fractions, i.e., dissolved OM (DOM) and particulate OM (POM). Historically, the threshold in size of the mesh or filter used to separate the different bacterial fractions was chosen arbitrary and still often varies among disciplines and between individual studies. Today, there is no general consensus, and the threshold can vary from 0.8 μm (e.g., Smith et al. 2013) to 5–10 μm (e.g., Bižić-Ionescu et al. 2014).

This methodological inconsistency might be responsible for the partly contradictory conclusions regarding the phylogenetic patterns in community composition of FL vs. PA bacteria. Some authors have tried to correlate the occurrence of typically copiotrophic (those found in environments rich in nutrients) and generalist bacterial groups such as *Gammaproteobacteria* and *Flavobacterium* with sources of POM, while oligotrophic-adapted bacteria (those which typically survive in environmental conditions that have much lower carbon concentrations) tend to primarily occur in the dissolved fraction (Biers et al. 2009; Ortega-Retuerta et al. 2013; Rieck et al.

2015). Contrariwise, other researchers have pointed to a more random bacterial colonization of the POM and, therefore, to the absence of a consistent pattern in FL vs. PA bacteria communities (Hollibaugh et al. 2000; Ghiglione et al. 2007). Although we agree with the "lottery" theory for the very initial colonization, which increases a certain degree of randomness (e.g., seasonality of FL bacteria which represent the bulk for the "lottery drawings"; Rösel and Grossart 2012), an increasing number of studies highlight marked genomic differences among copiotrophic and oligotrophic bacteria, which strongly support the existence of a consistent functional pattern in microbial distribution and community structure among the various microhabitats (Grossart 2010; Burke et al. 2011; Tang et al. 2014; Giovannoni et al. 2014).

However, in nature the distinction between DOM and POM is anything but sharp. In the literature it is described that OM in the environment forms a size continuum (Azam 1998; Verdugo et al. 2004). Molecules and single polymers in solution undergo a passive diffusion and can spontaneously coagulate or aggregate to form marine micro- and macro-gels. These tridimensional structures are formed continuously by purely physical and chemical processes. A physical gel is formed by polymer interconnections through tangles and cross-links (electrostatic interactions based on the polymers' charge density and interactions among hydrophilic and hydrophobic domains), which are characterized by low activation energy (<50 kJ mol^{-1}). Thus the formation of these gels is a fully reversible process with a high dynamic in the dispersion and self-assembly equilibrium that stretches from the polymers' dispersion phase to polymer aggregation in nanogels (0.1–0.2 μm) and eventually micro-gels (3–6 μm) (e.g., Verdugo 2012).

The dichotomic distinction between FL-PA bacteria and DOM-POM might not be enough as the natural distribution of the realistic size structure (both organisms and OM) within an ecosystem cannot be reduced and just simplified into two size classes. Therefore, a possible solution is to consider more precise size fractions, establishing the size threshold based on the true size of particles present within each specific environment. A more precise distinction of size classes (Mestre et al. 2017) or particle types with a defined sinking behavior (i.e., specific residence time in the water column) is necessary to achieve a meaningful ecological simplification of the real world. It also must be understood that particle size and type distribution within eutrophic environments are significantly different from that of ultra-oligotrophic environments (e.g., upwelling areas will have more and larger particles than the open ocean but presumably less and smaller particles than in estuarine areas).

In order to avoid confusion with the cited literature, in the following sections, we will keep the distinction between FL and PA bacteria based on the threshold adopted in the cited paper. However, we will consider the colloidal phase not as a homogenously diluted environment but rather as a heterogeneous puzzle in which some pieces are represented by micro-gel structures.

2.3 Microhabitat Descriptions

2.3.1 Diffusion-Controlled Water Phase (DifP) and Colloidal Phase (ColP)

2.3.1.1 Physiochemical Characterization and Major Dynamics

The aquatic environment surrounding the cell is an extreme diluted environment (Fig. 2.1a) where nutrients and substrates are relatively dispersed (McCarthy et al. 1996). The foraging distance might play a crucial role since at the cell scale, the aquatic medium is highly viscous (characterized by a low Reynolds number) which can involve diffusion limitations. Free-living bacteria can be roughly categorized as either immotile waiting for the food to come to them or choosing to hunt by swimming and thereby increase their encounter rate with the food particles. Being motile means that their food intake would increase only with the square root of the bacteria's velocity (Purcell 1977). As an alternative intermediate option, the bacteria can move toward patches with a higher nutrient concentration than the surrounding water, e.g., micro- and macro-gels, particles, aggregates, etc. It is worth pointing out that some cells and colonies due to their sizes or shapes with a Reynolds number > 1 can escape the diffusion limitation and experience the advantage of advection processes (such as catabolite dispersion) (KarpBoss et al. 1996).

Autotrophic bacteria mainly rely on the availability of inorganic nutrients (Agawin et al. 2000), and their suitable niches are constrained by the need for light. Inorganic nutrient concentrations are in turn affected by the metabolism of heterotrophic bacteria, which rely on the availability and composition of the OM (Hansell and Carlson 2002). However, a large fraction of bacteria in global superficial seawater have shown the capability to utilize both sunlight and OM circumventing conditions of energy and carbon limitation; these photo-heterotrophic groups are known as proteorhodopsin-containing bacteria (Béjà et al. 2000) and aerobic anoxygenic phototrophic (AAP) bacteria (Yurkov and Csotonyi 2009; Koblížek 2015). The consumption of OM by these organisms through enzymatic cleavage is coupled with the release of inorganic compounds (e.g., nitrogen and phosphate), in a process called mineralization of the OM. Thus the distribution and bioavailability of the OM have dramatic impacts on bacterial community structure and function. The DOM (which comprises the major fraction of compounds forming marine gels) represents the largest OM pool, being roughly ten times more abundant than POM in most aquatic systems, although the DOM present in the open ocean is predominantly recalcitrant and poorly bioavailable (Williams et al. 1969).

As argued in the previous section (Sect. 2.2), the distinction between DOM and POM is entirely dependent on methodological constraints since the polymers which make up the DOM undergo continuous processes of nano- and eventually micro-gel formation. The marine gels are hotspot of nutrients and OM of the ColP microhabitats, since they generally contain a solid/solvent ratio of $\sim$1% and thus higher OM concentrations than DifP microhabitats (i.e., $\sim$10 g L^{-1} against $\sim$10^{-3} g L^{-1} of the sea water). The gel formation process is greatly influenced by properties of the water mass (dielectric properties, osmolarity, pH, and ionic composition) and local environmental factors (temperature and pressure) but also by the quality of the polymers (length and chemical properties) and by their concentrations, which can also influence the dimension and the stability of the gels (Verdugo 2012 and references therein).

Input/output and transformation of the DOM have strong implications for the three-dimensional structure of the ColP microhabitats. The principal DOM input stems from algal exudates, sloppy feeding associated with zooplankton grazing, and cell lysis mediated by viral infection and by bacterial lysis (on phytoplankton or other bacteria) (Carlson 2002 and references therein). Moreover, DOM plumes created by bacterial POM degradation provide a rich, although ephemeral hotspot of polymers and nutrients which might be rapidly consumed by FL bacteria (Smith and Azam 1992; Kiørboe and Jackson 2001) or undergo DifP and ColP microhabitat formation. Biodegradation represents one of the major processes that affect the concentration and composition of the present DOM pool. Bacteria can rapidly degrade carbohydrates, proteins, lipids, and organic acids which represent the labile fraction of the DOM (although the composition varies according to DOM origin) but in turn also secrete new compounds and leave the more refractory fractions behind (Kirchman 2003). Similar to biodegradation, photolysis can greatly modify the DOM pool by breaking apart polymers and large molecules into small compounds (e.g., phenol, acetate, and propionate) which can in turn be degraded by the microbial community stimulating their growth and metabolism (Moran and Zepp 1997; Hutalle-Schmelzer et al. 2010). Refractory, high molecular weight compounds can also be produced by photooxidation processes devoting a fraction of DOM to biodegradation (Obernosterer et al. 1999). This physical process can also produce harmful compounds such as hydrogen peroxide which can induce modifications in the composition of the associated bacterial community selecting for species resistant to reactive oxygen species (Scully et al. 2003; Glaeser et al. 2010; Glaeser et al. 2014).

Gel formation has a critical importance for nutrient and metabolite uptake of bacterial cells since it reduces the dilution of the DOM pool in water. DOM represents a huge biomass ($\sim$700 gigatons, where 1 Gt $= 10^{15}$ g), but it can be evenly distributed with final molar concentrations in the order of 10^{-6} mol L^{-1}. In contrast, within a micro-gel a particle of 600 Da (upper threshold for microbial uptake) can reach OM concentrations of 10 g L^{-1}. As a result, the matrix of the gel limits the DOM dispersion of the enzymatically cleaved compounds and thus decreases the necessary foraging distance allowing bacteria to save energy when obtaining their nutrient supply (Verdugo et al. 2008) and to increase their growth yield (Fernandez et al. 2019).

2.3.1.2 Bacterial Adaptations: To Eat and Not Be Eaten

The two major concerns of each living organism are to eat and not to be eaten. Therefore, the FL bacteria have developed several strategies to exploit the available micro-niches paying attention to these constrains. As a consequence, FL bacteria tend to be streamlined both in cell and genome size to reduce their expenditures for maintenance energy and their vulnerability to grazing.

A small cell size offers several advantages to the organism such as a high surface-to-volume ratio, which maximizes their capabilities for nutrient/substrate uptake and thus provides a physiological advantage against larger cells in nutrient-limited environments (Kirchman 2012). Moreover, a reduced cell dimension is a useful strategy to avoid grazing losses. The freshwater *Actinobacteria*, for example, have cell dimensions of $\approx$0.1 μm in diameter (ultramicrobacteria), and their populations reach maximum abundances in the presence of heterotrophic nanoflagellates. This clearly points to a negative grazing selection on the basis of prey size (Jezbera et al. 2006; Grujčić et al. 2015). Other abundant clades of ultramicrobacteria include the marine *Alphaproteobacteria* SAR11 and the freshwater sister group LD12. The *Alphaproteobacteria* SAR11 has been defined as cryptic fugitive, being almost "invisible" to grazers due to its reduced biomass and slow growth rate (Yooseph et al. 2010). Other strategies adopted from FL bacteria to avoid predation are the formation of large colonies. This cellular organization confers additional buoyancy and circumvents the handling capability of smaller grazers (e.g., heterotrophic nanoflagellates). Filament formation is a common but diffuse trait among different phylogenetic groups such as *Cyanobacteria*, *Proteobacteria*, and *Bacteroidetes* (Hahn et al. 1999). Nevertheless, these bacterial colonies can be efficiently ingested by larger predators such as the zooplankton (e.g., Schauer et al. 2006) with remarkable effects on the carbon flux throughout the trophic food web. Despite these favorable traits and the fact that filamentous bacteria can account for up to 50% of the total bacterial biomass (Corno et al. 2008), filament-forming colonies do at times represent a low percentage of the total FL abundance, and their direct ingestion by the zooplankton represents an energetic shortcut toward the upper trophic levels (since each trophic level involves a loss of energy in the C-/biomass transfer). Another mechanism of defense against predation is the production of toxins, e.g., microcystins from *Cyanobacteria*, that are triggered by the presence of grazers (Jang et al. 2003; Ger et al. 2016).

Genome streamlining is typical in FL bacteria (Ghylin et al. 2014) (Fig. 2.2) and generally involves the reduction of regulatory genes (e.g., σ factor) and an increase in relative abundance of core genes (Giovannoni et al. 2014). However, some relevant exceptions have been reported, e.g., for the *Prochlorococcus* strains MIT9303 and MIT9313 (Kettler et al. 2007). Living in a highly diluted environment, FL bacteria encounter a reduced variety of exploitable compounds, and several ancillary genes might be poorly expressed. When a gene became "useless," because of the lack or the ephemeral nature of the triggering stimuli, the occurrence of a possible disruptive mutation will not cause any functional disadvantage to the

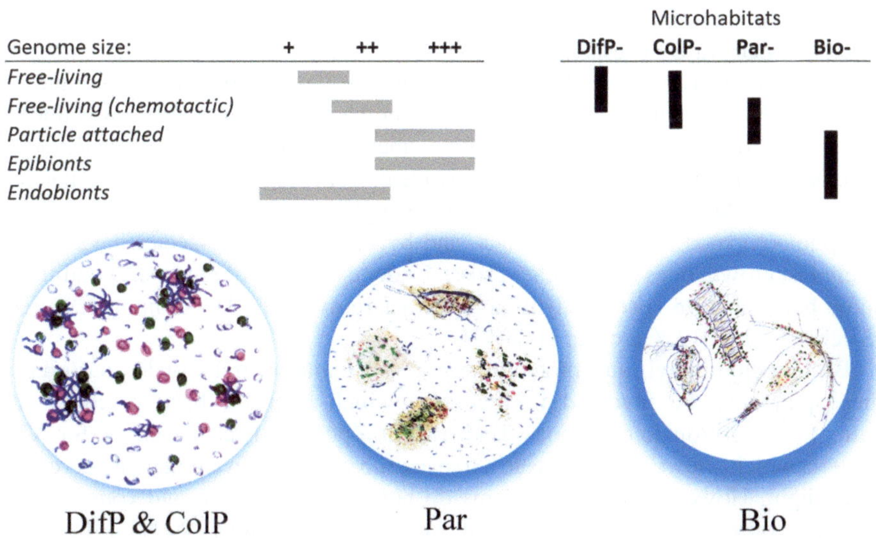

Fig. 2.2 The figure reports the relative genome size among different categories of bacteria (horizontal gray bars) and correlates their occurrence in the microhabitats (vertical black bars)

carrying organisms. The absence of a negative selection, therefore, will not decrease the fitness of these bacteria, leading eventually to loss of functional genes (Ochman and Davalos 2006 and references therein). Having a streamlined genome allows FL bacteria to reduce their growth and replication costs (Ochman and Davalos 2006), a feature that endorses for their dominance in such nutrient-depleted environments. A remarkable example is the *Alphaproteobacteria* SAR11, which dominates the bacterial communities in the oligotrophic ocean thanks to a streamlined genome, an optimal surface-to-volume ratio, and the presence of several high-affinity membrane transporters (Lauro et al. 2009). Sometimes genome streamlining is so pronounced that even core genes can be lost in the presence of well-established mutualistic interactions among different species. This process has been conceptualized as the Black Queen Hypothesis that explains positive selection of bacteria with disruptive mutations of core genes when these genes code for functions that are costly and leaky in the community (Morris et al. 2012). Vitamins and amino acids are some of these "public" goods whose production and availability become essential for several streamlined bacteria leading them to the so-called auxotrophy for these compounds (Garcia et al. 2015). Genome streamlining, however, is not a compulsory requirement for FL bacteria, and relevant differences in the ancillary-core gene ratio have been reported for the same clade due to long-term adaptation processes. Lauro et al. (2009) reported on how genomes of cyanobacteria dwelling in the open ocean have more oligotrophic traits than do cyanobacteria from freshwater ecosystems. In contrast, the copiotrophic gammaproteobacterium, *Photobacterium angustum*, has a broad genetic repertoire of genes to sense and react to environmental cues such as sudden nutrient influx or depletion; it is even able to switch its lifestyle and metabolic pathways in response to depletion of nutrient patches, e.g., by changing from a PA to a FL lifestyle (Lauro et al. 2009). Among FL bacterial communities, there are several other examples of copiotrophic bacteria often reported as abundant species in their ecosystem. The relative abundance of the *Betaproteobacteria Limnohabitans* and *Polynucleobacter* may result from their ability to exploit patches of low molecular weight compounds such as simple organic acids, monosaccharides, and photooxidated humic compounds, and by virtue of these species having relatively fast growth rates (see Salcher 2014). Under extremely oligotrophic conditions, some bacteria have specifically adapted to share available but limiting resources (Salcher et al. 2013; Garcia et al. 2015).

Bacteriochlorophyll and proteorhodopsin are common traits of FL bacteria as these pigments are involved in processes which help FL bacteria cope with extremely oligotrophic conditions. Bacteriochlorophyll A is characteristic for AAP bacteria which allows them to create a transmembrane reductive potential harvesting green light. However, these organisms do not contain Rubisco and consequently fully rely on organic carbon for growth (Yurkov and Csotonyi 2009). It has been estimated that the AAP bacteria represent about 24% of total bacteria in most oligotrophic aquatic environments (Lami et al. 2007). The coding potential for the proteorhodopsin, a transmembrane light-driven proton pump known to support some of the most costly cellular energetic processes such as the uptake of vitamin B1 (Gómez-Consarnau et al. 2016), has been detected in up to 75% of marine bacteria (especially, *Gammaproteobacteria* and *Bacteroidetes*).

Generally, FL bacteria present marked seasonal fluctuations in population size since they are directly exposed to environmental fluctuations (while particles and hosts represent relatively stable bacterial refuges). Whenever environmental conditions result in changes of the present micro-niches, the existing FL bacteria population experiences a drastic decrease in favor of those bacteria equipped with the respective metabolic machinery to efficiently exploit the new conditions (e.g., Fuhrman et al. 2006).

Among free-living bacteria, there is increasing evidence for the presence of organisms shifting from a passive to an active chemotactic lifestyle exploiting patches of OM and nutrients, and some otherwise free-living bacteria might also be involved in Par and Bio microhabitat colonization (Grossart 2010). It has to be kept in mind that the conditions allowing for this "microhabitat jump" are expected to be quite stringent since the metabolic potential of FL bacteria differs significantly from that of typical PA bacteria due to the high degree of specialization and also the genome streamlining of FL bacteria (Ghylin et al. 2014; Giovannoni et al. 2014). Perhaps the "traditional" concept of FL bacteria, which would suggest them to be mostly passive and waiting for their food particles to arrive, needs to be reconsidered. Genomes of FL oligotrophic-adapted bacteria are still poorly explored due to the historical cultivation limitations which recently have been overcome thanks to SAG approaches. SAG analyses are unveiling the extreme adaptation of these bacteria by revealing not only their substrate specialization (Rinke et al. 2013) but also some "unexpected" versatility; hereinafter we briefly highlight two interesting examples of specialized bacteria and one example of generalist FL bacteria. Genome analyses of the widespread FL *Actinobacteria* acI have described its preference for carbohydrate and N-rich organic compounds highlighting a further level of specialization for the clade acI-A and acI-B in substrate utilization (Ghylin et al. 2014). The *Alphaproteobacteria Caulobacter crescentus* has a dimorphic life cycle that separates a chemotactic behavior in the presence of nutrients and an immotile but reproductive stage. This species is consequently equipped with genes that mediate the environmental sensing of carbon and ammonium or amino acid starvation and then favor chemotactic behavior and genes that mediate adhesion to nutrient patches (Boutte and Crosson 2013). The proteomic insight of the *Alphaproteobacteria Ruegeria pomeroyi DSS-3* revealed that 30% of its genes encode for basal living function (housekeeping genes), while 20% are responsible for physiological responses to the induced environmental perturbations within the "natural" ranges of variability for these factors (salinity, temperature, UV, aromatic compounds). Interestingly, 50% of its proteome was never expressed during experimental treatments suggesting that very specific physiological responses are required to exploit different substrates and thus express a generalist lifestyle (Christie-Oleza et al. 2012).

2.3.2 *Marine and Freshwater Particles (Par)*

2.3.2.1 Physiochemical Characterization and Major Dynamics

The Par microhabitats are extremely heterogeneous environments as particle composition can be significantly different depending on their origin, chemical composition, and age (Fig. 2.1b). Marine habitats are mainly characterized by the occurrence of autochthonous POM (produced by organisms inhabiting that ecosystem), while freshwater systems can experience a significant input of allochthonous terrestrial POM depending on the degree of the aquatic-terrestrial coupling. Furthermore, autochthonous OM in aquatic systems differs in bioavailability and turnover depending on its mode of production: (1) direct (excretion and exudates) or (2) indirect (body fragment, carcasses).

Importantly, Par microhabitats are quickly and intensively colonized by bacteria whose fraction can represent 25% of the total bacterial abundance in the epipelagic zone, showing a tight correlation with algal abundance and a fast decrease with depth (PA bacteria have a relative abundance <4% in mesopelagic layers; Mével et al. 2008). Immediately after the attachment, bacteria are able to activate a transcriptional response and increase their enzymatic activities (Grossart et al. 2007). Although they do not represent the majority of the bacterial community in most aquatic environments, PA bacteria usually account for >50% of the total bacterial activity (Ghiglione et al. 2007; Grossart et al. 2007; Rösel and Grossart 2012; Stocker 2012), and extremely high values were recorded in areas characterized by elevated trophic status (e.g., up to 80% of the total bacterial activity in a mesotrophic area of the NE Mediterranean Sea; Ghiglione et al. 2007). The reason for these high fractions of PA bacteria lies in the tight coupling with the available substrate in the Par microhabitats and the likely metabolic cooperation among PA bacteria (Wahl et al. 2012).

Marine particles are mainly represented by autochthonous POM, and its composition is, therefore, strictly linked to the local organisms producing it. Three major categories of autochthonous POM can be distinguished: (1) exudates, (2) excretions, and (3) detritus and carcasses. Exudates (1) are mainly composed of polymers such as carbohydrates, proteins, and glycoproteins which are continuously released by algae (e.g., diatoms, cyanobacteria) and that can reach remarkable concentrations in the surface waters during algal bloom events. The chemical composition of these substrates is related to both algal physiology and metabolism, which can vary for the same species living in different environments (Becker et al. 2014). The composition of the exudates eventually shapes the bacterial community by favoring that guild of bacteria which can most efficiently exploit and degrade the aggregates; a clear example is the tight coupling between the algae and bacteria successions during bloom events (Bunse et al. 2016; Needham and Fuhrman 2016). In turn, some bacteria are capable of interacting with the algae and deeply influence the bloom dynamics controlling both bloom development and aggregation processes (Grossart et al. 2006; Amin et al. 2015). Organic matter excretions (2) represent almost one third of the ingested food of marine organisms such as zooplankton (e.g., copepods and salps) and fish. Particles such as fecal pellets are rapidly colonized by PA bacteria but also present relevant abundances of enteric bacteria which were released from the digestive tracts of the host (Bio microhabitats, see below). These Par microhabitats are characterized by several factors such as a different physical structure and origin (e.g., copepod excretions are enveloped by a peritrophic membrane which delays their remineralization process) that affect their sinking rates (i.e., small-size excretions are mostly degraded by bacteria in the water column, while larger and heavier excretions are involved in sedimentation processes). This largely influences OM colonization and degradation (Turner 2002; Tang et al. 2010). Detritus and carcasses (3), e.g., of phyto- and zooplanktonic organisms share the same fate of the excretions (e.g., Tang et al. 2014) and due to their high density are involved in the biological carbon pump (Jiao et al. 2010) as well as in the dispersion process of attached bacteria. Particulate-associated bacteria can cover longer distances and spread across physical barriers otherwise impenetrable for FL bacteria (e.g., thermocline and water masses fronts) (Grossart et al. 2010).

In addition to a pronounced autochthonous fraction in limnetic habitats, freshwater particles generally contain significant portions of terrestrial POM, which is often recalcitrant, being hard to degrade due to the incorporation of minerals (e.g., Ortega-Retuerta et al. 2013), lignocellulose compounds, and humic matter. Lignocellulose degradation on land is carried out mostly by fungi; however, their function is not as well known in the aquatic environments where they appear to be abundant but less diverse and of similar importance for OM cycling as are heterotrophic bacteria (Crowther and Grossart 2015; Grossart et al. 2019). Humic matter represents a class of molecules with a high degree of chemical complexity; hence their degradation understandably might be slow, energetically demanding, requiring the presence of specific metabolic pathways and likely favoring the co-occurrence of specific bacterial species (Habe and Omori 2003; Grossart 2010). A remarkable boost in the degradation of these refractory substrates in the presence of more labile OM can be related to co-metabolism processes. This co-metabolism is known as a "priming effect" (Horvath 1972) and is represented by mainly three hypotheses: (1) extracellular enzymes produced to degrade the labile fraction can also cleave a part of the refractory OM; (2) by-products formed by the degradation of labile OM support the energetic costs of refractory OM degradation conducted by more specific bacteria; and (3) bacteria degrading labile OM acquire enough energy to produce specialized enzymes and thus are able to cleave the refractory OM (Guenet et al. 2010). Although the theory of priming is still largely debated, it can explain the consumption of recalcitrant OM discharged by a river in the presence of exudates produced by a phytoplankton bloom and possibly also the remineralization of refractory OM originating from the deep sea in upwelling zones (Bianchi 2011). Throughout time, all Par microhabitats evolve and change their tridimensional structures, chemical composition, and internal gradients, along with the microbial degradation processes that play a major role in particle turnover. The new architecture in turn can trigger a succession of PA bacteria communities on particles with a sequential selection for species with the most suitable metabolic capabilities that further degrade and evolve in a continuous, mutualistic, and dynamic process. Lecleir et al. (2014) found a rapid change—within 24 h from sample collection—in the PA microbial community retrieved through sediment traps, and Fontanez et al. (2015) with a similar approach report on a shift from DOM- to POM-associated bacterial groups (*Alteromonadales* and *Flavobacteriales*) to pathogens and saprophyte groups (*Vibrionales* and *Campylobacterales*). In addition, fungi may have a relevant impact on OM turnover, in particular on larger particles (Grossart et al. 2019).

Initial microhabitat colonization processes on particulates might be random and dependent on variables that include seasonal and abiotic factors which control the composition of the pool of potential PA bacteria waiting in DifP and ColP microhabitats and also be affected by the population inhabiting other Par microhabitats (serving as seeding hotspots). Once settled onto a particulate, PA bacteria can have energetic advantages that allow them to prevent the settlement of competitors through activities such as antibiotic production (Long and Azam 2001; Wahl et al. 2012). These dynamics might explain why clear patterns in the composition of the PA bacteria community are still missing, although we suggest that this variability more likely can be due to the high heterogeneity of particles as highlighted by Ortega-Retuerta et al. (2013) who demonstrated that the bacterial community composition is more correlated with the particle quality than quantity.

Their remarkable associated bacterial activity and the relatively high substrate bioavailability make Par microhabitats pivotal hotspots for processes affecting biogeochemical cycles, in particular OM export and remineralization of nutrients (Simon et al. 2002). The need for a better and more comprehensive knowledge of particle diversity and bacterial enzymatic capacity is evident, as emphasized by demands for new, more focused investigations. The application of novel methodologies such as "omics", NMR, and Fourier transform ion cyclotron resonance (FTICR) mass spectrometry coupled with an improved sampling design which takes into account the heterogeneity of Par microhabitats will refine the description of global biogeochemical cycles.

2.3.2.2 Bacterial Adaptations

In the past years, the number of studies addressing differences between PA and FL bacteria communities has enormously increased, and the high-throughput metabarcoding approach is providing a more comprehensive and detailed description of their in-depth composition. Two important obstacles for generalizing the obtained results are cases of organisms being able to efficiently commute between FL and PA life strategies (Lauro et al. 2009) and the unrepresentative size fractioning for FL and PA bacteria neglecting the true size structure (see Sect. 2.2). However, based on the most recent literature, we suggest the existence of a large number of guilds of copiotrophic organisms whose genomes have evolved and been selected for a PA lifestyle. Even though these species are considered generalists, each guild has developed specific adaptations providing a higher fitness which, therefore, leads to their dominance on different kinds of Par microhabitats as has been described above. In order to test for the existence of these specific bacterial guilds, we reviewed the results of several studies evaluating the PA bacteria community composition across different Par microhabitats, and here, we try to highlight the underlying phylogenetic and functional patterns.

Rieck et al. (2015) analyzed the PA bacteria diversity in the Baltic Sea which represents one of the largest brackish basins in the world due to a limited water exchange with the open ocean and high freshwater loads from the Scandinavian rivers entering the system in the north. The authors highlighted the dominance of *Planctomycetes* and *Betaproteobacteria* on freshwater particles (oligohaline) and a dominance of flavobacteria on marine particles. Among flavobacteria, *Polaribacter*, *Fluviicola*, and *Formosa* genera are known to be associated with particles rich in OM as they have several enzymes for polysaccharide cleavage (Walsh et al. 2013; Bižić-Ionescu et al. 2014), and *Formosa* and *Planctomycetes* are specifically involved in the remineralization and degradation processes of algal biomass (Pizzetti et al. 2011; Teeling et al. 2012). These findings agree with the general seasonal patterns of Par microhabitats. Whereas during summer autochthonous particles are produced by phytoplanktonic blooms, allochthonous particles are introduced in the oligohaline area by high riverine runoff during the rainy seasons in fall/winter. The dominance of PA flavobacteria in marine areas is usually shared with other bacterial groups such as *Verrucomicrobia* (Ortega-Retuerta et al. 2013) and especially *Gammaproteobacteria* (*Alteromonadales*, *Oceanospirillales*, and *Vibrionales*) (Bižić-Ionescu et al. 2014; Fontanez et al. 2015). Limnetic systems have a different PA community composition revealing a dominance of *Alphaproteobacteria* and *Betaproteobacteria* over *Gammaproteobacteria*, *Flavobacterium*, and *Planctomycetes*. Moreover, limnetic

PA bacterial communities show a higher degree of specialization than do marine PA communities (Bižić-Ionescu et al. 2014).

In the marine environment, an increasing water depth is usually associated with strong gradients in nutrient availability, oxygen concentration, and temperature greatly affecting bacterial abundance and community composition. However, the diversity of PA bacteria seems to be highly correlated with particle composition rather than with abiotic variables of the surrounding water. Studies on sinking particles are still rare, but there is increasing evidence for their enrichment in *Epsilonproteobacteria* carrying out sulfur oxidation (*Sulfurimonas*, *Sulfurovum*, and *Sulfuricurvum*) as well as sulfur and nitrogen reduction processes (*Sulfurospirillum*) in the absence of molecular oxygen (Fontanez et al. 2015). Some *Epsilonproteobacteria* are associated with metazoan surfaces and their digestive tracts (*Arcobacter*, *Wolinella*, and *Campylobacter*) (Gugliandolo et al. 2008) explaining their presence with the death and sinking of the host. Of further significance is the fact that particles represent micro-niches with oxygen concentrations that often are lower than in the surrounding water. Gradients of decreasing oxygen concentrations were shown to be correlated with the high microbial activity taking place on Par microhabitats that is even sufficient to maintain anoxic condition within large particles (e.g., fecal pellets or copepod carcasses; Alldredge and Cohen 1987; Ploug et al. 2008). The resulting micro-niches can thus host ephemeral loci for anoxygenic processes such as nitrogen fixation, even in oxic zones of the water column (Riemann et al. 2010; Glud et al. 2015). Indeed, Ganesh et al. (2014) reported for Par microhabitats collected at the oxygen minimum zone an enrichment in genes involved in the last steps of the denitrification process.

Based on these findings, we speculate on the existence of phylogenetic base guilds of generalistic copiotrophic PA bacteria well-adapted to degrade and exploit particles characterized by different chemical composition and architectures. The taxonomic resolution provided for these guilds, however, is still low (phyla to order); thus, it is difficult to describe specific functional and metabolic patterns (some examples are given by Ganesh et al. 2014; Fontanez et al. 2015). Here, we identified two major problematic issues: (1) methodological biases in the size fractioning procedures which prevent a clear characterization of FL and PA bacterial communities (particle disruption, wrong mesh size, etc.) and (2) high heterogeneity of Par microhabitats which generate even greater differences between micro-niches hosting a huge diversity of associated bacteria (e.g., Amin et al. 2012).

Despite these limitations, it is possible to identify some essential traits of PA bacteria which indicate their capability to exploit the complex mixtures of polymers forming the Par microhabitats. What is considered essential for PA bacteria are the motility and chemotactic behaviors allowing the organisms not only to find the Par microhabitat but also to dynamically attach onto and detach from surfaces in accordance with their physiochemical properties (Grossart et al. 2001; Fenchel 2002; Raina et al. 2019). Once settled onto the particles, many bacteria show a high production of antibiotic substances that act as deterrent against possible competitors (Long and Azam 2001; Grossart et al. 2004) although they still experience the top-down control by many specialized grazers (Kiørboe et al. 2004; Corno et al.

2013) and of viral infection (Suttle 2007). The genomes of PA-associated bacteria are generally larger than the genomes of FL bacteria (Giovannoni et al. 2014), presumably representing the fact that PA-associated bacteria often have a higher number of genes related to their greater capability for polymeric substrate degradation. Furthermore, PA bacteria have a wide repertoire of genes coding for extracellular and membrane-attached enzymes such as those for the digestion of phytoplankton exopolysaccharides (Smith et al. 2013) and other carbohydrates (α-mannosidase, α-L-fucosidase, and L-fucose permease; Teeling et al. 2012; Kappelmann et al. 2019). The PA bacteria can also express a wide range of transporters (e.g., TonB-dependent transporters) for the uptake of amino acids, nucleotides, and coenzymes cleaved from the particles (Smith et al. 2013). A broad set of PA bacteria-associated genes mediates bacterial anchoring to the substrate and allows the microbes to colonize different types of particles (as well as organisms and hosts) (Ganesh et al. 2014). Among the large fractions of unclassified genes recognized for PA bacteria are several putative sequences for the mobilization of DNA expression for signaling-related pathways (Allen et al. 2013). Although the synthesis of proteorhodopsins and actinorhodopsins represents a typical trait of FL bacteria, the AAP are also abundant among PA bacteria (e.g., Lami et al. 2009; Simon et al. 2014). This trait has deep implications for the bacterial metabolism, particle turnover ratio, and eventually carbon flux as the APP organisms are capable of doubling their efficiency of organic carbon assimilation in comparison to strict heterotrophic bacteria (Yurkov and Cssotonyi 2009). Finally, the high density of bacterial cells which colonize Par microhabitats (as well as Bio microhabitats, see below) allows bacteria to stay in close contact, a condition that is quite rare in DifP and ColP microhabitats. The proximity of bacterial cells favors processes like conjugation and lateral gene exchange (Paul 1999) rendering particles as possible seeding spots for genetic exchange which further increases the variability and functionality of PA bacteria (Ganesh et al. 2014).

Overall, Par microhabitats have been often overlooked in the past, and so far, the effect of PA bacteria on microbial diversity and functionality in aquatic ecosystems has been greatly neglected. Nowadays, the increasing utilization of flow cytometry to estimate bacterial abundances might represent a possible new source of bias; while this technique provides unquestionable advantages in terms of resolution and analysis time, it completely ignores PA bacteria that have to be counted "manually" which many scientists avoid doing. Generally, the PA fraction of the microbial community (including also PA fungi and protists) exhibits higher metabolic activities than do microbes inhabiting the DifP and ColP microhabitats. Therefore, PA microbes have a significant impact on the balance of several biogeochemical cycles (e.g., carbon and nitrogen). As briefly reported above, some studies have started to shed light on the dynamics of biological processes associated with Par microhabitats although it will be crucial to further focus our research efforts on the functioning and efficiency of these processes in order to provide more accurate and reliable estimations on the role of aquatic systems in global C, N, and P fluxes.

2.3.3 *Living Biosphere (Bio)*

2.3.3.1 Physiochemical Characterization and Major Dynamics

Bio microhabitats represent the most diverse and dynamic range of micro-niches (Fig. 2.1c). Their high complexity is determined by the variable nature of the "substrate" exploited by the bacteria, i.e., the host organism. These consist mainly of eukaryotes ranging from protists to whales (although relevant associations have been reported among bacteria, cyanobacteria, and archaea, e.g., Huber et al. 2002; Aharonovich and Sher 2016). The host's size scales up the size and diversity of the associated bacterial community and introduces a different time scale for changes in the microhabitat conditions (especially for the bigger and long-lived eukaryotes). Each host species has its own physiology, behavior, and life stage development that eventually affect bacterial colonization via changes in surface properties (external or internal) for such variables as nutrient concentration and chemical gradients (e.g., Grossart et al. 2010; Scavotto et al. 2015).

The associated bacteria can be found at the host's surface (epibionts) and within some of its internal organs (endobionts). The epibiont-endobiont distinction is pivotal since it has severe implications in terms of bacterial traits and ecology and allows us to compare bacterial characteristics associated with the microhabitats described in this chapter. Epibionts are accounted as PA bacteria because they mainly exploit OM released by the host, while endobionts share several characteristics more typically associated with FL bacteria since they live in a more homeostatic environment leading to relevant genome streamlining.

Bacteria inhabiting Bio microhabitats undergo the same dynamics of random colonization reported for PA bacteria (see Sect. 2.3.2) although for some tight and specific associations, a recurrent pattern of the associated bacterial community has been shown (e.g., Sharp et al. 2007). Thereby, the recruitment process mainly accounts for epibiotic communities which change and evolve in composition over time with continuous addition (and attempts of addition) from the surrounding environment. In contrast, endobiotic communities tend to be more stable (Wahl et al. 2012). The distinction between epibiotic and endobiotic characteristics can be related to the specific conditions of micro-niches that favor bacterial species which have evolved to maximize their fitness at the given conditions.

Host and bacteria can directly interact with each other by releasing attracting chemicals and inhibitory substances among the latter being antibiotics and toxic compounds (e.g., Tang et al. 2010; Amin et al. 2012). Host-bacteria interactions can span over a wide range of symbiosis types, from positive such as mutualism and commensalism to negative such as parasitism and saprophytism. These types of interactions differ in their specificity and evolutionary history (e.g., mutualism presumably requires a long coevolution and is more species-specific as compared to commensalism; Kirchman 2012).

In a highly diluted oligotrophic environment, the host represents an exceptionally dense and diverse hotspot of OM and nutrients supporting increased bacterial growth rates and specific microbial processes absent in the surrounding water. Consequently, OM-associated bacterial communities can reach extremely high volume-specific densities, especially on zooplankton (up to 10^{11} cells ml^{-1}; Tang et al. 2010). Such high bacterial densities often lead to intra- and interspecies competitions with severe functional consequences. The associated effects of bacterial metabolism may have important implications for OM nutrient cycling and hence influence trophic transfer efficiency as well as the well-being of the host. Understanding the capability of bacteria to sense their population density through so-called quorum sensing mechanisms (QS) is crucial to understanding the dynamics of host-bacteria interactions and their ecological consequences (Miller and Bassler 2001). This type of cell-to-cell communication relies primarily on the detection of small molecules, so-called autoinducers (AI) (Dobretsov et al. 2009), excreted into the environment by each bacterial cell, reaching a certain threshold concentration when a critical bacterial cell concentration has been reached. At this threshold of saturation, a signaling cascade is promoted by the bacterial cell detector mechanisms that eventually evoke a genetically coded response (Case et al. 2008). Examples of bacterial strategies adopted for competition with each other are (1) the responses to the reduction of available anchoring surface (Reid et al. 2001), (2) the depletion of a specific substrate or nutrient essential for the metabolism of a competitor, (3) the alteration of the chemical environment (pH, oxygen), and (4) the production of antagonistic substances (e.g., antibiotics; Wannamaker 1980). Bacterial antagonistic compounds thereby produced can also affect the host (Steinberg et al. 2011).

Phytoplankton provide several different OM compounds via exudation, thus supporting the energetic requirements of the associated bacteria, and this represents one of the most important host contributions to the complex biotic interactions (Seymour et al. 2017). One of the most abundant phytoplankton exudates consists of transparent exopolymeric polysaccharides (TEP), which have been shown to be rapidly colonized and degraded by bacteria (Grossart et al. 2006). This class of OM compounds is released especially during exponential algal growth (e.g., blooms), and these compounds have been suggested as a way in which phytoplankton attract and promote bacterial settlement (Amin et al. 2012); *Thalassiosira pseudonana*, for example, releases a sulfonate compound whose catabolic genes are poorly distributed in the ocean, but it can be utilized by a co-occurring *Roseobacter* bacterium (Durham et al. 2015). Bacterial colonization of phytoplankton grants permanent substrate supply in the phycosphere, the boundary layer surrounding the phytoplankton cell. On the other hand, bacteria can also mediate this process by releasing proteins or polysaccharides which create a tridimensional anchoring structure (Rinta-Kanto et al. 2012). Recently, host-bacteria interactions within the phycosphere have been shown to be finely tuned by complex mechanisms of communication. Recognition from the host of the approaching bacteria and vice versa is essential for these interactions (Teplitski et al. 2011). Such inter-kingdom communication is made possible by the evolution of small signaling molecules (e.g., the AI) and of specific binding receptors on the association active surfaces. Each host and bacteria have developed different signaling compounds during their evolution; however, they all share some chemical features, and in order to limit the dispersion, these molecules are typically hydrophobic and lipid-based. Moreover, allochthonous signaling receptors have been discovered on different organisms involved in inter-kingdom interaction, supporting the idea of long coevolution processes between the host and specific bacterial strains (see Amin et al. 2012).

Similar to phytoplankton, zooplankton hosts can provide nutrient-rich microhabitats to the bacteria but have greatly been neglected by aquatic microbial ecologists. As predators, they can accumulate high OM concentrations which can be well exploited by both endobiotic and epibiotic bacterial communities. The latter have a convenient proximity to the source of nutrients originating from either sloppy feeding or of excreted matter (Møller et al. 2003), and, for instance, nitrogen-rich excretions have been shown to increase the growth of epibiotic bacteria (Carman and Dobbs 1997). Moreover, the exoskeleton itself can be considered a favorable substrate since the inability of many microbes to effectively store carbohydrates leads to low C:N and C:P ratios (Beers 1966). A substantial amount of bacteria are more or less passively ingested by zooplankton, and some of these bacteria can well survive the ingestion-digestion process (see Tang et al. 2010) effectively rendering those ingested microbes as endobiotic commensals. This specific microenvironment is characterized by several remarkable constrains that determine the composition of the endobiotic bacterial community such as rapid changes in pH, oxygen, and OM concentrations, as well as a short gut passage time. These properties require essential bacterial adaptations for successful bacterial settlement within the host enabling a rapid effective response to external stress and require an efficient anchoring strategy. Their particular lifestyle is reflected in their extraordinary resistance to standard disinfection procedures which may even result in a serious threat for public health (Bichai et al. 2008). Eventually, some of the endobiotic bacteria are excreted via fecal pellets and consequently have to survive a shift from Bio to Par microhabitats. This abrupt change of microhabitat conditions may negatively select for specialist bacterial with streamlined genomes, which do not alter the fitness of transient PA bacteria.

An extra degree of complexity of Bio microhabitats is represented by the physiology of the organismic hosts, both phyto- and zooplankton. Responses to fluctuations in environmental factors or other stressors (e.g., pollutants) as well as different life stages and physiological conditions of the host can imply remarkable changes in microhabitat conditions (see Wahl et al. 2012). These changes favor different bacterial species and induce shifts in the composition of the associated bacterial community; some species might also change their behavior and thus render different kinds of trophic interactions (e.g., from commensalism to parasitism). As an example, in response to warming stress (temperature and UV), the red macroalga *Delisea pulchra* loses its capability to quench the QS mechanism of associated bacteria via the production of furanones; therefore the bacteria evolve a pathogenic behavior inducing tissue damage and bleaching (Steinberg et al. 2011). Dispersal and differentiation of the associated bacteria can be induced by the presence of nitric oxide (NO), a typical biologic signal released by diatoms (Steinberg et al. 2011). While the presence of NO in the phycosphere can act as repellent for most bacteria, it will create micro-niches for those bacteria which are capable of detoxifying it. During stress conditions the release of NO increases induction of the detachment of some associated bacteria while attracting scavenger bacteria capable of sensing NO concentrations (DeGroote and Fang 1995).

In general, Bio microhabitats have been largely overlooked by microbial ecologists. "Environmental" properties of the micro-niches can be well defined for most of the hosts based on information about physiology, life cycles, and effects of external stressors, although we still have relatively little information for some host groups, e.g., protists. Ecological effects induced by changes associated with these host properties on bacterial community are still largely neglected, and thus future investigations should tackle possible host-mediated consequences for bacterial species composition and functionality. We already know that the host can affect bacterial dispersion by acting as a dispersal courier in a similar way as do the sinking particles in the Par microhabitats. However, organismic hosts provide an increased dissemination possibility for the associated bacteria by not only sedimenting downward but also by active movement in the water column (e.g., Grossart et al. 2010).

2.3.3.2 Bacterial Adaptations

Similar to PA bacteria, bacteria associated to living organisms are usually sorted out during sampling (by size fractionation) or are underrepresented in the sample (as the volume of this specific micro-niche is usually only a few mL; Grossart 2010). Most of the available data on the host-associated bacteria stems from investigations that targeted sponges (e.g., Sharp et al. 2007) although some information is also available for zooplankton and algal hosts (Møller et al. 2007; Tang et al. 2010; Egan et al. 2013; Behringer et al. 2018). Bacteria associated with these organisms are involved in a long-lasting coevolution process with their host which can be still seen in the host genomes (see Amin et al. 2012). Our main knowledge gap, however, is that we still do not know or poorly understand the metabolic functioning of the host-associated bacteria (Wahl et al. 2012; Durham et al. 2015; Amin et al. 2015).

Generally, epibiotic bacteria are characterized by mechanisms which are QS dependent, and the most relevant trait is the production of antibiotic compounds. Bacteria associated to the host surface represent a main producer of antibiotics in seawater (Wahl et al. 2012), but they also produce secondary metabolites, e.g., acids from sugar fermentation to stave off potential competing bacteria by inducing negative chemotaxis (Madigan and Martinko 2006) or inducing permanent binding of the competitor cells to the substrates (Boyd et al. 1999). The capability of producing antibiotic compounds is not limited to specific phylotypes or bioregions but rather represents a widely spread trait, and many bacteria can modulate the production of various compounds when necessary (Bode et al. 2002). Generalist-epibiotic bacteria such as *Pseudoalteromonas* and *Streptomyces* tend to produce active molecules affecting a broad range of species as they are mostly associated with crowded microhabitats, while specialized-endobiotic bacteria might produce less antibiotics and differently target specific competitor species (Hibbing et al. 2010). An alternative hypothesis relates the production of antibiotics to their functioning as signaling molecules (Sengupta et al. 2013) rather than as being harmful compounds since their production cost is usually elevated and the targeted species can quickly evolve to overcome this inhibition (Hibbing et al. 2010).

Epibiotic bacteria can actively find and colonize their hosts thanks to chemotactic behavior which allows them to follow chemical gradients and also possible hydrodynamic cues (Carman and Dobbs 1997; Raina et al. 2019). The major phyla associated with phytoplanktonic hosts are *Bacteroidetes* and *Proteobacteria* with some genera having been particularly found associated with diatoms such as *Sulfitobacter*, *Roseobacter*, *Alteromonas*, and *Flavobacterium* (see Amin et al. 2012). Interestingly, while consistent patterns in microalgal-associated bacterial communities are found at high taxonomic levels (e.g., phylum), no similarities are observed at lower ranks (i.e., genera or species) (Egan et al. 2013); this can be related to methodological limitation but might also be correlated with specific and selective host-bacteria interactions. Some studies provide evidences that the degrees of variability of bacterial communities associated with the same algal species are also reflected in community functionality rather than composition (Burke et al. 2011). Aharonovich and Sher (2016) have shown that associations of *Alteromonas macleodii* with different strains of the cyanobacteria *Prochlorococcus* can lead to completely different interactions. For example, the presence of the heterotrophic bacteria reduces the cyanobacterial expression of several core genes in the strain MED4, e.g., those involved in the biosynthesis of amino acids, purines, pyrimidines, vitamin B12, fatty acids, and phospholipids as well as of genes responsible for DNA replication and cell division. MED4, however, showed an increasing growth rate in the presence of *Alteromonas* denoting a synergic relationship with the heterotrophic bacterium providing available basic substrates to the associate cyanobacterium. On the contrary, the *Prochlorococcus* strain MIT9313 showed a competitive relationship with *Alteromonas* which included an increasing expression of the gene for uptake of small compounds and also the production of antimicrobial or signaling compounds by *Prochlorococcus* (Aharonovich and Sher 2016).

In some cases, FL bacteria also can be associated with dinoflagellate and diatom blooms, e.g., *Roseobacter* (*Alphaproteobacteria*), which can take up and utilize some of the algal osmolytes, especially dimethylsulfoniopropionate (DMSP) (see Egan et al. 2013).

Several studies have tried to characterize the community composition of bacteria associated with zooplankton although it is not a trivial task to separate epibionts and endobionts. According to the theory proposed by Harris (1993), endobionts are more affected by the feeding activity, and thus the resident bacteria can be distinguished from transient ones by analyzing the stability of the community composition. This can be obtained by feeding the zooplankton with an axenic culture and subsequently tracking the decreasing richness of the associated bacterial community composition until it is stable. However, feeding experiments with xenic prey cultures may indicate the presence of several transient bacterial groups which are capable of anchoring within the organism's stomach exploiting the rich OM pool, and these bacteria mainly belong to the groups of *Pseudoalteromonas*, *Sulfitobacter*, and *Roseobacter* (Tang et al. 2009). The bacterial community composition of crustacean zooplankton in lakes has been observed mainly to be composed of 36 different phylogenetic groups belonging to 6 major bacterial taxa: *Actinobacteria*, *Firmicutes*, *Bacteroidetes*, *Alphaproteobacteria*, *Betaproteobacteria*, and *Gammaproteobacteria* (Grossart et al. 2009). The community composition of the sole endobionts appears less rich and is represented by *Bacteroidetes*, *Alphaproteobacteria*, *Betaproteobacteria*, and *Gammaproteobacteria* (mainly *Pseudoalteromonas*) (Peter and Sommaruga 2008). As for phytoplankton, a generalization of the bacterial community composition on zooplankton is only possible at high taxonomic levels since (as pointed out above) the environmental conditions of the colonizing surface can drastically change between host species.

Zooplankton endobionts can rely on high numbers of nutrient particles which allow them to thrive on OM substrates which are otherwise extremely diluted in seawater, e.g., DMSP-consuming bacteria (Diaz et al. 1992). Environmental DMSP concentrations can be particularly high during the senescence of phytoplanktonic blooms, but release of DMSP by active phytoplankton is almost negligible. Compounds such as DMSP can reach μM or mM concentration in the zooplankton guts, being released from the digestion of the phytoplankton prey (see Tang et al. 2010). The *Betaproteobacteria Limnohabitans* exploiting the nutrient-rich conditions within the gut of *Daphnia magna* was found responsible for 8% of the leucine and 9% of the *N*-acetyl-D-glucosamine uptake of the total bacterioplankton community demonstrating the high metabolic activity of endobiotic bacteria. Although these rates might be not astonishing, they have severe implications for ecosystem dynamics since by ingesting a *Daphnia*, a predator will transfer also the carbon fixed by *Limnohabitans* (Eckert and Pernthaler 2014). This "shortcut" in the food web increases the efficiency of the carbon flux and eventually the energy budget of the ecosystem. Due to the particular microenvironmental conditions within the guts, specific processes can take place, which otherwise cannot occur in the surrounding seawater. For example, the anoxic environment of full copepod guts allows for nitrogen fixation by the cyanobacteria UCYN-A, while gammaproteobacterial nifH sequences (gene candidate to identify nitrogen fixation in environmental samples) were highly represented in starved copepods. The *Gammaproteobacteria* might be represented by *Vibrio* spp. which often is associated with the exoskeleton and gut lining of living copepods (Rawlings et al. 2007) from where it can retrieve the energy necessary to support the energetically demanding nitrogenase activity (Scavotto et al. 2015). Another example is the occurrence of denitrification in zooplankton carcasses in an otherwise fully oxic water column with profound implications for aquatic nitrogen cycling (Glud et al. 2015).

In summary, bacterial endobiont organisms are capable of penetrating the host tissue and may survive the host's digestion tract thus often establishing a long-lasting relationship with their host. This process has been described several times for different hosts (Kikuchi 2009; Archibald 2015) demonstrating its apparent evolutionary advantage (although there are similar cases of associations with pathogens and saprophytes). A general characteristic shared by all endobionts (resident species) is their extreme genome streamlining (Fig. 2.2); this process is even more emphasized than in FL bacteria (genome of 5–10 Mb) and leads to the typically small genomes of pathogenic species (2–5 Mb) and often even smaller genome sizes for symbionts (0.5–1.5 Mb). Symbionts have a stable genome organization without lateral gene transfer (Ochman and Davalos 2006), and their minimal genomic size is a direct effect of the extreme stability of the environment where several core functions and metabolites are provided by the host. In exchange, symbionts may provide their hosts either the capability to thrive on formerly unusable energy sources (diet shift) or to cope with otherwise nutritionally deficient diets. Terrestrial bacteria-host associations have shown loss of essential genes (information processing, energy metabolism, cell envelope synthesis) which possibly provides evidence for phylogenetic specific coevolution processes (Bennett et al. 2014). Driven by the technological improvements, there is an increasing awareness of the critical importance of the microbiome and its implication for the host metabolome (Fischer et al. 2017). The unexpected diversity of bacterial species present in several Bio microhabitats such as the human body (Eloe-Fadrosh and Rasko 2013) raises questions on their possible interaction with the host. These studies reveal that bacteria can significantly influence host processes, even being entirely responsible for some host processes, and can have deep implications on the dynamics of interactions among environment, host, and bacteria (Leonard et al. 2015).

It is worth remembering that bacteria inhabiting Bio microhabitats can be also FL bacteria since they need to maintain their population in the absence of a host. Epibionts can be involved in the exploitation of carcasses or fecal pellets of their host and thus extend the life-span of their microhabitat beyond the lifetime of the host; these bacteria can also actively attach and detach from the host thus being able to survive also as FL bacteria (Møller et al. 2007). The endobionts, however, experience drastic changes in their micro-niches upon the host's death, and some hosts can also release the associated bacteria into the surrounding seawater, which is the case of bacterial associations between the squid *Euprymna scolopes* and the bacterium *Vibrio fischeri* (Nyholm 2016). Therefore, these bacteria need to adopt ad hoc strategies in order to rapidly colonize another host. Genome streamlining of the endobionts is a relevant obstacle to their survival outside of the host, and the vertical transmission to the host's progeny can be a valid strategy.

2.4 Human-Mediated Implications

Following a size-based gradient, we would like to highlight some major concerns related to human impacts on the structure and complexity of aquatic microhabitats as earlier described in this chapter. We tried to provide some clues for the present and future transformation dictated from human activities onto marine microhabitats. The DifP and ColP microhabitats are mostly affected by the continuous discharge of DOM, heavy metals, and toxic compounds (antibiotics, drugs, new chemicals) (Fig. 2.3a); these products enter aquatic ecosystems mainly through drainage into rivers, lakes, and coastal areas which represent the most affected biomes. While

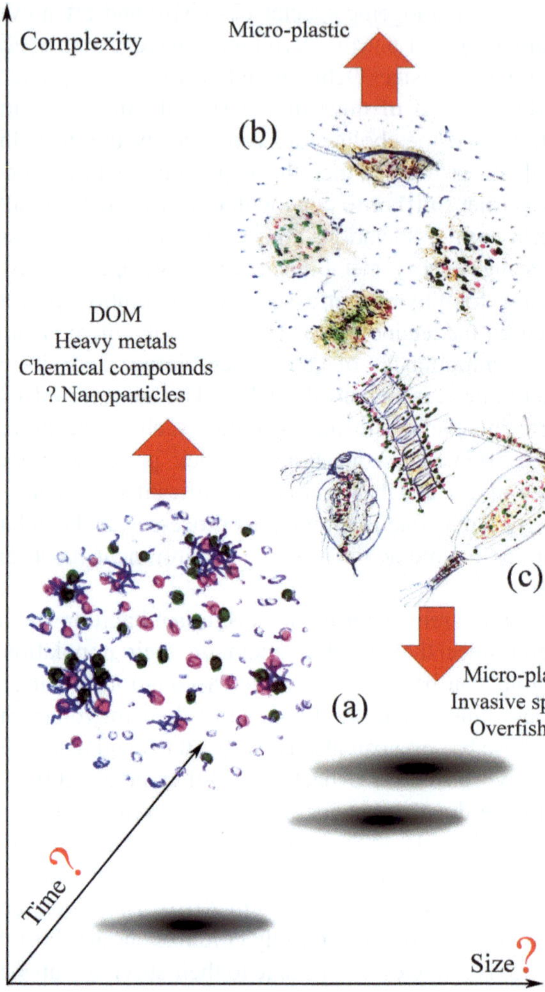

Fig. 2.3 Effects of the major anthropogenic stressors on the complexity of the aquatic microhabitats: the diffusion-controlled water phase (a), colloidal phase (b), freshwater and marine particles (c), and living biosphere (d). Question marks are added to size and time axes because the future effects are still unpredictable

these pollutants have often lethal or weakening effects on large organisms, they might represent an exploitable resource for bacteria and archaea. In particular, DOM, heavy metals, and chemical compounds can increase the chemical and physical complexity of DifP and ColP microhabitats, thus affecting the spatial heterogeneity and selecting for different bacterial species and guilds (e.g., increasing sulfur-reducing *Verrucomicrobia* species in the presence of Hg and MeHg (methylmercury); Vishnivetskaya et al. 2011). During the last decade, we also started to feed these microhabitats with a novel category of compounds, the nanoparticles. One of the most used nanomaterials consists of silver nanoparticles which have been shown to release Ag ions as consequence of the interactions with humic acids. Silver nanoparticles and dissolved Ag ions have inhibitory effects on bacterial growth (a potential oxidative stressor) and represent a critical toxic risk when they are discharged into natural environments (Fabrega et al. 2009). Although the full environmental consequences of several types of nanoparticles remain unclear, these compounds are capable of affecting bacterial community functions especially in DifP and ColP microhabitats where the bacteria can hardly escape from dissolved toxic compounds.

The major anthropic impacts on Par microhabitats are represented by the tremendous number of micro-plastic particles (Fig. 2.3b) which are released into the environment. The accumulation of this marine litter in the ocean is dramatically increasing in terms of diversity, and the abundances of these microhabitats provide a significant quantity of new colonizable surfaces. The photodegradation process converts plastic wastes trapped inside the principal oceanic gyres into micro- and nanoscale pieces which might represent formidable surfaces for microbial biofilm development. Microbial associations with micro-plastic particles are characterized by extremely high cell densities and high rates of metabolic activities (see Sect. 2.3.2); therefore, these particles have the potential to boost microbially mediated processes like OM remineralization. The micro-plastic itself can represent a source of energy for some bacteria which are capable of degrading it (Skariyachan et al. 2016; Urbanek et al. 2018), and the elevated abundance of micro-plastic might strongly shape the bacterial community composition selecting for specific guilds (Zettler et al. 2013; Arias-Andres et al. 2018) which can eventually overwhelm and potentially even replace the natural microbial community in some areas of the open oceans. However, while for bacteria micro-plastic can represent favorable resources and colonization sites, an increasing number of studies are suggesting negative effects of these particles upon macrofauna. Plastic microparticles will negatively affect many Bio microhabitats (Fig. 2.3c). Zooplankton organisms as well as fish can ingest these compounds and consequently experience decreased growth rates and fitness, energetic depletion over time, behavior alterations, and reduced survival rates (Cole et al. 2015; Lönnstedt and Eklöv 2016). In addition, as a result of micro-plastic ingestion, these particles end up in fecal pellets and can be rapidly exported to deeper water layers (Cole et al. 2016). This process further increases the heterogeneity of Par microhabitats but might also significantly affect carbon fluxes and the energetic budget of carbon-limited bacterial communities—typically for the deep ocean. Furthermore, the microbial niches associated with Par microhabitats are likely the most impacted by eutrophication and soil erosion which are mainly connected with agricultural activities, deforestation, and changes in land use. While the increase in nutrient load affects all microhabitats in coastal and freshwater environments (e.g., Smith 2003), the concomitant input of terrestrial particles abruptly shifts quantity and quality of Par microhabitats. The allochthonous particles have been shown to reshape bacterial communities by providing new exploitable surfaces and substrates (Traving et al. 2017) but also potentially seeding the local communities with allochthonous microorganisms. The problem of alien species is further increased by anthropogenic water pollution, e.g., by ballast water input (Williams et al. 1988), directly adding and selecting for species which in a pristine environment are not common or even not present. Thereby, bacteria and other microorganisms preferentially survive different ballast water treatments on particles (Tang et al.

2011), which can make them invasive and alien species within profound but so far unstudied consequences for microbial diversity and functioning.

Bio microhabitat complexity is further threatened by several stressors. The introduction of invasive species usually implies a reduction of the diversity of local communities since the successful establishment of an alien tends to overrule local species. As a result, a drastic reduction of Bio microhabitat diversity can be observed with tremendous consequential effects on the endobiotic bacteria. The generalist-epibiotic bacteria likely can colonize the new resident species, but most of the specialized-endobiotic bacteria will perish or became ecologically inactive as resisting stage. Shifts in functioning of the host-associated bacteria are largely unpredictable since they are largely dependent on specific characteristics of the new hosts. Moreover, additional large-scale and unpredictable effects that are triggered by overfishing result in the removal of specific groups of organisms with severe implications for food web dynamics, e.g., due to alterations of inter- and intraspecies guild competition, top-down trophic cascades (Hessen and Kaartvedt 2014), and reduction in diversity and abundance of Bio microhabitats. The least but not the last stressor is the human-mediated increase of pathogen dispersal such as viruses, bacteria, fungi, and other protists. The impact of a new pathogen in the environment can be predicted for its designated host; however, similar to what is discussed above for alien species, the overall consequences at an ecosystem or regional level can be hardly conceived (e.g., cascade effects, species-barrier crossing, etc.) (García-Vásquez et al. 2017).

2.5 Scaling Up to the Biomes

At a first glance, if one looks into lake or seawater, one might think that these systems represent exclusively homogeneous habitats in which everything is diluted and no relevant differences exist between them. However, each single aquatic environment is composed of specific microhabitats as described above, and some major differences can already be ascribed by the comparison of freshwater and marine areas. In this last section, we will try to highlight a gradient in microhabitat composition among different aquatic systems and upscale the implications of these microhabitats for ecosystem functioning as a whole.

Small lakes experience the strongest impacts from land inputs of OM due to their high ratio between land/water interface area and the relatively smaller volume of the basin. The trophic state of these biomes is subjected to abrupt changes in their physical and chemical features as their relatively small volume grants only a low buffering capacity against environmental fluctuations (some might be even too small to undergo thermal stratification such as polymictic lakes). The terrestrial-particle load exerts strong effects on the ecosystem by triggering shifts in both the bacterial community composition and the community's associated functionalities (e.g., Bižić-Ionescu et al. 2014). Small lakes thus tend to be mostly composed by Par microhabitats of terrestrial origin whose degradation often leads to anoxic events in the deeper layers due to oxygen consumption associated with OM settlement. Part of the OM recycled at the surface also supports the growth of primary producers increasing the Bio microhabitats, especially during periods of high rainfall (Tranvick and Jansson 2002); a remarkable enrichment in inorganic nutrient concentrations can lead to microbial bloom events that eventually sink and thus further increase the oxygen consumption at the lake bottom.

The coastlines of large lakes are still highly influenced by terrestrial input; however, due to the much higher volume of these basins, their terrestrially related processes contribute less extensively toward the overall microhabitat composition of the ecosystem. Likewise, marine coastal areas also benefit from allochthonous input of Par microhabitats. In the sunlit layers of both biomes, POM remineralization, coupled with nutrient input from the deep-water layers (mixing events in spring and fall seasons, upwelling), supports the growth of phytoplankton populations which in turn enrich the environment with new autochthonous Par microhabitats (exudates, fecal pellets, and later on of carcasses) and Bio microhabitats (phytoplankton itself and zooplankton grazers). These marine and freshwater Par microhabitats, however, are colonized by different bacterial species since they differ in chemical composition and physical properties; for instance, in lakes there exist specific bacterial guilds, e.g., for the degradation of humic compounds, while in marine areas species adapted to diatom and cyanobacteria, OM degradation are favored. The different lability of bacterial substrates implies different microbial turnover rates and, therefore, different OM fluxes. Moreover, marine and freshwater systems select for different phytoplankton and zooplankton communities increasing the dissimilarity of Bio microhabitats between the two environments. Abnormal dynamics occur in special water bodies such as estuaries and lagoons where direct contact between freshwater and seawater takes place. Mixture of both water types leads to the formation of Par microhabitats ("turbidity maximum") with different origins that imply the presence of bacterial communities with exceptionally different composition and functionalities and thus might lead to peculiar dynamics such as the priming effect (Treignier et al. 2006).

The inner part of big lakes, which in terms of volume is the dominant portion of these habitats, generally shares the same dynamics as open sea environments. The surface layers of lakes (epilimnion) and of the open sea (epipelagic) are dominated by DifP and ColP microhabitats as the consequence of persistent oligotrophic conditions. These microhabitats are pervaded by rare environmental fluctuations and thus are more stable, characterized by long-term seasonal variations contrary to Par and, especially, Bio microhabitats that can experience extremely high dynamics (i.e., process rates measured in hours rather than days). The relatively small size and low bacterial activity of the DifP and ColP microhabitats in big lakes and open sea, however, can pair and even overcome the bio-process rates occurring in the larger and more active Par and Bio microhabitats. The reason for this lies in the huge extension of DifP-ColP-dominated lake and open ocean biomes which represent more than 80% of all aquatic habitats accounting for ca. 40% of total primary producer biomass and 70% of net primary production (Valiela 1995).

Similarly, the deep layer of large lakes (meta- and hypolimnion) and marine areas (meso- and bathypelagic layers) is dominated by DifP and ColP microhabitats although the ColP microhabitats typically represent mostly unproductive and recalcitrant spots (Williams et al. 1969). Both Par and Bio microhabitats are mostly absent or extremely rare (diluted) in these deep layers, with the occurrence of microhabitats largely dependent upon distance from the coast, the above presence of productive epipelagic areas, and the oceanographic dynamics related to circulation patterns. Offshore deep regions are thus unlikely to experience the presence of nutrient-rich microhabitats since the sunlit layers of oligotrophic open oceans are populated by small-size primary producers (mainly *Synechococcus* and *Prochlorococcus*) whose allochthonous exports are rapidly consumed in the upper part of the water column. Constrained in allochthonous inputs and in autochthonous OM production, the deep ocean represents the most oligotrophic biome. The bacterial activity and biogeochemical processes in this biome are, therefore, limited although relevant in a global perspective due to the tremendous volume of the deep sea (Arístegui et al. 2009). Recent studies indicate that these microbial communities are not inactive as previously thought and that the gap between energy requirements and OM inputs in these biomes might be covered by autochthonous production of OM by chemoautotrophic bacteria (Swan et al. 2011). Anyhow, these environments are dominated by DifP and refractory ColP microhabitats harboring little diverse and active microbial communities.

2.6 Future Perspectives

We believe that in the near future, several exciting advances will be achieved regarding comprehension of the extreme and vibrant heterogeneity of bacterial microhabitats. Technological improvements coupled with a more representative experimental design likely will provide high-resolution data useful to upscale our knowledge regarding ongoing biological dynamics from the nano- and microscales to the macro- and global scales. Technologies such as SAG, MS-MS, and DESI-MS are providing us the capability to unveil the genetic potential and metabolically characterize even individual cells. Our core perspectives will be refined with the development of future instrumental and methodological improvements. A special effort should be invested to cover two major critical knowledge gaps: (1) the missing coupling between biochemistry and molecular techniques for fine-scale analyses of the environmental-biological diversity and (2) the development of biogeochemical models which can take into account the bacterial heterogeneity of microhabitats and scale up to ecosystem and global levels. In this way it will be possible to eliminate some of our present uncertainty which nullifies the study of processes occurring at extremely limited spatial and temporal scales; these methodological errors magnify when being scaled up to broader ecosystem levels and thus still keep our estimation of global biogeochemical fluxes far from realistic.

Compliance with Ethical Standards

Funding This study was funded by HFSP (grant number RGB 0020/2016).

Conflict of Interest Luca Zoccarato declares that he has no conflict of interest. Peter Grossart declares that he has no conflict of interest.

Ethical Approval This article does not contain any studies with human participants or animals performed by any of the authors.

References

Agawin NSR, Duarte CM, Agustí S (2000) Nutrient and temperature control of the contribution of picoplankton to phytoplankton biomass and production. Limnol Oceanogr 45:591–600. https://doi.org/10.4319/lo.2000.45.3.0591

Aharonovich D, Sher D (2016) Transcriptional response of prochlorococcus to co-culture with a marine Alteromonas: differences between strains and the involvement of putative infochemicals. ISME J 10:2892–2906. https://doi.org/10.1038/ismej.2016.70

Alldredge AL, Cohen Y (1987) Can microscale chemical patches persist in the sea? Microelectrode study of marine snow, fecal pellets. Science 235:689–691. https://doi.org/10.1126/science.235.4789.689

Allen AE, Allen LZ, McCrow JP (2013) Lineage specific gene family enrichment at the microscale in marine systems. Curr Opin Microbiol 16:605–617

Amin SA, Parker MS, Armbrust EV (2012) Interactions between diatoms and bacteria. Microbiol Mol Biol Rev 76:667–684. https://doi.org/10.1128/MMBR.00007-12

Amin SA, Hmelo LR, van Tol HM et al (2015) Interaction and signalling between a cosmopolitan phytoplankton and associated bacteria. Nature 522:98–101. https://doi.org/10.1038/nature14488

Archibald JM (2015) Endosymbiosis and eukaryotic cell evolution. Curr Biol 25:911–921

Arias-Andres M, Klümper U, Rojas-Jimenez K, Grossart H-P (2018) Microplastic pollution increases gene exchange in aquatic ecosystems. Environ Pollut 237:253–261

Arístegui J, Gasol JM, Duarte CM, Herndl GJ (2009) Microbial oceanography of the dark ocean's pelagic realm. Limnol Oceanogr 54:1501–1529. https://doi.org/10.4319/lo.2009.54.5.1501

Azam F (1998) Microbial control of oceanic carbon flux: the plot thickens. Science 280:694–696. https://doi.org/10.1126/science.280.5364.694

Becker JW, Berube PM, Follett CL et al (2014) Closely related phytoplankton species produce similar suites of dissolved organic matter. Front Microbiol 5:111. https://doi.org/10.3389/fmicb.2014.00111

Beers JR (1966) Studies on the chemical composition of the major zooplankton groups in the Sargasso Sea off Bermuda. Limnol Oceanogr 11:520–528. https://doi.org/10.4319/lo.1966.11.4.0520

Behringer G, Ochsenkühn MA, Fei C, Fanning J, Koester JA, Amin SA (2018) Bacterial communities of diatoms display strong conservation across strains and time. Front Microbiol 9:659. https://doi.org/10.3389/fmicb.2018.00659

Béjà O, Aravind L, Koonin EV et al (2000) Bacterial rhodopsin: evidence for a new type of phototrophy in the sea. Science 289:1902–1906. https://doi.org/10.1126/science.289.5486.1902

Bennett GM, McCutcheon JP, MacDonald BR et al (2014) Differential genome evolution between companion symbionts in an insect-bacterial symbiosis. mBio 5:1–12. https://doi.org/10.1128/mBio.01697-14

Bianchi TS (2011) The role of terrestrially derived organic carbon in the coastal ocean: a changing paradigm and the priming effect. Proc Natl Acad Sci USA 108:19473–19481. https://doi.org/10.1073/pnas.1017982108

Bichai F, Payment P, Barbeau B (2008) Protection of waterborne pathogens by higher organisms in drinking water: a review. Can J Microbiol 54:509–524. https://doi.org/10.1139/W08-039

Biers EJ, Sun S, Howard EC (2009) Prokaryotic genomes and diversity in surface ocean waters: interrogating the global ocean sampling metagenome. Appl Environ Microbiol 75:2221–2229. https://doi.org/10.1128/AEM.02118-08

Bižić-Ionescu M, Zeder M, Ionescu D et al (2014) Comparison of bacterial communities on limnic versus coastal marine particles reveals profound differences in colonization. Environ Microbiol 17:1–36. https://doi.org/10.1111/1462-2920.12466

Bode HB, Bethe B, Höfs R, Zeeck A (2002) Big effects from small changes: possible ways to explore nature's chemical diversity. ChemBioChem 3:619–627

Boutte CC, Crosson S (2013) Bacterial lifestyle shapes stringent response activation. Trends Microbiol 21:174–180. https://doi.org/10.1016/j.tim.2013.01.002

Boyd KG, Adams DR, Burgess JG (1999) Antibacterial and repellent activities of marine bacteria associated with algal surfaces. Biofouling 14:227–236. https://doi.org/10.1080/08927019909378414

Bunse C, Bertos-Fortis M, Sassenhagen I et al (2016) Spatio-temporal interdependence of bacteria and phytoplankton during a Baltic Sea spring bloom. Front Microbiol 7:517. https://doi.org/10.3389/fmicb.2016.00517

Burghardt KT, Oswald JS (2015) Trophic ecology: bottom-up and top-down interactions across aquatic and terrestrial systems. Cambridge University Press, Cambridge

Burke C, Thomas T, Lewis M et al (2011) Composition, uniqueness and variability of the epiphytic bacterial community of the green alga *Ulva australis*. ISME J 5:590–600. https://doi.org/10.1038/ismej.2010.164

Carlson CA (2002) Production and removal processes. In: Biogeochemistry of marine dissolved organic matter. Academic Press, Elsevier, San Diego, pp 139–151

Carman KR, Dobbs FC (1997) Epibiotic microorganisms on copepods and other marine crustaceans. Microsc Res Tech 37:116–135. https://doi.org/10.1002/(SICI)1097-0029(19970415)37:2<116::AID-JEMT2>3.0.CO;2-M

Case RJ, Labbate M, Kjelleberg S (2008) AHL-driven quorum-sensing circuits: their frequency and function among the proteobacteria. ISME J 2:345–349. https://doi.org/10.1038/ismej.2008.13

Christie-Oleza JA, Fernandez B, Nogales B et al (2012) Proteomic insights into the lifestyle of an environmentally relevant marine bacterium. ISME J 6:124–135. https://doi.org/10.1038/ismej.2011.86

Church MJ (2008) Resource control of bacterial dynamics in the sea. In: Kirchman DL (ed) Microbial ecology of the oceans. Wiley, New York, pp 335–382

Cole M, Lindeque PK, Fileman ES et al (2015) The impact of microplastics on feeding, function and fecundity in an abundant marine copepod. Environ Sci Technol 49:1130–1137

Cole M, Lindeque PK, Fileman E et al (2016) Microplastics alter the properties and sinking rates of zooplankton faecal pellets. Environ Sci Technol 50:3239–3246. https://doi.org/10.1021/acs.est.5b05905

Corno G, Caravati E, Callieri C, Bertoni R (2008) Effects of predation pressure on bacterial abundance, diversity, and size-structure distribution in an oligotrophic system. J Limnol 67:107–119. https://doi.org/10.4081/jlimnol.2008.107

Corno G, Villiger J, Pernthaler J (2013) Coaggregation in a microbial predator-prey system affects competition and trophic transfer efficiency. Ecology 94:870–881. https://doi.org/10.1890/12-1652.1

Crowther TW, Grossart H-P (2015) The role of bottom-up and top-down interactions in determining microbial and fungal diversity and function. In: Hanley TC, La Pierre KJ (eds) Trophic ecology: bottom-up and top-down interactions across aquatic and terrestrial systems. Cambridge University Press, Cambridge, pp 260–287

DeGroote M, Fang F (1995) NO inhibitions: antimicrobial properties of nitric oxide. Clin Infect Dis 21:162–165. https://doi.org/10.1007/0-306-46816-6_12

Diaz M, Visscher P, Taylor B (1992) Metabolism of dimethylsulfoniopropionate and glycine betaine by a marine bacterium. FEMS Microbiol Lett 96:61–66. https://doi.org/10.1111/j.1574-6968.1992.tb05394.x

Dobretsov S, Teplitski M, Paul V (2009) Mini-review: Quorum sensing in the marine environment and its relationship to biofouling. Biofouling 25:413–427. https://doi.org/10.1080/08927010902853516

Durham BP, Sharma S, Luo H et al (2015) Cryptic carbon and sulfur cycling between surface ocean plankton. Proc Natl Acad Sci 112:453–457. https://doi.org/10.1073/pnas.1413137112

Eckert EM, Pernthaler J (2014) Bacterial epibionts of Daphnia: a potential route for the transfer of dissolved organic carbon in freshwater food webs. ISME J 8:1–12. https://doi.org/10.1038/ismej.2014.39

Egan S, Harder T, Burke C et al (2013) The seaweed holobiont: understanding seaweed-bacteria interactions. FEMS Microbiol Rev 37:462–476. https://doi.org/10.1111/1574-6976.12011

Eloe-Fadrosh EA, Rasko DA (2013) The human microbiome: from symbiosis to pathogenesis. Annu Rev Med 64:145–163. https://doi.org/10.1146/annurev-med-010312-133513

Fabrega J, Fawcett SR, Renshaw JC, Lead JR (2009) Silver nanoparticle impact on bacterial growth: effect of pH, concentration, and organic matter. Environ Sci Technol 43:7285–7290. https://doi.org/10.1021/es803259g

Fenchel T (2002) Microbial behavior in a heterogeneous world. Science 296:1068–1071. https://doi.org/10.1126/science.1070118

Fernandez VI, Yawata Y, Stocker R (2019) A foraging Mandala for aquatic microorganisms. ISME J 13(3):563–575

Fischer CN, Trautman EP, Crawford JM, Stabb EV, Jo H, Broderick NA (2017) Metabolite exchange between microbiome members produces compounds that influence Drosophila behavior. eLife 6

Fontanez KM, Eppley JM, Samo TJ et al (2015) Microbial community structure and function on sinking particles in the North Pacific Subtropical Gyre. Front Microbiol 6:469. https://doi.org/10.3389/fmicb.2015.00469

Fuhrman JA, Hewson I, Schwalbach MS et al (2006) Annually reoccurring bacterial communities are predictable from ocean conditions. Proc Natl Acad Sci USA 103:13104–13109. https://doi.org/10.1073/pnas.0602399103

Ganesh S, Parris DJ, DeLong EF, Stewart FJ (2014) Metagenomic analysis of size-fractionated picoplankton in a marine oxygen minimum zone. ISME J 8:187–211. https://doi.org/10.1038/ismej.2013.144

Garcia SL, Buck M, McMahon KD et al (2015) Auxotrophy and intrapopulation complementary in the interactome of a cultivated freshwater model community. Mol Ecol 24:4449–4459. https://doi.org/10.1111/mec.13319

García-Vásquez A, Razo-Mendivil U, Rubio-Godoy M (2017) Triple trouble? Invasive poeciliid fishes carry the introduced tilapia pathogen *Gyrodactylus cichlidarum* in the Mexican highlands. Vet Parasitol 235:37–40

Gärdes A, Iversen MH, Grossart H-P et al (2011) Diatom-associated bacteria are required for aggregation of *Thalassiosira weissflogii*. ISME J 5:436–445. https://doi.org/10.1038/ismej.2010.145

Ger KA, Faassen EJ, Pennino MG, Lürling M (2016) Effect of the toxin (microcystin) content of microcystis on copepod grazing. Harmful Algae 52:34–45. https://doi.org/10.1016/j.hal.2015.12.008

Ghiglione JF, Mevel G, Pujo-Pay M et al (2007) Diel and seasonal variations in abundance, activity, and community structure of particle-attached and free-living bacteria in NW Mediterranean Sea. Microb Ecol 54:217–231. https://doi.org/10.1007/s00248-006-9189-7

Ghylin TW, Garcia SL, Moya F et al (2014) Comparative single-cell genomics reveals potential ecological niches for the freshwater acI Actinobacteria lineage. ISME J 8:2503–2516. https://doi.org/10.1038/ismej.2014.135

Giovannoni SJ, Cameron Thrash J, Temperton B (2014) Implications of streamlining theory for microbial ecology. ISME J 8:1553–1565. https://doi.org/10.1038/ismej.2014.60

Glaeser SP, Grossart H-P, Glaeser J (2010) Singlet oxygen, a neglected but important environmental factor: short-term and long-term effects on bacterioplankton composition in a humic lake. Environ Microbiol 12:3124–3136. https://doi.org/10.1111/j.1462-2920.2010.02285.x

Glaeser SP, Berghoff BA, Stratmann V et al (2014) Contrasting effects of singlet oxygen and hydrogen peroxide on bacterial community composition in a humic lake. PLoS One 9:e92518. https://doi.org/10.1371/journal.pone.0092518

Glasl B, Herndl GJ, Frade PR (2016) The microbiome of coral surface mucus has a key role in mediating holobiont health and survival upon disturbance. ISME J 10:2280–2292. https://doi.org/10.1038/ismej.2016.9

Glud RN, Grossart H-P, Larsen M et al (2015) Copepod carcasses as microbial hot spots for pelagic denitrification. Limnol Oceanogr 60:2026–2036. https://doi.org/10.1002/lno.10149

Gómez-Consarnau L, González JM, Riedel T et al (2016) Proteorhodopsin light-enhanced growth linked to vitamin-B1 acquisition in marine Flavobacteria. ISME J 10:1102–1112. https://doi.org/10.1038/ismej.2015.196

Grossart H-P (1999) Interactions between marine bacteria and axenic diatoms (*Cylindrotheca fusiformis*, Nitzschia laevis, and *Thalassiosira weissflogii*) incubated under various conditions in the lab. Aquat Microb Ecol 19:1–11. https://doi.org/10.3354/ame019001

Grossart H-P (2010) Ecological consequences of bacterioplankton lifestyles: changes in concepts are needed. Environ Microbiol Rep 2:706–714. https://doi.org/10.1111/j.1758-2229.2010.00179.x

Grossart H-P, Riemann L, Azam F (2001) Bacterial motility in the sea and its ecological implications. Aquat Microb Ecol 25:247–258. https://doi.org/10.3354/ame025247

Grossart H-P, Schlingloff A, Bernhard M et al (2004) Antagonistic activity of bacteria isolated from organic aggregates of the German Wadden Sea. FEMS Microbiol Ecol 47:387–396. https://doi.org/10.1016/S0168-6496(03)00305-2

Grossart H-P, Levold F, Allgaier M et al (2005) Marine diatom species harbour distinct bacterial communities. Environ Microbiol 7:860–873. https://doi.org/10.1111/j.1462-2920.2005.00759.x

Grossart H-P, Czub G, Simon M (2006) Algae-bacteria interactions and their effects on aggregation and organic matter flux in the sea. Environ Microbiol 8:1074–1084. https://doi.org/10.1111/j.1462-2920.2006.00999.x

Grossart H-P, Tang KW, Kiørboe T, Ploug H (2007) Comparison of cell-specific activity between free-living and attached bacteria using isolates and natural assemblages. FEMS Microbiol Lett 266:194–200. https://doi.org/10.1111/j.1574-6968.2006.00520.x

Grossart H-P, Dziallas C, Tang KW (2009) Bacterial diversity associated with freshwater zooplankton. Environ Microbiol Rep 1:50–55. https://doi.org/10.1111/j.1758-2229.2008.00003.x

Grossart H-P, Dziallas C, Leunert F, Tang KW (2010) Bacteria dispersal by hitchhiking on zooplankton. Proc Natl Acad Sci USA 107:11959–11964. https://doi.org/10.1073/pnas.1000668107

Grossart H-P, Van den Wyngaert S, Kagami M, Wurzbacher C, Cunliffe M, Rojas-Jimenez K (2019) Fungi in aquatic ecosystems. Nat Rev Microbiol. https://doi.org/10.1038/s41579-019-0175-8

Grujčić V, Kasalický V, Šimek K (2015) Prey-specific growth responses of freshwater flagellate communities induced by morphologically distinct bacteria from the genus Limnohabitans. Appl Environ Microbiol 81:4993–5002. https://doi.org/10.1128/AEM.00396-15

Guenet B, Danger M, Abbadie L, Lacroix G (2010) Priming effect: bridging the gap between terrestrial and aquatic ecology. Ecology 91:2850–2861. https://doi.org/10.1890/09-1968.1

Gugliandolo C, Irrera GP, Lentini V, Maugeri TL (2008) Pathogenic Vibrio, Aeromonas and Arcobacter spp. associated with copepods in the Straits of Messina (Italy). Mar Pollut Bull 56:600–606. https://doi.org/10.1016/j.marpolbul.2007.12.001

Habe H, Omori T (2003) Genetics of polycyclic aromatic hydrocarbon metabolism in diverse aerobic bacteria. Biosci Biotechnol Biochem 67:225–243. https://doi.org/10.1271/bbb.67.225

Hahn MW, Moore ERB, Hüfle MG (1999) Bacterial filament formation, a defense mechanism against flagellate grazing, is growth rate controlled in bacteria of different phyla. Appl Environ Microbiol 65:25–35

Hansell DA, Carlson CA (2002) Biogeochemistry of marine dissolved organic matter. Academic Press, Delft

Harris JM (1993) The presence, nature, and role of gut microflora in aquatic invertebrates: a synthesis. Microb Ecol 25:195–231. https://doi.org/10.1007/BF00171889

Hessen DO, Kaartvedt S (2014) Top-down cascades in lakes and oceans: different perspectives but same story? J Plankton Res 36:914–924. https://doi.org/10.1093/plankt/fbu040

Hibbing ME, Fuqua C, Parsek MR, Peterson SB (2010) Bacterial competition: surviving and thriving in the microbial jungle. Nat Rev Microbiol 8:15–25. https://doi.org/10.1038/nrmicro2259

Hollibaugh JT, Wong PS, Murrell MC (2000) Similarity of particle-associated and free-living bacterial communities in northern San Francisco Bay, California. Aquat Microb Ecol 21:103–114. https://doi.org/10.3354/ame021103

Horvath RS (1972) Microbial co-metabolism and the degradation of organic compounds in nature. Bacteriol Rev 36:146–155

Huber H, Hohn MJ, Rachel R et al (2002) A new phylum of Archaea represented by a nanosized hyperthermophilic symbiont. Nature 417:63–67. https://doi.org/10.1038/417063a

Hutalle-Schmelzer KML, Zwirnmann E, Krüger A, Grossart H-P (2010) Enrichment and cultivation of pelagic bacteria from a humic lake using phenol and humic matter additions. FEMS Microbiol Ecol 72:58–73. https://doi.org/10.1111/j.1574-6941.2009.00831.x

Hutchinson GE (1961) The paradox of the plankton author. Am Nat 95:137–145. https://doi.org/10.1086/282171

Jang MH, Ha K, Joo GJ, Takamura N (2003) Toxin production of cyanobacteria is increased by exposure to zooplankton. Freshw Biol 48:1540–1550. https://doi.org/10.1046/j.1365-2427.2003.01107.x

Jezbera J, Horňák K, Šimek K (2006) Prey selectivity of bacterivorous protists in different size fractions of reservoir water amended with nutrients. Environ Microbiol 8:1330–1339. https://doi.org/10.1111/j.1462-2920.2006.01026.x

Jiao N, Herndl GJ, Hansell DA et al (2010) Microbial production of recalcitrant dissolved organic matter: long-term carbon storage in the global ocean. Nat Rev Microbiol 8:593–599. https://doi.org/10.1038/nrmicro2386

Kappelmann L, Krüger K, Hehemann J-H, Harder J, Markert S, Unfried F, Becher D, Shapiro N, Schweder T, Amann RI, Teeling H (2019) Polysaccharide utilization loci of North Sea Flavobacteriia as basis for using SusC/D-protein expression for predicting major phytoplankton glycans. ISME J 13(1):76–91

KarpBoss L, Boss E, Jumars PA (1996) Nutrient fluxes to planktonic osmotrophs in the presence of fluid motion. Oceanogr Mar Biol 34(34):71–107

Kettler GC, Martiny AC, Huang K et al (2007) Patterns and implications of gene gain and loss in the evolution of Prochlorococcus. PLoS Genet 3:2515–2528. https://doi.org/10.1371/journal.pgen.0030231

Kikuchi Y (2009) Endosymbiotic bacteria in insects: their diversity and culturability. Microbes Environ 24:195–204. https://doi.org/10.1264/jsme2.ME09140S

Kiørboe T, Jackson GA (2001) Marine snow, organic solute plumes, and optimal chemosensory behavior of bacteria. Limnol Oceanogr 46:1309–1318. https://doi.org/10.4319/lo.2001.46.6.1309

Kiørboe T, Thygesen UH (2001) Fluid motion and solute distribution around sinking aggregates. II. Implications for remote detection by colonizing zooplankters. Mar Ecol Prog Ser 211:15–25. https://doi.org/10.3354/meps211015

Kiørboe T, Grossart H-P, Ploug H, Tang KW (2002) Mechanisms and rates of colonisation of sinking aggregates. Appl Environ Microbiol 68:3996–4006. https://doi.org/10.1128/AEM.68.8.3996

Kiørboe T, Grossart H-P, Ploug H et al (2004) Particle-associated flagellates: swimming patterns, colonization rates, and grazing on attached bacteria. Aquat Microb Ecol 35:141–152. https://doi.org/10.3354/ame035141

Kirchman DL (2003) The contribution of monomers and other low-molecular weight compounds to the flux of dissolved organic matter in aquatic ecosystems. In: Findlay SEG, Sinsabaugh RL (eds) Aquatic ecosystems: interactivity of dissolved organic matter. Elsevier Science, New York, pp 217–241

Kirchman DL (2012) Processes in microbial ecology. Oxford University Press, Oxford

Koblížek M (2015) Ecology of aerobic anoxygenic phototrophs in aquatic environments. FEMS Microbiol Rev 39:854–870

Lami R, Cottrell MT, Ras J et al (2007) High abundances of aerobic anoxygenic photosynthetic bacteria in the South Pacific Ocean. Appl Environ Microbiol 73:4198–4205. https://doi.org/10.1128/AEM.02652-06

Lami R, Čuperová Z, Ras J et al (2009) Distribution of free-living and particle-attached aerobic anoxygenic phototrophic bacteria in marine environments. Aquat Microb Ecol 55:31–38. https://doi.org/10.3354/ame01282

Lauro FM, McDougald D, Thomas T et al (2009) The genomic basis of trophic strategy in marine bacteria. Proc Natl Acad Sci USA 106:15527–15533. https://doi.org/10.1073/pnas.0903507106

Lecleir GR, Debruyn JM, Maas EW et al (2014) Temporal changes in particle-associated microbial communities after interception by nonlethal sediment traps. FEMS Microbiol Ecol 87:153–163. https://doi.org/10.1111/1574-6941.12213

Leonard MM, Camhi S, Huedo-Medina TB, Fasano A (2015) Celiac disease genomic, environmental, microbiome, and metabolomic (CDGEMM) study design: approach to the future of personalized prevention of celiac disease. Nutrients 7:9325–9336. https://doi.org/10.3390/nu7115470

Long RA, Azam F (2001) Antagonistic interactions among marine bacteria. Appl Environ Microbiol 67:4875–4983. https://doi.org/10.1128/AEM.67.11.4975-4983.2001

Lönnstedt OM, Eklöv P (2016) Environmentally relevant concentrations of microplastic particles influence larval fish ecology. Science 352:1213–1216. https://doi.org/10.1126/science.aad8828

Madigan M, Martinko J (2006) Brock biology of microorganisms. Pearson Education, Boston

McCarthy M, Hedges J, Benner R (1996) Major biochemical composition of dissolved high molecular weight organic matter in seawater. Mar Chem 55:281–297. https://doi.org/10.1016/S0304-4203(96)00041-2

Mestre M, Borrull E, Mm S, Gasol JM (2017) Patterns of bacterial diversity in the marine planktonic particulate matter continuum. ISME J 11:999–1010. https://doi.org/10.1038/ismej.2016.166

Mével G, Vernet M, Goutx M, Ghiglione JF (2008) Seasonal to hour variation scales in abundance and production of total and particle-attached bacteria in the open NW Mediterranean Sea (0–1000 m). Biogeosciences 5:1573–1586. https://doi.org/10.5194/bg-5-1573-2008

Miller MB, Bassler BL (2001) Quorum sensing in bacteria. Annu Rev Microbiol 55:165–199. https://doi.org/10.1146/annurev.micro.55.1.165

Møller EF, Thor P, Nielsen TG (2003) Production of DOC by *Calanus finmarchicus*, *C. glacialis* and *C. hyperboreus* through sloppy feeding and leakage from fecal pellets. Mar Ecol Prog Ser 262:185–191. https://doi.org/10.3354/meps262185

Møller EF, Riemann L, Søndergaard M (2007) Bacteria associated with copepods: abundance, activity and community composition. Aquat Microb Ecol 47:99–106. https://doi.org/10.3354/ame047099

Moran MA, Zepp RG (1997) Role of photoreactions in the formation of biologically labile compounds from dissolved organic matter. Limnol Oceanogr 42:1307–1316. https://doi.org/10.4319/lo.1997.42.6.1307

Moran MA, Satinsky B, Gifford SM et al (2013) Sizing up metatranscriptomics. ISME J 7:237–243. https://doi.org/10.1038/ismej.2012.94

Morris J, Lenski RE, Zinser ER et al (2012) The black queen hypothesis: evolution of Dependencies through adaptive gene loss. MBio 3:e00036. https://doi.org/10.1128/mBio.00036-12.Updated

Needham DM, Fuhrman JA (2016) Pronounced daily succession of phytoplankton, archaea and bacteria following a spring bloom. Nat Microbiol 1:1–7. https://doi.org/10.1038/NMICROBIOL.2016.5

Nyholm SV (2016) Fiat Lux: the Squid–Vibrio association as a model for understanding host–microbe associations. In: Hurst CJ (ed) The mechanistic benefits of microbial symbionts. Springer, New York, pp 295–315

Obernosterer I, Reitner B, Herndl GJ (1999) Contrasting effects of solar radiation on dissolved organic matter and its bioavailability to marine bacterioplankton. Limnol Oceanogr 44:1645–1654. https://doi.org/10.4319/lo.1999.44.7.1645

Ochman H, Davalos L (2006) The nature and dynamics of bacterial genomes. Science 311:1730–1734. https://doi.org/10.1126/science.1119966

Ortega-Retuerta E, Joux F, Jeffrey WH, Ghiglione JF (2013) Spatial variability of particle-attached and free-living bacterial diversity in surface waters from the Mackenzie River to the Beaufort Sea (Canadian Arctic). Biogeosciences 10:2747–2759. https://doi.org/10.5194/bg-10-2747-2013

Paul JH (1999) Microbial gene transfer: an ecological perspective. J Mol Microbiol Biotechnol 1:45–50

Paul C, Pohnert G (2011) Interactions of the algicidal bacterium Kordia algicida with diatoms: regulated protease excretion for specific algal lysis. PLoS One 6:e21032. https://doi.org/10.1371/journal.pone.0021032

Peter H, Sommaruga R (2008) An evaluation of methods to study the gut bacterial community composition of freshwater zooplankton. J Plankton Res 30:997–1006. https://doi.org/10.1093/plankt/fbn061

Pizzetti I, Gobet A, Fuchs BM et al (2011) Abundance and diversity of Planctomycetes in a Tyrrhenian coastal system of central Italy. Aquat Microb Ecol 65:129–141. https://doi.org/10.3354/ame01535

Ploug H, Iversen MH, Koski M, Buitenhuis ET (2008) Production, oxygen respiration rates, and sinking velocity of copepod fecal pellets: direct measurements of ballasting by opal and calcite. Limnol Oceanogr 53:469–476. https://doi.org/10.4319/lo.2008.53.2.0469

Purcell EM (1977) Life at low Reynolds number. Am J Phys 45:3. https://doi.org/10.1063/1.1761474

Raina J-B, Fernandez V, Lambert B, Stocker R, Seymour JR (2019) The role of microbial motility and chemotaxis in symbiosis. Nat Rev Microbiol. https://doi.org/10.1038/s41579-019-0182-9

Rawlings TK, Ruiz GM, Colwell RR (2007) Association of *Vibrio cholerae* O1 El Tor and O139 Bengal with the copepods *Acartia tonsa* and *Eurytemora affinis*. Appl Environ Microbiol 73:7926–7933. https://doi.org/10.1128/AEM.01238-07

Reid G, Howard J, Gan BS (2001) Can bacterial interference prevent infection? Trends Microbiol 9:424–428. https://doi.org/10.1016/S0966-842X(01)02132-1

Rieck A, Herlemann DPR, Jürgens K, Grossart H-P (2015) Particle-associated differ from free-living bacteria in surface waters of the baltic sea. Front Microbiol 6:1297. https://doi.org/10.3389/fmicb.2015.01297

Riemann L, Farnelid H, Steward GF (2010) Nitrogenase genes in non-cyanobacterial plankton: prevalence, diversity and regulation in marine waters. Aquat Microb Ecol 61:235–247

Rinke C, Schwientek P, Sczyrba A et al (2013) Insights into the phylogeny and coding potential of microbial dark matter. Nature 499:431–437. https://doi.org/10.1038/nature12352

Rinta-Kanto JM, Sun S, Sharma S et al (2012) Bacterial community transcription patterns during a marine phytoplankton bloom. Environ Microbiol 14:228–239. https://doi.org/10.1111/j.1462-2920.2011.02602.x

Rösel S, Grossart H-P (2012) Contrasting dynamics in activity and community composition of free-living and particle-associated bacteria in spring. Aquat Microb Ecol 66:169–181. https://doi.org/10.3354/ame01568

Salcher MM (2014) Same same but different: ecological niche partitioning of planktonic freshwater prokaryotes. J Limnol 73:74–87. https://doi.org/10.4081/jlimnol.2014.808

Salcher MM, Posch T, Pernthaler J (2013) In situ substrate preferences of abundant bacterioplankton populations in a prealpine freshwater lake. ISME J 7:896–907. https://doi.org/10.1038/ismej.2012.162

Scavotto RE, Dziallas C, Bentzon-Tilia M et al (2015) Nitrogen-fixing bacteria associated with copepods in coastal waters of the North Atlantic Ocean. Environ Microbiol 17:3754–3765. https://doi.org/10.1111/1462-2920.12777

Schauer M, Jiang J, Hahn MW (2006) Recurrent seasonal variations in abundance and composition of filamentous SOL cluster bacteria (Saprospiraceae, Bacteroidetes) in oligomesotrophic Lake Mondsee (Austria). Appl Environ Microbiol 72:4704–4712. https://doi.org/10.1128/AEM.02935-05

Scully NM, Tranvik LJ, Cooper WJ (2003) Photochemical effects on the interaction of enzymes and dissolved organic matter in natural waters. Limnol Oceanogr 48:1818–1824. https://doi.org/10.4319/lo.2003.48.5.1818

Sengupta S, Chattopadhyay MK, Grossart H-P (2013) The multifaceted roles of antibiotics and antibiotic resistance in nature. Front Microbiol 4:47. https://doi.org/10.3389/fmicb.2013.00047

Seymour JR, Amin SA, Raina J-B, Stocker R (2017) Zooming in on the phycosphere: the ecological interface for phytoplankton–bacteria relationships. Nat Microbiol 2:17065. https://doi.org/10.1038/nmicrobiol.2017.65

Sharp KH, Eam B, John Faulkner D, Haygood MG (2007) Vertical transmission of diverse microbes in the tropical sponge *Corticium* sp. Appl Environ Microbiol 73:622–629. https://doi.org/10.1128/AEM.01493-06

Shendure J, Ji H (2008) Next-generation DNA sequencing. Nat Biotechnol 26:1135–1145. https://doi.org/10.1038/nbt1486

Simon M, Grossart H-P, Schweitzer B, Ploug H (2002) Microbial ecology of organic aggregates in aquatic ecosystems. Aquat Microb Ecol 28:175–211. https://doi.org/10.3354/ame028175

Simon HM, Smith MW, Herfort L (2014) Metagenomic insights into particles and their associated microbiota in a coastal margin ecosystem. Front Microbiol 5:466. https://doi.org/10.3389/fmicb.2014.00466

Skariyachan S, Manjunatha V, Sultana S et al (2016) Novel bacterial consortia isolated from plastic garbage processing areas demonstrated enhanced degradation for low density polyethylene. Environ Sci Pollut Res 23:18307–18319. https://doi.org/10.1007/s11356-016-7000-y

Smith V (2003) Eutrophication of freshwater and coastal marine ecosystems a global problem. Environ Sci Pollut Res 10:126–139. https://doi.org/10.1065/espr2002.12.142

Smith DC, Azam F (1992) A simple, economical method for measuring bacterial protein synthesis rates in seawater using ^{3}H-leucine. Mar Biol Res 6:107–114

Smith MW, Allen LZ, Allen AE et al (2013) Contrasting genomic properties of free-living and particle-attached microbial assemblages within a coastal ecosystem. Front Microbiol 4:120. https://doi.org/10.3389/fmicb.2013.00120

Steinberg PD, Rice SA, Campbell AH et al (2011) Interfaces between bacterial and eukaryotic "neuroecology". Integr Comp Biol 51:794–806. https://doi.org/10.1093/icb/icr115

Stocker R (2012) Marine microbes see a sea of gradients. Science 338:628–633. https://doi.org/10.1126/science.1208929

Stocker R, Seymour JR, Samadani A et al (2008) Rapid chemotactic response enables marine bacteria to exploit ephemeral microscale nutrient patches. Proc Natl Acad Sci USA 105:4209–4214. https://doi.org/10.1073/pnas.0709765105

Suttle CA (2007) Marine viruses—major players in the global ecosystem. Nat Rev Microbiol 5:801–812. https://doi.org/10.1038/nrmicro1750

Swan BK, Martinez-Garcia M, Preston CM et al (2011) Potential for chemolithoautotrophy among ubiquitous bacteria lineages in the dark ocean. Science 333:1296–1300. https://doi.org/10.1126/science.1203690

Tang KW, Dziallas C, Hutalle-Schmelzer K, Grossart H-P (2009) Effects of food on bacterial community composition associated with the copepod *Acartia tonsa* Dana. Biol Lett 5:549–553. https://doi.org/10.1098/rsbl.2009.0076

Tang KW, Turk V, Grossart H-P (2010) Linkage between crustacean zooplankton and aquatic bacteria. Aquat Microb Ecol 61:261–277. https://doi.org/10.3354/ame01424

Tang KW, Dziallas C, Grossart H-P (2011) Zooplankton and aggregates as refuge for aquatic bacteria: protection from UV, heat and ozone stresses used for water treatment. Environ Microbiol 13:378–390. https://doi.org/10.1111/j.1462-2920.2010.02335.x

Tang KW, Gladyshev MI, Dubovskaya OP et al (2014) Zooplankton carcasses and non-predatory mortality in freshwater and inland sea environments. J Plankton Res 36:597–612. https://doi.org/10.1093/plankt/fbu014

Teeling H, Fuchs BM, Becher D et al (2012) Substrate-controlled succession of marine bacterioplankton populations induced by a phytoplankton bloom. Science 336:608–611. https://doi.org/10.1126/science.1218344

Teplitski M, Mathesius U, Rumbaugh KP (2011) Perception and degradation of N-Acyl homoserine lactone quorum sensing signals by mammalian and plant cells. Chem Rev 111:100–116. https://doi.org/10.1021/cr100045m

Tranvick LJ, Jansson M (2002) Climate change (communication arising): terrestrial export of organic carbon. Nature 415:861–862. https://doi.org/10.1038/415862a

Traving SJ, Rowe O, Jakobsen NM et al (2017) The effect of increased loads of dissolved organic matter on estuarine microbial community composition and function. Front Microbiol 8:351. https://doi.org/10.3389/fmicb.2017.00351

Treignier C, Derenne S, Saliot A (2006) Terrestrial and marine n-alcohol inputs and degradation processes relating to a sudden turbidity current in the Zaire canyon. Org Geochem 37:1170–1184. https://doi.org/10.1016/j.orggeochem.2006.03.010

Turner JT (2002) Zooplankton fecal pellets, marine snow and sinking phytoplankton blooms. Aquat Microb Ecol 27:57–102. https://doi.org/10.3354/Ame027057

Urbanek AK, Rymowicz W, Mirończuk AM (2018) Degradation of plastics and plastic-degrading bacteria in cold marine habitats. Appl Microbiol Biotechnol 102(18):7669–7678

Valiela I (1995) Marine ecological processes. Springer, New York

van Tol HM, Amin SA, Armbrust EV (2016) Ubiquitous marine bacterium inhibits diatom cell division. ISME J 11:31–42. https://doi.org/10.1038/ismej.2016.112

Verdugo P (2012) Marine microgels. Annu Rev Mar Sci 4:375–400. https://doi.org/10.1146/annurev-marine-120709-142759

Verdugo P, Alldredge AL, Azam F et al (2004) The oceanic gel phase: a bridge in the DOM-POM continuum. Mar Chem 92:67–85. https://doi.org/10.1029/2002GL016046

Verdugo P, Orellana MV, Chin W-C et al (2008) Marine biopolymer self-assembly: implications for carbon cycling in the ocean. Faraday Discuss 139:393–398. https://doi.org/10.1039/b800149a

Vezzulli L, Brettar I, Pezzati E et al (2012) Long-term effects of ocean warming on the prokaryotic community: evidence from the vibrios. ISME J 6:21–30. https://doi.org/10.1038/ismej.2011.89

Vishnivetskaya TA, Mosher JJ, Palumbo AV et al (2011) Mercury and other heavy metals influence bacterial community structure in contaminated Tennessee streams. Appl Environ Microbiol 77:302–311. https://doi.org/10.1128/AEM.01715-10

Volk T, Hoffert MI (1985) Ocean carbon pumps: analysis of relative strengths and efficiencies in ocean-driven atmospheric CO_2 changes. In: Sundquist ET, Broecker WS (eds) The carbon cycle and atmospheric CO2: natural variations Archaean to present. American Geophysical Union; Geophysical Monograph 32, Washington, DC, pp 99–110

Wahl M, Goecke F, Labes A et al (2012) The second skin: ecological role of epibiotic biofilms on marine organisms. Front Microbiol 3:292. https://doi.org/10.3389/fmicb.2012.00292
Walsh DA, Lafontaine J, Grossart H-P (2013) On the eco-evolutionary relationships of fresh and salt water bacteria and the role of gene transfer in their adaptation. In: Gophna U (ed) Lateral gene transfer in evolution. Springer, New York, pp 55–77
Wannamaker LW (1980) Bacterial interference and competition. Scand J Infect Dis Suppl Suppl 24:82–85
Whitman WB, Coleman DC, Wiebe WJ (1998) Prokaryotes: the unseen majority. Proc Natl Acad Sci USA 95:6578–6583. https://doi.org/10.1073/pnas.95.12.6578
Williams PM, Oeschger H, Kinney P (1969) Natural radiocarbon activity of the dissolved organic carbon in the north-east Pacific Ocean. Nature 224:256–258. https://doi.org/10.1038/224256a0
Williams RJ, Griffiths FB, Van der Wal EJ, Kelly J (1988) Cargo vessel ballast water as a vector for the transport of non-indigenous marine species. Estuar Coast Shelf Sci 26:409–420. https://doi.org/10.1016/0272-7714(88)90021-2
Yooseph S, Nealson KH, Rusch DB et al (2010) Genomic and functional adaptation in surface ocean planktonic prokaryotes. Nature 468:60–66. https://doi.org/10.1038/nature09530
Yurkov VV, Csotonyi JT (2009) New light on aerobic anoxygenic phototrophs. In: Hunter CN, Daldal F, Thurnauer MC, Beatty JT (eds) The purple phototrophic bacteria. Springer, Dordrecht, Netherlands, pp 31–55
Zettler ER, Mincer TJ, Amaral-Zettler LA (2013) Life in the "plastisphere": microbial communities on plastic marine debris. Environ Sci Technol 47:7137–7146. https://doi.org/10.1021/es401288x

Chapter 3
Biofilms: Besieged Cities or Thriving Ports?

Otini Kroukamp, Elanna Bester, and Gideon M. Wolfaardt

Abstract From a humble beginning, with less than 50 articles published per year with the term "biofilm" in the title prior to 1990, research output on this topic has grown dramatically with concurrent improved understanding of this form of microbial existence. While remarkable advances in molecular techniques perhaps enabled the major share of this growing knowledge base, we argue in this chapter that due consideration of the biofilms' physical environment in our experimental design, measurements and interpretation of results, is needed. For instance, the effect of flow, and its effect on nutrient and metabolite flux, is markedly different for cells attached to the surface (impacted by both the hydrodynamic effect and physicochemical properties of the surface) compared to those in the quiescent zone of low flow close to the surface (impacted by only the hydrodynamic effect of the surface) and the moving bulk fluid further away (little or no effect of the surface). The quiescent zone bordering a biofilm presents an area where planktonic cells can remain, roam around the attached biomass, and increase in density due to reduced flow resulting in dilution rates that are lower than cell growth rate, thus potentially playing an important role in both immigration into and emigration from the biofilm. This may yield a notably different progression of biofilm development and maintenance and thus a higher degree of fluidity than the discrete stages as depicted by the classical view of biofilm development.

3.1 Introduction

The recognition that biofilms, defined as surface-associated communities of microorganisms, are the prevalent mode in which microbes exist has, in addition to the myriad of areas where biofilms intersect with human interests, resulted in an

O. Kroukamp (✉) · G. M. Wolfaardt
Department of Chemistry and Biology, Ryerson University, Toronto, ON, Canada
e-mail: mkroukam@ryerson.ca

E. Bester
Department of Microbiology, University of Stellenbosch, Matieland, Stellenbosch, South Africa

C. J. Hurst (ed.), *The Structure and Function of Aquatic Microbial Communities*,
Advances in Environmental Microbiology 7,
https://doi.org/10.1007/978-3-030-16775-2_3

exponential increase in research efforts. A recent search of Elsevier's Scopus database (www.scopus.com) for articles published between 1976 and 2017 with the term "biofilm" in the title yielded 16,286 documents. Whereas 50 articles were published in 1990, this number grew consistently in subsequent years to 182 (2000), 504 (2005), 974 (2010), and 1790 (2016). As could be expected, the main subject areas represented by these publications were immunology and microbiology (36.1%), medicine (32.7%), biochemistry, genetics, and biology (29.7%), followed by environmental science (22.5%) and agricultural and biological sciences (17.9%). The remainder consisted of publications in fields as diverse as dentistry, chemical engineering, physics and astronomy, materials science, and mathematics, to name but a few. While this illustrates the multidisciplinary nature of biofilm-related research, it should also provide a sobering reality for newcomers to the field; the incorporation of fundamental principles from multiple, unrelated disciplines typically is required to adequately investigate biofilms. This naturally implies that investigators must not only be aware of but also become conversant in subjects that may fall outside of their primary training. Furthermore, by design and necessity, microbiological research remains firmly entrenched in homogenous, pure culture studies involving planktonic suspensions of microbes, despite the fact that spatially and temporally heterogeneous, surface-associated aggregates, consisting of multiple microbial species are more representative of prokaryote existence outside of the laboratory. In this chapter, we aim to illustrate how critical it is to foster an adequate awareness and understanding of the impact of the environment on biofilms. To achieve this, we focus on key physical aspects of aqueous environment and their influence on both individual bacterial cells and biofilms. Judicious use of analogies is advocated as tools to aid in the simplification of complex concepts to, firstly, convey information, secondly, facilitate understanding, and, thirdly, allow selection of the most crucial parameters to incorporate in experimental studies.

3.2 Rationale

3.2.1 Living Beings Interact with the Environment

Living beings are influenced by their environment and impact their environment in return. These interactions and resultant reciprocal changes may be minuscule or significant. In the field of biological sciences, we usually try to gain an increased understanding of a focus area by taking measurements during a carefully designed experiment. These collective measurements are then repackaged into descriptions, models, theories, and representations to (often incrementally) answer the questions of "what" and "how" that inspired our curiosity in the first place. For example, Quinn and Keough (2002) show in their first chapter that the scientific method involves stating a research hypothesis (as opposed to statistical hypothesis testing for inferential statistics) and how, with proper experimental design involving the measurement of copepod (predator) mortality, it is possible to postulate a model formulating a probable reason why dinoflagellates (prey) luminesce when the water is stirred.

The particular model mentioned by Quinn and Keough (2002) was referred to as a "burglar alarm" where the experimenters predicted that dinoflagellate bioluminescence would attract fish (copepod predators) leading to increased copepod mortality and indirectly to greater dinoflagellate survival. In this example, the experimenters seemingly considered sufficient role players (fish, copepods, luminescent and nonluminescent dinoflagellates) to address their research hypothesis after measuring copepod mortality. What was inherent, but perhaps invisible and therefore not considered in the experiment, was the particular environment where everything took place.

David Foster Wallace (2008) opened his Kenyon College commencement address of 2005 as follows:

> There are these two young fish swimming along and they happen to meet an older fish swimming the other way, who nods at them and says "Morning, boys. How's the water?" And the two young fish swim on for a bit, and then eventually one of them looks over at the other and goes "What the hell is water?"

As experimenters we remain vulnerable to let the obvious go unnoticed. In addition, we may also be lacking either in the ability to measure some (or all) environmental parameters, or being unable to accurately simulate these in an experiment. The latter could be because we do not fully understand which parameters, or interactions between parameters, are most relevant or because it is either physically impossible or too expensive to adequately re-create those relevant conditions within the laboratory. Naturally, the particular environment under which we conducted the experiment may not have influenced our design and eventual conclusion—but even in such cases it is best practice to describe the environment as thoroughly as possible rather than to appear ignorant about its influence. Since the influence from the environment is a given, the question is therefore simply whether the magnitude of influence is relevant or not. For small organisms such as microbes, a particular environment will affect the metabolism, growth, motility, gene expression, and behavior. This is true for single microbes as well as extracellular polymeric substance (EPS) matrix-encased biofilms which exist in close proximity as well as attached to surfaces. In this chapter we argue why the physical environment of aqueous biofilms should be considered in our experimental design, measurements, and interpretation of results.

3.2.2 Points of Consideration for the Next Generation of Biofilm Researchers

Biofilm research occurs and is perceived from many vantage points, including engineering, physics, and biological sciences, with each field making a unique and important contribution to the whole. Novice researchers entering the field of biofilm research will bring along with them an understanding and vocabulary native to their field. They will describe and interpret their findings through the lenses of their experience. While engineers and physicists are trained with an acute awareness of physical and chemical environmental contributors, biologists often have minimal such awareness, and it is precisely this limitation that may hinder progress in the field.

When encountering any new field of study, students often are encouraged to use analogies as a method and tool for simplifying complex scenarios. This tool has deservedly stood the test of time as both a communication aid and a simplified means of fostering understanding. However, despite our best intentions, the potential exists that the similarities and correspondence with the original may break down at some point or that our intuition may fail us.

In this chapter we will indeed introduce analogies, while attempting to be mindful of our assumptions, to gain insight into the physical environment encountered by microbes in and around a biofilm. We hope to achieve this by first assembling a list of key aspects which will be considered parameters relating to the aqueous biofilm environment with special emphasis on those aspects most affecting of microbes. In the final section, these parameters will be applied to an example where awareness of the environment may aid in experimental design and interpretation of experimental results, thus leading to an increased understanding of the world around us.

3.3 Background

3.3.1 Microbiology Legacy

Human beings are inextricably intertwined with microbes—perhaps in no way more evidently than the fact that each living human body houses more microbial genes and cells than it does those of the human host (Qin et al. 2010). On any given day, the human path intersects with microbes or microbial functions that directly or indirectly influence the quality of life in areas as diverse as human health, enjoyment (food), and environmental sustainability (nutrient cycles). It is no wonder then that humans are motivated to understand and control microbial growth, death, and function. Although the collective effects of microbes have been observed for millennia, their existence went largely undetected until the invention of the microscope. Since microbes were originally interesting to humans by virtue of their effects (disease, fermentation, etc.), and measurable (noticeable) effects usually require large numbers of microbes, these organisms were and still are grown to sufficient quantities to satisfy the resolution of the experimental techniques of the day. For example, the well-mixed, pure culture batch flask allowed early researchers to learn more about a particular microbial species under specific growth conditions (e.g., at a specific time point) by assuming that the measured parameter of a homogenous collective was a representative average of each individual member. Even though the population's metabolism may change over time, for example, when switching from one carbon source to another (diauxic shift) or from one electron acceptor to another, the microbes will all display similar behavior on average. In contrast, systems that are not well-mixed such as static batch cultures and biofilms (more about the latter in upcoming sections) and the responses of their constituent microbes to gradients of various forms (e.g., nutrients, redox potential) are expected to develop over time (Stewart 2003). It should be noted, however, that it may be hard to distinguish whether gradients develop first, followed by changes in microbial physiology, or whether microbial activity leads to gradients of environmental parameters (or a combination of both processes occurring simultaneously). Since each microbe is acutely affected by its immediate environment (e.g., temperature, solute concentration, viscosity, etc.), the assumption of spatial homogeneity is a very convenient one from an experimental perspective, especially given that an empirical result can be reproduced fairly predictably under similar conditions. It is therefore not surprising that the legacy of studying microbiology from a spatially homogenous viewpoint still remains in textbooks and undergraduate training. Outside the laboratory, however, microbiology occurs in spatially heterogeneous systems such as biofilms (Geesey et al. 1978).

3.3.2 *Spatial and Temporal Heterogeneity Associated with Biofilms*

Despite the observation that biofilms formed by various bacterial species proceed through similar steps and exhibit comparable properties, researchers have been unable to determine whether a biofilm-specific genetic program is in operation or if the resultant properties represent the culmination of the response and adaptation of individual cells to local environmental change (Kjelleberg and Givskov 2007). The inability to identify a biofilm-specific program may in fact be due to the acknowledged spatial heterogeneity of biofilms (Stewart and Franklin 2008). The physical, chemical, and biological properties of microenvironments within biofilms fluctuate continually due to a combination of various factors, which may include nutrient and oxygen diffusion gradients, sloughing, predation, etc. Individual cells respond and adapt to the prevailing conditions, leading to the establishment of physiologically differentiated subpopulations within a biofilm.

Due to its nature, a biofilm (even a pure culture biofilm) will be spatially heterogeneous in terms of:

- Chemical environment, e.g.:
  - Nutrients (both electron acceptor and donor)
  - Excreted products
  - Hydrogen ion (pH) concentrations
- Physiological manifestation
  - Gene expression
  - Mode of metabolism
  - Motility

The resulting spatial heterogeneity complicates a study since the emergent properties of the interactions encountered may be difficult or impossible to foresee or delineate. While this spatial heterogeneity is evident even in pure culture biofilms, it is not a big leap to conclude that the complexity may be much more pronounced in a multispecies environment. As a result, the strategy to research biofilm cells and their behavior has had to be adapted, when necessary, to address this lack of experimental resolution. Depending on whether viewed from an engineering (applied) or fundamental perspective, one could focus on global effects (overall mass balances or functional measurements) or conversely shift toward experimental techniques adapted to microscale to study local subpopulations within the biofilm with regard to their gene expression and immediate surroundings (De Beer et al. 1996; Rani et al. 2007). Fortunately, researchers are starting to recognize how heterogeneity stemming from single cell behavior affects collective function (Martins and Locke 2015).

Even more so than is the case for planktonic operation, explicit experimental conditions [e.g., experiments conducted under static flow conditions at solid-air interface (i.e., colony biofilms) or solid-liquid interface (i.e., microtiter plate biofilm assays), or flow conditions (i.e., flow cells or moving bed biofilm reactor)] should be taken into account when interpreting any biofilm-related experimental results given the added complexity resulting from granular spatial heterogeneity. Lewandowski and Beyenal (2013) argue for the need to consider the different levels of focus, whether it involves the very close scrutiny of microelectrodes or comparing the efficiency of different biofilm carriers in wastewater treatment.

While all of this might seem daunting to new researchers in the field, we hope to show that familiarization with a few key concepts will aid substantially in both understanding and communicating results and interpretations. In the next section, we will address certain areas related to how microbes respond to their physical environment, with a focus on aqueous systems, and how mastery of these building blocks can act as implements in a toolbox where the researcher can use some and neglect others to make sense of real-life observations.

3.4 Appropriate Analogies and Rubrics

A preferred means to convey new and complex concepts is by way of analogies. Analogies are based on prior knowledge, thereby making it easier for the reader (or listener) to categorize, classify, and position the new knowledge. In addition to facilitating understanding, analogies have the added advantage that complex concepts can be communicated parsimoniously, e.g., a bean being described as kidney-shaped and vice versa.

Once these analogies have been related, it can be assembled into rubrics or lenses for the researcher (or reader) to communicate. Given that the environment affects microbial behavior, it is important to avoid drawing conclusions from generalized inductive reasoning without regard for the experimental conditions that contribute toward the measured microbial behavior. Conversely, an undiscerning, exhaustive listing of experimental conditions may be counterproductive in that it may limit understanding or be too cumbersome for the reader to wade through. The benefit of these rubrics is that areas of observation (whether on the scale of a research study or a measured parameter) can be evaluated rapidly and effectively allow for the capturing of the relevant important contributing factors without having to be exhaustive. For example, Flemming et al. (2007) describes the extracellular polymeric substances (EPS) as "The house of biofilm cells." With our understanding of a house, this mental picture makes it easier to conceptualize and explain different biofilm behaviors resulting from the presence of EPS such as structural stability, exchange of genetic material (communication and transfer of knowledge), and retention of nutrients and crucial enzymes. Similarly, Watnick and Kolter (2000) use the analogy of "biofilms being cities of microbes" to explain biofilm development and spatial arrangement by correlating it to geographical settling patterns and zoning laws found in a city.

Analogies can be of great benefit during the process of gaining understanding in an unknown or related field. However, caution should always be exercised in that one should remain aware of terminology use as well as the explicit and inherent assumptions—especially those that are valid in the progenitor field or system—that can perhaps indiscriminately be carried over to the new endeavor. For example, when we attempt to explain microbial behavior by using anthropomorphic terminologies such as altruism, it is often challenging to remain neutral about the human components associated with this term such as choice, compassion, goodwill, willing sacrifice, etc. Referring to microbial altruism and cooperation, West et al. (2007) stated:

> However, progress is often hindered by poor communication between scientists, with different people using the same term to mean different things, or different terms to mean the same thing. This can obscure what is biologically important, and what is not. The potential for such semantic confusion is greatest with interdisciplinary research.

With this in mind, an analogy to describe biofilm development under aqueous conditions is presented in the next section. Our aim is to provide a framework to evaluate biofilm experimental results by focusing on the differences in environmental conditions that result in many of the discrepancies observed between different biofilm studies conducted in aqueous environments. We hope to show that by simply being aware of whether or not the biofilm under consideration developed under flow versus stagnant conditions, the reader will become cognizant of how these differences influence viscosity, nutrient concentration, motility, and association with a surface and a biofilm, to name but a few parameters.

3.4.1 Thriving Port City or Besieged City?

According to this analogy, an aqueous biofilm surrounded by flowing conditions can be envisaged as a port city with its harbor. Pertinent to this analogy, one can consider three distinct spatial zones related to the flow-surrounded biofilm and discuss how these zones interact via space and time to influence characteristic biofilm behavior. The three zones under consideration include the free-flowing bulk liquid, the fluid adjacent to the surface or biofilm, and the gel-like biofilm itself which consists of microbial cells and the EPS matrix. The characteristics of a port city naturally lends itself as a focal point for the initiation of colonization, but even long after the establishment of thriving communities, the same "gateway" character of these cities maintains their strategic importance. For example, when considering an established coastal country, we are usually aware of the importance of port cities especially as it relates to world commerce (as well as immigration and emigration during the time prior to air travel). But it is sometimes easy to forget about the strategic and crucial early roles that these port cities played in the establishment of the particular country or nation—e.g.:

- Regional trade between coast and hinterland
- Transit points for goods and people
- Possible entry points for invasions
- Immigration and emigration
- World commerce (Gilbert 2006)

During the days of naval exploration, *natural harbors* were prized as access points connecting the motherland (via the ocean) to the new land to be explored. The sheltered nature of the natural harbor provided safety and rest after being exposed and vulnerable during the ocean voyage, as well as access to the potential renewal of sustenance available on land. As previously mentioned, the comparison of a biofilm to a port city and its associated harbor will involve the recognition of three spatially differentiated zones. Each of these zones will be discussed in terms of physical characteristics (e.g., viscosity, diffusion, and advection), the impact of these characteristics on microbial motility, and how both the aspects of spatial zones and physical characteristics relate to emigrating from and immigration to microbial life on surfaces.

While port cities epitomize the concept of flow and facilitation of maintenance and growth, both in terms of resources and residents, a besieged city demonstrates the opposite. The absence of flow (or severely hindered movement) engenders conditions of stress, starvation, stagnation, refuse buildup, dormancy, and death. Compared to the flow system, its stagnant counterpart will conceptually be divided into two zones: the biofilm attached to a solid surface and the overlaying stagnant bulk fluid. Continuing with our analogy, the stagnant system could be considered a city in a swamp where all inhabitants and food will enter or exit at a much slower rate compared to a flow system, to the extent that stress and starvation conditions will be the eventual outcome if population growth is not controlled.

It should be noted that the two analogies of port cities and besieged cities were arbitrarily chosen and are by no means all-encompassing descriptions of biofilm environments but rather very commonly occurring scenarios. Prior to continuing with the above analogy and describing how the physical conditions in each of these zones impact on various aspects of microbial life at or near surfaces, it is essential to first look at how a microbe experiences a watery environment. This may seem unnecessary if we inadvertently assume that a microbe will behave in a manner analogous to humans or another macroscopic organism's swimming behavior. However, in the course of the following sections, the relevance of doing so should become evident.

The selection of microbial swimming as a focus area is intended to demonstrate how this seemingly simple, often observed bacterial parameter can be influenced by its environment.

3.5 Key Physical Parameters

3.5.1 Microbes in Aqueous Solutions

General biology training usually does not include a comprehensive study of fluids either at rest or in motion (fluid mechanics), and even when included in the curriculum, it must be recognized that applying fluid mechanics to biological systems requires facing unique assumptions and challenges. For example, not all "rule of thumb" numbers such as transition values for Reynolds numbers and boundary layer thickness are applicable, as reviewed by Alexander (2016), and therefore this subject matter might be better taught with specific applications and constraints in mind (Loudon 1999) rather than as broad generalities only.

Here, we do not attempt to act as an authoritative text on fluid mechanics but will rather point the reader to worthwhile resources while at the same time hoping to raise an increased awareness of key concepts (Persat et al. 2015). As an example, due to the microscopic size of a microorganism, the impact of a watery environment is quite distinct from that experienced by a macroscopic organism (e.g., a fish or a human). The ability of a bacterium to propel itself relative to its surroundings when submersed in an aqueous solution is equivalent to a human being's ability to move around in honey. This is a prime example of an instance where attempting to interpret a scenario through the lens of a known experience, obtained either through observation or personal experience (e.g., swimming), could result in erroneous interpretation of experimental results and thus reaching an incorrect conclusion.

In the next few sections, the dimensionless Reynolds number will provide a lens to understand the challenges encountered by microbial swimmers.

3.5.2 The Reynolds Number

Physicists, mathematicians, and engineers have long attempted to understand their objects of study and translate this understanding into models, theories, or laws that could be encoded into a mathematical formula or theorem. Ideally, it would be possible to fully describe every object or process in the natural world with elegant mathematical formulas for which all parameters and variables are known or could be measured. In practice, however, in many instances scientific efforts lead to the development of unwieldy equations with mathematical operators that require advanced training to understand and solve. To overcome these limitations, assumptions are made that allow simplification of the formulas to address relevant questions related to a particular aspect of the object of study. The choice of which parameters to include or exclude in experimental observations and measurements is usually dictated by some characteristic that will satisfy a practical need for the investigator, for example, excluding parameters that are not deemed to contribute significantly.

In the case presented here, the object of observation is either a fluid moving relative to a rigid object such as a fluid moving in a conduit such as a pipe or a swimming body moving relative to its surrounding fluid. The question that we wish to answer relates to whether the characteristics of the fluid and the characteristics of the rigid object influence the movement of one another (if at all). The Reynolds number, which describes the ratio of inertial to viscous forces of a fluid under defined conditions, can be applied to answer this question.

3.5.2.1 Inertial Force of a Fluid

When a fluid element (i.e., a small volume or unit of water) moves through space, an inertial force is ascribed to it which will resist a change in momentum; the latter could be either a change in direction or velocity. This inertial force is directly proportional to the density of the fluid and the square of the characteristic velocity meaning that when the fluid element has higher mass (as represented by the density term) or higher velocity, greater force will be required to stop it. The inertial force (F_{inertia}) is derived from Newton's second law of motion:

$$F_{\text{inertia}} = ma \tag{3.1}$$

$$m = \rho V = \rho l^3 \tag{3.2}$$

$$a = \frac{U}{t} = \frac{Ul}{tl} = \frac{U^2}{l} \tag{3.3}$$

$$F_{\text{inertia}} = \rho U^2 l^2 \tag{3.4}$$

where m is mass, a is acceleration, ρ is the liquid density, V is the volume, l is the characteristic length, U is the velocity of the fluid, and t is time.

3.5.2.2 Viscosity of a Fluid

Another force that may play a role in the movement of a fluid element is that of viscosity (F_{viscous}). Fluid sticks to itself, so if one pictures adjacent water elements as sheets of paper, the viscosity would be an indication of the stickiness between these sheets or the "glueyness" of the fluid (Vogel 1994). Viscosity can also be viewed as fluid friction, with friction being an indication of the resistance generated upon the relative movement of two solid objects that are in contact with each other. Key to understanding viscosity as fluid friction is the realization that neighboring fluid elements are moving relative to each other at different velocities. For example, a sheet of water moving at a faster velocity will be slowed down by an adjacent, slower moving sheet. Conversely, the faster moving sheet will tend to drag the slower moving sheet at an increased velocity and if the adjacent sheets of water are moving at the same velocity, the viscous force between these will disappear.

With regard to viscosity, we have thus far only considered fluid moving relative to itself; however these observations can be extended to the case where a fluid moves relative to a solid object. In addition to fluid elements sticking to one another, the fluid will also stick to a solid object that it is in contact with. In fact, the extent of this affinity is so great that this phenomenon is referred to as the "no-slip" condition. This means that this layer of fluid is considered to have zero velocity relative to the surface of the object (i.e., both are moving at the same velocity). By implication, we can infer that a liquid velocity gradient will extend from the solid surface into the bulk liquid phase. A simple example to demonstrate this is by observing debris floating down a slow flowing river that travel slower when close to the bank compared to those floating in the middle of the stream. From this we can conclude that solid surfaces in contact with a fluid, whether a static pipe wall or a solid swimming body immersed in a fluid, will always result in viscous forces being present in the layers close to the surface where these velocity gradients exist.

The derivation for the equation for viscous force (F_{viscous}) can be found in most fluid mechanic textbooks and is given by

$$F_{\text{viscous}} = \frac{\mu U l^2}{l} \tag{3.5}$$

where μ is the dynamic viscosity of the fluid.

Terminology can sometimes be confusing, especially when a word might have one meaning in everyday speech but a different or nuanced meaning in a certain research field or when particular background knowledge is required to avoid misinterpretation. An example of the latter is found in fluid mechanics literature, where the fluid layer close to a surface is referred to as the viscous layer or the "region where viscous flow occurs" as opposed to inviscid flow further away from the surface. A person familiar with the field will realize that the dynamic viscosity for an incompressible Newtonian fluid such as water is indeed a constant (as is evident from the derivation of the shear stress equation—see any fluid mechanics textbook for an example), but a non-expert may assume that the viscosity decreases with increasing distance away from the surface.

When considering the movement of a fluid element, the analogies from Vogel (1994) are very useful: he describes inertial forces as reflecting the "individuality" of these fluid elements, while the viscous force reveals their "groupiness." In practice, this can be viewed as follows: when inertial forces dominate, fluid movement can be described by the progress of a milling crowd, while the movement resembles that of a disciplined march when viscous forces prevail (Vogel 1994).

3.5.2.3 Reynolds Number Equation

For specific flow conditions and geometries, it is therefore possible to comment on whether the flow conditions are dominated by inertial or viscous forces by considering the ratio of these forces acting on a fluid element. This ratio, represented by the dimensionless Reynolds number (Re), is indeed frequently used in the field of fluid mechanics to make a distinction between inertia-dominated flow (turbulent flow) and viscosity-dominated flow (laminar flow).

$$\mathrm{Re} = \frac{F_{\text{inertia}}}{F_{\text{viscocity}}} = \frac{\rho l^2 U^2}{\mu l^2 U/l} = \frac{\rho l U}{\mu} \tag{3.6}$$

with the characteristic length, l, and characteristic velocity, U, being specific to different flow geometries and conditions.

3.5.2.4 Points for Consideration When Applying the Reynolds Number

In reference to the ideal scenario of having appropriate mathematical formulas to assist our understanding of the physical environment, Vogel (1994) praises the usefulness of the abovementioned ratio by calling the:

> Peculiarly powerful Reynolds number the centrepiece of biological fluid mechanics.

And Vogel goes on to say that this “almost magical variable” is “the nearest thing we have to a completely general guide to what is likely to happen when solid and fluid move with respect to each other.”

After reading these commendations from an expert in the field of fluid mechanics, a biologist might conclude that he or she has encountered some sort of equational jackpot—which it indeed can be—but this is only true after careful consideration of how the Reynolds number should be understood and used.

In our case, we described the Reynolds number as the ratio of forces acting on a fluid element. However, it should be noted that there are various physical interpretations of the Reynolds number, four of which are mentioned by Lauga and Powers (2009). For the purpose described in this chapter, the main function of the Reynolds number is to provide an indication under which flow conditions inertial (turbulent) or viscous (laminar) forces will dominate, given the characteristics (e.g., density, viscosity, velocity, etc.) of the fluid under consideration. Rather than attaching importance to the specific numeric value of the Reynolds number, close attention should be paid to the critical Reynolds number, which indicates when flow transitions from laminar to turbulent. For example, the transition Reynolds number for flow in a circular pipe starts at approximately 2000, and turbulent flow can be expected at a Reynolds number of 4000 (Brading et al. 1995). It is important to note that the transition Reynolds number is dependent on the geometry and flow conditions. For example, each geometry results in a different definition for the characteristic length, l, and the characteristic velocity, U; e.g., for a circular pipe, the characteristic length is equal to the inner diameter of the pipe, whereas the characteristic velocity is the average fluid velocity. In essence, the Reynolds number indicates the character of the flow regardless which of the contributing variables cause the change. A tenfold reduction in characteristic length will increase relative viscous effects by a factor of 10, similar to what would happen if the viscosity increased tenfold (Vogel 1994).

3.5.2.5 Reynolds Number for a Swimming Body Such as a Bacterium

Osborne Reynolds (after which the Reynolds number was named) investigated the transition from laminar to turbulent flow in pipes, which represent a fluid moving past a stationary surface. However, the same reasoning regarding the ratio of inertial forces to viscous forces can be applied to a rigid body moving through stationary fluid, given that the appropriate characteristic length and characteristic viscosity are defined. For a bacterial cell moving through a fluid, the characteristic length is taken as the greatest length of the swimming body in the direction of flow, and the characteristic velocity is the difference in velocity between the swimmer and the velocity of the bulk fluid. In stationary fluid, the bulk fluid velocity will be reduced to zero and the characteristic velocity used to calculate the Reynolds number will be equal to the velocity of the swimmer.

The Reynolds number for a swimming microorganism can therefore be determined by using the following approximate values (Lauga and Powers 2009); the density and dynamic viscosity of water can be given by $\rho \approx 10^3$ kg m^{-3} and $\mu \approx 10^{-3}$ Pa s, respectively, the characteristic length of microbes ranges between 1 and 10 µm, and the characteristic speed is $\approx$10–30 µm s^{-1}. These values will yield a Reynolds number of approximately 10^{-5} to 10^{-4}; these very low numbers indicate that a microorganism swimming in water will experience strong viscous resistance. In comparison, a 2-m-tall human, swimming at 1–2 m s^{-1} in water, would be equivalent to a Reynolds number in the order of 10^5 to 10^4, thus indicating a dominance of inertial forces. Only if a human would attempt to swim in honey, with a dynamic viscosity roughly 10,000 times that of water, would this person experience viscosity-dominated laminar swimming conditions similar to what bacteria experience when moving in water.

Related to whether viscous forces will dominate or not, we can use the same equations and reasoning to determine what distance a swimmer will coast (i.e., float) after all activity has stopped. Lauga and Powers (2009) derived equations to show that the coasting distance, d, under inertia-dominated flow conditions is given by

$$d \approx l \frac{\rho_{\text{swimmer}}}{\rho_{\text{fluid}}} \tag{3.7}$$

where l is the characteristic length of the swimmer and ρ the density of the swimmer and fluid, respectively. This equation indicates that a human swimmer will coast for a few meters in the water before coming to a standstill.

Under conditions where viscous forces dominate, the coasting distance is given by

$$d \approx l\text{Re}\frac{\rho_{\text{swimmer}}}{\rho_{\text{fluid}}} \tag{3.8}$$

where l is again the characteristic length and Re is the Reynolds number calculated as before, indicating that a bacterium will slow down after about 0.1 nm, within a time frame in the order of microseconds (Purcell 1977).

3.5.3 Reduced Flow Velocity Near a Surface

The liquid flowing near a wall, such as the inner surface of a pipe, will be dominated by viscous forces under both laminar and turbulent conditions. This region is associated with a gradient in flow velocity, with increasing fluid velocity observed further away from a surface. While the velocity profile in this region is strictly speaking parabolic, the assumption of linearity is acceptable if only a small distance is considered (Rao 2010). Engineers refer to this region as the hydrodynamic boundary layer, and conceptually it separates this fluid from the zones further away from the surface where viscous forces are negligible compared to inertial forces (Brading et al. 1995). Under turbulent flow conditions, the flow close to the surface is referred to as the "laminar sublayer" or "viscous sublayer" (Brading et al. 1995). From a biological perspective (e.g., biofilms growing on a pipe surface), the region close to the surface can be viewed as a "surface microenvironment" (Caldwell and Lawrence 1988; Watnick and Kolter 2000) as opposed to the macro-environment encountered in the bulk fluid flow further away from the surface.

For smooth surfaces a conservative thickness, δ, for this sublayer can experimentally be determined as

$$\delta = \frac{5v}{\sqrt{\tau_{\text{W}}/\rho}} \tag{3.9}$$

where $v(v = \mu/\rho)$ is the kinematic viscosity of the fluid, ρ is the density of the fluid, and τ_{w} is the shear stress at the wall (Rao 2010; Agrawal 2012). The thickness of δ is determined experimentally from velocity profile graphs (the regions where these profiles dominate and where they intersect) portraying the different layers that occur under turbulent flow conditions. Depending on which portions of a particular equation or specific intersection of two profiles are used, some textbooks will provide a δ with a different constant (Allan 1995; Rubin and Atkinson 2001), e.g.:

$$\delta = \frac{11.6\nu}{\sqrt{\tau_W/\rho}} \tag{3.10}$$

More detailed explanations for this discrepancy can further be explored elsewhere (White 2010). Despite the use of these different constants, an approximation of the thickness of the viscous sublayer can be garnered from reported values, although scarce in the current literature. Rough approximations for industrially (Rubin and Atkinson 2001) and environmentally relevant conditions (Kumarasamy and Maharaj 2015) estimate this thickness to be in the order of 10^2 μm. This would suggest that the viscous sublayer may provide sufficiently retarded flow close to the surface for microbes to attach and form biofilms even under turbulent flow conditions. The latter has been reported for *Listeria innocua* (Perni et al. 2006), mixed species biofilms (Percival et al. 1999), and *E. coli* (Teodósio et al. 2011).

3.5.4 Flow in Non-Newtonian Fluids Such as a Biofilm

Up to now we have mostly considered Newtonian fluids (e.g., water) where the viscosity does not change when the rate of shear strain (change of strain or deformation with respect to time) is increased or decreased. However, the behavior of Newtonian fluids can change drastically when small particles are suspended in the liquid or when macromolecules are dissolved in the liquid (Brown and Jaeger 2011). The focus now turns to two aspects related to this nonlinear fluid behavior as it relates to biofilms, i.e., shear strain rate-induced nonlinearities (shear-thickening and shear-thinning) and viscoelastic behavior.

Some suspensions will be liquid-like when slightly perturbed, but once exposed to higher impact, an increase in viscosity will harden the liquid to such an extent that it could support a person running across it without sinking into it (Waitukaitis and Jaeger 2012). This phenomenon is termed shear-thickening, whereas a decrease in viscosity upon an increased rate of shear strain is called shear-thinning. Since a particular fluid may change its behavior depending on the shear rate applied, a fluid cannot be unequivocally classified as either shear-thinning or shear-thickening. These terms should therefore rather be considered as dependent on the flow conditions, as opposed to a characteristic of the fluid itself (Mewis 2012). Furthermore, the definitions of shear-thinning and shear-thickening (often portrayed as a graph of shear stress vs strain rate) represent another example where the terminology may be misleading since these terms refer to *shear*, while the definitions require strain rate to be the controlling factor (Mewis 2012). Biofilms or EPS matrix encapsulated microbial cells (Flemming and Wingender 2010) have been described as a complex fluid (Wilking et al. 2011) and recorded to display shear-thinning behavior (Houari et al. 2008; Billings et al. 2015; Patsios et al. 2015).

Some sources in the literature may also refer to "strain hardening" or "shear stiffening" (functional definitions will be mentioned later in this paragraph) (Barai et al. 2016), but these terms should not be confused with shear-thickening or shear-thinning (Mewis 2012). Rather, the former terms are used to indicate the occurrence of viscoelasticity, which is normally determined from a plot of shear stress vs strain (Fabbri and Stoodley 2016), as opposed to using plots of shear stress vs strain rate to visualize shear-thickening or shear-thinning behavior (Stoodley et al. 1999a, 2002a; Klapper et al. 2002; Peterson et al. 2015). A viscoelastic material will display both viscous and elastic properties, meaning that it deforms when placed under stress. Once the stress is removed, the material returns to its previous state, which may seem similar, but not necessarily identical, to the original state (Peterson et al. 2015; Fabbri and Stoodley 2016).

3.6 Bacterial Locomotion

In the previous section, key interactions between the movement of fluids and solid surfaces or bodies relative to one another were discussed. In the following section, focus is shifted to bacterial motility with the aim of understanding how microbes reach and interact with surfaces, since this is a crucial step in biofilm development. It should be noted that while bacterial motility plays a role in surface association, not all microbes are capable of locomotion and furthermore that other factors, such as fluid hydrodynamics, do play a role in translocating microbes to surfaces.

Prokaryotes can propel themselves in different ways through liquids, including swimming, swarming, gliding, twitching, or floating, among others as reviewed by Jarrell and McBride (2008). Those authors detailed various mechanisms for movement, including the use of surface appendages such as flagella and pili, or internal structures such as the cytoskeleton and gas vesicles. The method of propulsion and mechanism does not always correspond; for example, swimming with or without flagella or gliding with or without pili is possible. Since the scope of this chapter is limited to aqueous environments, the main focus of this section will be flagellar swimming. Most bacterial flagella are thin ($\sim$20–50 nm diameter) helical structures (Berg 2003) that can extend from the cell body for several cell lengths. Information regarding flagellar structure, assembly (Macnab 2003), and regulatory genes has been reviewed by Jarrell and McBride (2008).

3.6.1 Single Cell Swimming

While flagellar-mediated swimming is a common means of microbial propulsion, there are a few subclassifications of this mode of locomotion that become relevant to the discussion of near-surface swimming. In particular, we will consider differences in flagellar number, location or arrangement on the cell body, and swimming method.

The number and location of prokaryotic flagella vary among different species (Merino et al. 2006). Examples include single (monotrichous) or multiple (lophotrichous) polar flagella as found in *Pseudomonas aeruginosa* and *Pseudomonas putida*, respectively (Theves et al. 2013), and multiple flagella distributed uniformly over the cell body (peritrichous) such as seen for *Escherichia coli* and *Salmonella enterica* (Berg 2003; Merino et al. 2006). The specific swimming method strongly influences microbial behavior when an organism is interacting with its environment, such as when it encounters surfaces or nutrient gradients. Two important components of the method of swimming are speed and the ability to change direction, with the mechanism of the latter varying substantially among different flagellar arrangements.

Bacterial swimming speeds vary widely; e.g., *Bdellovibrio bacteriovorus* with a single polar flagellum (Lambert et al. 2006) was observed to swim at 160 $\mu m\ s^{-1}$, whereas *P. putida* with multiple polar flagella was able to swim at speed of 75 $\mu m\ s^{-1}$ (Harwood et al. 1989), and *E. coli* was capable of reaching speed of 35 $\mu m\ s^{-1}$ with peritrichous flagella (McCarter 2005).

Microbes with a single polar flagellum, such as *P. aeruginosa*, propel themselves forward in a pusher motion by counterclockwise (CCW) rotation of their flagella or reverse their direction in a straight trajectory by clockwise (CW) rotation (puller motion; Cai et al. 2016). Peritrichous bacteria, such as *E. coli*, bundle their flagella, and CCW rotation similarly drives the cells forward (called a "run"), while clockwise rotation disrupts the flagellar bundle and results in a "tumble" (Jarrell and McBride 2008; Cai et al. 2016) or "twiddle" as the original authors pointed out (Berg and Brown 1972). It should be noted that the direction of motion resulting from CW or CCW rotation depends on the handedness of the flagella of a particular species. CCW rotation of the left-handed filaments of *E. coli* drives the cell forward, whereas the single, right-handed helical filament of the bacterium *Caulobacter crescentus* move the cells forward by rotating CW and reverse the cell by CCW rotation (Lele et al. 2015). Thus, a switch between CCW and CW flagellar rotation brings about both propulsion and direction changes for peritrichous bacteria, while alternating between CCW and CW rotation for bacteria with a single flagellum will merely move the cells forward or backward in a straight trajectory (in the absence of a nearby surface).

While not a focus area in this chapter, eukaryotic flagella are considered to be distinct from prokaryotic flagella (Berg 2003) in terms of location, structure, function, and swimming patterns (Polin et al. 2009; Guasto et al. 2012; Moran et al. 2014). Among the eukaryotes, *Chlamydomonas* (algal unicellular flagellates) are capable of beating their flagella in two different ways: a breaststroke and an undulatory (waves travelling down the flagellum) stroke (Tam and Hosoi 2011).

3.6.2 Collective Swimming

While only single cell swimming has been considered thus far, collective swimming by numerous cells results in interesting deviations from the norm and may influence physical fluid parameters such as viscosity and diffusion. For example, Lushi et al. (2014) investigated self-organizing patterns exhibited by dense suspensions of confined bacteria. However, the interactions between microbes and fluid parameters are of greater relevance to the present discussion and will remain the focus area.

Sokolov and Aranson (2009) measured the shear viscosity in thin films resulting from swimming *Bacillus subtilis* using two complementary experimental approaches [shear viscosity can be determined experimentally as the ratio of the tangential shear stress and the shear rate (Sunthar 2010)]. The first approach involved measuring the deterioration of a large vortex, whereas the second determined the viscous torque on a rotating magnetic particle in the presence of various concentrations of *B. subtilis* cells. The authors report a decrease in liquid viscosity of up to a factor of 7; this decrease was relative to the viscosity of a fluid containing non-motile or no bacteria, and it was dependent on the bacterial concentration and swimming speed (Sokolov and Aranson 2009). It seems, however, that the above observation was particular to pusher-type swimmers such as *B. subtilis* since Rafaï et al. (2010) showed experimentally that puller-type swimming by the unicellular microalgae, *Chlamydomonas reinhardtii*, resulted in an increase in fluid viscosity compared to the same concentration of dead cells.

Ishikawa (2009) discussed how puller-type swimmers will repel each other when swimming side by side, while pusher-type swimmers will attract each other. The author goes on to describe how suspensions of swimming microbes can affect both fluid rheology and diffusive properties (enhancing the diffusion of dissolved chemicals in the suspension) by influencing these fluid properties on a mesoscale (between microscopic and macroscopic; Ishikawa 2009). Underhill et al. (2008) expand on the latter as it relates to the mode of swimming and showed that the pusher motion enhanced the diffusion of tracer molecules more so than did puller-type swimming.

Experimental simulations of a bath containing swimming *E. coli* indicated a significantly larger effective diffusion coefficient for a tracer molecule suspended in the fluid, compared to the thermal effective diffusion coefficient exhibited by a fluid containing only passive particles (Morozov and Marenduzzo 2014). The authors concluded that this increase in effective diffusion is dependent on the concentration of bacteria, their swimming speed, as well as a characteristic velocity field created by a single swimming bacterium (Morozov and Marenduzzo 2014). Wolgemuth (2008) investigated the formation of transient jet and vortex patterns in the presence of high densities of *B. subtilis* using a theoretic model. According to the simulations performed, active swimming by the bacteria resulted in turbulent mixing of the fluid, with fluid velocities exceeding the maximum swimming speed set for individual bacteria.

3.6.3 Surface Interaction

3.6.3.1 Swimming Close to a Surface

The tendency for swimming cells to remain in close proximity to surfaces in a time frame ranging from seconds to minutes has been ascribed to hydrodynamic effects (Vigeant et al. 2002; DiLuzio et al. 2005; Conrad 2012), as was shown for non-tumbling *E. coli* (Berke et al. 2008).

Confirmation of the role of bacterial locomotion in surface association was provided by Galajda et al. (2007). The authors established that swimming rather than non-swimming microorganisms were trapped along a wall with micro-fabricated funnel-shaped openings. Pusher cells strongly migrate toward nearby boundaries or surfaces when few in number. However, at higher cell concentrations, this coherence is disrupted by large-scale interaction between swimming cells (Hernandez-Ortiz et al. 2005).

The accumulation of cells at a surface has also been demonstrated to depend on swimming speed and cell length, with faster and longer cells accumulating to a greater degree than slower and shorter ones (Li et al. 2011). In addition, the tumbling frequency of peritrichous bacteria was reduced by 50% within 20 μm of a surface and has been proposed to contribute to near-surface cell trapping (Molaei et al. 2014).

The proximity of a surface is known to influence microbial swimming behavior such as a change in the microbe's trajectory or a reorientation of a cell. For example, *E. coli* swims in a CW, circular motion near a solid boundary; rotation of *E. coli*'s left-handed helix will cause the cells to turn to the right (CW) (Lauga and Powers 2009). The circular motion observed when monotrichous bacteria swim in reverse near a surface has been described as "run-and-arc" swimming (Karimi et al. 2015) and has been reported for a number of monotrichous bacteria including *Caulobacter crescents* and *P. aeruginosa*. Pusher cells have been found to reorientate themselves parallel to a surface, while pullers will orientate themselves at a right angle with respect to the wall, thereby appearing to swim into it (Lauga and Powers 2009). While the singly flagellated *Vibrio alginolyticus* usually swims backward and forward by alternating the rotation direction of its flagella, it behaves differently near a surface. Its forward and backward swimming speeds have been observed to differ significantly and the swimming motion to trace a more circular trajectory. The latter effect was explained by fluid dynamic interactions between the cell and the rigid boundary (Goto et al. 2005).

3.6.3.2 Surface Sensing

The question may arise as to how a bacterium senses that it is near a surface. Cells are able to sense direct contact with a surface or experience surface-associated changes in hydrodynamic conditions via different mechanisms, one of which involves the restriction in flagellar rotation (Harapanahalli et al. 2015). Belas (2014) investigated the role of rotating flagella in surface sensing and found that rotating flagella are used as mechanosensors to detect subtle changes in the operation of their motors when they near a surface (Belas 2014), specifically via the motor stators (Lele et al. 2013). Obstruction of flagellar rotation may trigger adhesion and surface-associated motion (Ellison and Brun 2015). Cairns et al. (2013) discussed how the inhibition of flagellar rotation triggers a signal transduction cascade in *B. subtilis*. In addition to the differential expression of genes linked to motion, the association with surfaces and the related mechanical interactions may influence virulence factor-related gene expression. The latter was shown by Siryaporn et al. (2014) for *P. aeruginosa* after attachment to various, chemically distinct surfaces and is worth considering in the study of biofilms' role in infection.

However, it should be noted that not all surface-sensing pathways involve flagella, such as the case of non-motile bacteria (McClaine and Ford 2002; Belas 2014). Some bacteria sense a surface when experiencing adhesion force-induced deformation of the cell wall (Harapanahalli et al. 2015).

3.6.3.3 Shear Trapping

In the absence of flow, the density of a dilute suspension of non-tumbling, peritrichous *B. subtilis* was found to be uniform, whereas the introduction of flow led to the accumulation of cells at surfaces (Rusconi et al. 2014). An increase in flow rate or shear corresponded to an increase in the concentration of cells at the surface and was termed "shear trapping" (Rusconi et al. 2014; Rusconi and Stocker 2015). This phenomenon required motile cells, since dead cells did not show any variation in spatial density (similar to the instance of no-flow) (Rusconi et al. 2014). Shear trapping not only increases surface attachment but also suppresses chemotaxis (Rusconi et al. 2014) and the ability of a cell to respond to external stimuli (Bearon and Hazel 2015).

Similar results were observed for *P. aeruginosa* (Lecuyer et al. 2011) where the time that the bacteria remained adhered to the surface increased nearly linearly as a result of an increase in wall shear stress from zero to $\sim$3.5 Pa. The authors furthermore observed that while the absence of type I pili, type IV pili, the flagellum, or EPS building block production affected the frequency with which cells could become either adhered or detached, none of these factors could account for the average increase in adhesion time correspondent with shear (Lecuyer et al. 2011). Since the same result was observed for tumbling *B. subtilis* and the monotrichous *P. aeruginosa*, it appears plausible that this behavior occurs irrespective of the swimming motion. Early reports by Molaei et al. (2014) indicated an increased incidence of shear trapping under conditions without flow (Molaei et al. 2014). In a subsequent study, the group (Molaei and Sheng 2016) used digital holographic microscopy to track swimming *E. coli* near surfaces under shear. Under these conditions, the authors found that shear mitigated the tumbling inhibition, which in turn reduced cell trapping close to the surface.

3.6.3.4 Swimming Upstream

Contrary to what may be assumed, microorganisms are able to swim and translocate over surfaces against the prevailing direction of flow (Hill et al. 2007; Rusconi and Stocker 2015). For example, *P. aeruginosa* cells are initially orientated along the direction of flow, after which the retraction of the polar type IV pili allow upstream twitching (Shen et al. 2012).

Kaya and Koser (2012) observed that *E. coli* can swim upstream under flow conditions at speeds exceeding 20 μm s^{-1}. In a previous section, it was mentioned that *E. coli* swims in circles when encountering a surface under quiescent conditions. In this mode the cell experiences an increased hydrodynamic drag which rotates and dips the front of the cell body, thereby keeping the bacterium pointed toward the surface with the flagella oriented away from the surface (Kaya and Koser 2012). Under flow conditions the drag on the flagella oriented further away from the surface than the cell body will rotate the cell to face upstream. The authors concluded that this upstream swimming behavior is not unique to the propulsion mechanisms of *E. coli* but merely requires a swimming microorganism moving freely over a surface under moderate flow conditions (Kaya and Koser 2012). Korber et al. (1989) compared surface colonization of motile and non-motile *P. fluorescens* under two flow velocities, 8 and 120 μm s^{-1}. They found that flagellated cells not only attached more rapidly than did nonflagellated cells under both velocities but also migrated upstream against the laminar flow.

3.7 Dissolved Nutrients

In this chapter the main focus is on how microbes interact with their physical aqueous environment. As a result, the question of how microbes interact with their chemical environment, especially as it pertains to breathables (electron acceptors) and edibles (electron donors) (Nealson 2003), the gradients of these chemicals and how it relates to chemotaxis will only be discussed briefly—its importance, however, should always be considered when interpreting biofilm behavior.

For biofilms growing on surfaces, a mass transport or diffusion boundary layer exists within the hydrodynamic boundary layer (Lewandowski and Beyenal 2013). This mass transport boundary layer can be demonstrated by the nutrient concentration gradient increasing with increasing distance from the surface (assuming that nutrients found in the bulk is being consumed as it approaches the biofilm on the surface); at distances close to the surface where fluid velocity is low, mass transport is diffusion dominated, whereas the higher flow velocities prevailing at greater distances from the surface result in convective mass transport.

3.7.1 Diffusion and Advection

If one considers a stationary, nutrient-filled fluid surrounding a spherical stationary microbial cell that assimilates all nutrients reaching its surface, a nutrient concentration of zero will result at the cell's surface (Dusenbery 2009). The difference in nutrient concentration between the cell surface and the bulk fluid will generate a concentration gradient which in turn will be the driving force for diffusion from the bulk to the cell surface. Given that the concentration of the particular nutrient is zero (at the cell surface) under these conditions, the question is whether the microbe would be able to increase its rate of nutrient uptake by swimming? To attempt an answer to the above question, some rudimentary calculations will provide insight into the relative contributions of the two main mechanisms of solute transport through a solution, namely, flow and diffusion. Under conditions of flow (advective transport) at velocity U, the time, t_a, required for a solute to move a distance l can be approximated (Purcell 1977; Dusenbery 2009) by

$$t_{\mathrm{a}} \approx \frac{l}{U} \tag{3.11}$$

For diffusion the time, t_d (Purcell 1977; Dusenbery 2009), for transport can be approximated as

$$t_{\mathrm{d}} \approx \frac{l^2}{D} \tag{3.12}$$

where D is the diffusion coefficient. One consequence of time being proportional to the squared distance is that diffusion times are very short over small distances; a small molecule will diffuse across a bacterial cell (~1 μm) in approximately a millisecond (approximating the diffusion coefficient in water as 10^{-9} m^2 s^{-1}). Both of the above approximations of transport time are derived from the one dimensional advective diffusion equation for an incompressible fluid where variable sizes have been estimated on the level of orders of magnitude rather than precise values. If the ratio of diffusion transport time to that of flow (advection) transport time is taken, we arrive at the dimensionless Péclet number (Vogel 2004; Dusenbery 2009):

$$Pe = \frac{t_{\mathrm{d}}}{t_{\mathrm{a}}} = \frac{Ul}{D} \tag{3.13}$$

Similar to the Reynolds number discussed earlier, the Péclet number is used to comment on the occurrence of certain regimes of flow (transport dominated by advection or diffusion) and exact values are of lesser importance, especially given that the Péclet number is not taking any geometrical details into account and has been derived from order of magnitude assumptions rather than exact variables (Vogel 2004).

From our earlier discussions into how an individual swimming microbe experiences a watery environment, it is evident that "at low Reynolds numbers you cannot shake off your environment" (Purcell 1977). Purcell (1977) and Vogel (2004) approximated a Péclet number of 10^{-2} for a microbe swimming in water, thereby indicating that swimming will not increase nutrient transport to the cell. On a practical level this implies that a swimming microbe carries a layer of fluid around with it, and the only means whereby nutrients can cross the fluid is via diffusion. This phenomenon is succinctly summarized by Dusenbery (2009) as follows: "they carry a halo of fluid depleted of nutrient around with them."

3.7.2 Chemotaxis

Since tumbling by an *E. coli* cell will result in a random change of direction, the question arises on how this microbe manages to navigate in a desired direction, e.g., toward a higher concentration of a food source or another chemoattractant. Chemotaxis is facilitated by regulating the run length, which in effect means that the runs in between tumbling events will be longer when moving in the direction of a chemoattractant (Cai et al. 2016). In contrast, the presence of a chemorepellent will increase the tumbling frequency resulting in a net change in overall swimming direction. The chemotactic ability of monotrichous bacteria such as *P. aeruginosa*, especially as it relates to directional changes, has been ascribed to a number of different strategies. Firstly, a change in direction has been attributed to random derailing due to Brownian motion of the fluid (Li et al. 2008). Secondly, *P. aeruginosa* can increase the likelihood of moving toward a chemoattractant by prolonging movement in a particular direction (Cai et al. 2016). It was found that *P. aeruginosa* can use either the forward or reverse motion to move up a chemoattractant gradient (Cai et al. 2016). The switch between CCW and CW rotation is separated by a pause, with the length of the latter positively correlated with the size of the turning angle (Qian et al. 2013). Thirdly, some uniflagellate bacteria such as *Vibrio alginolyticus* can turn or "flick" (Son et al. 2013) by utilizing a strain-induced collapse in its flagellum, which the authors call "an example in nature of a biological function stemming from a controlled mechanical failure."

Taylor and Stocker (2012) argue that microbes should derive sufficient nutritional benefit given the energy expenditure required for chemotactic swimming motility (speed). Maintaining optimal swimming speeds to balance this trade-off will be particularly important in nutrient-poor environments such as ocean water (Stocker et al. 2008). The authors found that the monotrichous (Gauthier et al. 1995) marine bacterium *Pseudoalteromonas haloplanktis* was able to engage in a chemotactic response that was >10 times as rapid as *E. coli*. This swift response was mainly attributed to the faster swimming speed of *P. haloplanktis* ($\sim$68 μm s^{-1}). Stocker (2011) mentions that the chemotactic motility pattern of monotrichous bacteria (a hybrid of forward and reverse movement with flicking) outperforms the chemotactic response of the run-and-tumble moving pattern exhibited by, for example, *E. coli*. Interestingly, of ~600 motile marine bacteria isolated, approximately 90% had a single polar flagellum (Leifson et al. 1964). The ability to respond rapidly may result in a competitive advantage given the transient microscale nutrient patches in the ocean.

3.8 Evaluating the Port City Analogy

To integrate the preceding discussions of the interactions between microbes and surfaces in a dynamic fluid environment, especially as it relates to bacterial motility, we will restrict the evaluation of our port city analogy to initial adhesion to a surface or biofilm (immigration), the association of motile cells with an existing biofilm, and detachment or dispersion from a biofilm (emigration).

When considering a port city with its harbor, we could recognize three zones, namely, the city, the harbor, and the ocean. These regions can in turn be related to the three zones of a biofilm in flowing conditions: the surface-attached biofilm, the quiescent zone of low flow close to the surface, and the moving bulk fluid further away from the surface (Fig. 3.1). Vigeant et al. (2002) similarly defined three zones but based these on the distance from the solid surface, namely, the near-surface constrained region, the near-surface bulk region, and the bulk fluid. The near-surface constrained zone is impacted by both the hydrodynamic effect and physicochemical

Fig. 3.1 A diagram of the proposed zones associated with a biofilm under flowing conditions (images taken from Bester et al. (2013) and not drawn to scale). According to the port city analogy, the biofilm corresponds to the city (purple cells and brown EPS matrix), whereas the region with reduced flow rates equates to the harbor (white background). In this zone, the reduced flow rates allow for freely motile cells to swim upstream and around the biomass. The zone furthest away from the surface (blue background) represents the ocean or bulk liquid flow where surface-associated hydrodynamics have no influence on the characteristics of the flow

properties of the surface, the near-surface bulk phase experiences only the hydrodynamic effect of the surface, while in the bulk fluid, no effect from the surface is evident.

According to the classical dogma, the formation of a biofilm is considered to be a developmental process which proceeds through four or five stages from attachment to maturation and detachment or dispersion (Sauer et al. 2002; Stoodley et al. 2002b). While this model provides a useful framework, it does not capture the complexity of surface-associated microbial growth. The particular focus of this model on the attached, sessile cells has the potential to restrict our perspective and understanding of surface-associated microbial communities.

For example, this model does not take into account the "harbor" region of the port city analogy where planktonic or freely motile cells can remain, meander around biofilm biomass, and multiply indefinitely due to reduced flow rates. This "low-flow" zone may be critically important in both immigration into and emigration from the biofilm. By extension, this may yield a notably different progression of biofilm development with a higher degree of fluidity than the discrete stages as depicted by the classical biofilm model.

An additional example of the inherent, but perhaps unrecognized bias of the biofilm developmental model is evident from an investigation into *Legionella pneumophila* biofilm formation. Mampel et al. (2006) found that continued planktonic cell replication (rather than sessile cell division) was necessary for biofilm formation under static conditions. *L. pneumophila* was furthermore unable to form robust biofilms under conditions of flow, but effectively associated with existing biofilms comprised of certain environmental bacterial species (*Microbacteria* and *Acinetobacter baumannii*), while other biofilms consisting of *Klebsiella pneumoniae*, *Pseudomonas* spp., or *Corynebacterium glutamicum* resisted *L. pneumophila* colonization (Mampel et al. 2006).

3.8.1 Biofilm Formation or Association: Immigration

Flow conditions can impact nutrient uptake and the rate of cell-to-cell encounters, while shear in particular can influence the attachment of bacteria to the surface as a first step in the process of microcolony and later biofilm formation (Brumley et al. 2015). Lemon et al. (2007) compared wild-type, flagellum-minus, and paralyzed-flagellum mutants of *Listeria monocytogenes* and found that flagellar motility was essential for initial surface attachment and biofilm formation in this species.

Vigeant et al. (2002) comments on the different forces at play when considering the classic biofilm development dogma. According to this theory, reversibly adhering cells originate from planktonic cells trapped by hydrodynamic forces after swimming close to the surface for extended periods. The subsequent, irreversible attachment of these cells can be explained by DLVO (Derjaguin, Landau, Verwey, and Overbeek) electrostatic interactions.

Much of the boundary effect theory is based on the no-slip wall condition which considers that at a solid boundary the fluid velocity relative to the boundary will be zero. Hu et al. (2015) predicted that *E. coli* can sense nanoscale surface slip as opposed to the usual no-slip assumption occurring at boundaries. Surface slip occurs at surfaces modified by, e.g., adsorbents and hydrophobic compounds (Lauga et al. 2007; Rothstein 2010). An increasing slip length, as could be achieved experimentally by coating a surface with the EPS polymer alginate, implies CCW circles of decreasing radius which may lead to increased surface adsorption due to a prolonged local residence time (Hu et al. 2015).

Reports such as the one by Barken et al. (2008) investigated whether the mere presence of certain organelles versus the presence of functional organelles influenced the outcome of a study deserve further consideration. The authors showed that both type IV pili and flagella are required for the development of mature multicellular structures in *P. aeruginosa* biofilms. Whereas motile, and thus functional flagella were necessary for this process, type IV pili were only required to be present but not necessarily functional, perhaps indicating a structural role in adhesion (Barken et al. 2008). Wood et al. (2006) reported similar findings in *E. coli* where the poorest biofilm formation, with respect to both architecture and thickness, was observed for strains with impaired motility.

Once a swimming cell adheres to a surface and biofilm formation is initiated, one could ask what happens to the appendages involved in motility. While a comparison of global gene expression between biofilm and planktonic *P. aeruginosa* only revealed a 1% difference, the expression of genes involved in pili and flagella synthesis were repressed in biofilms (Whiteley et al. 2001). In contrast Domka et al. (2007) showed that 20 motility-related genes are induced throughout all stages of *E. coli* biofilm development, ranging from early attachment stages to fully mature biofilms. Kalmokoff et al. (2006) demonstrated via proteomic analysis that the flagella motility complex plays an important role in initial attachment of *Campylobacter jejuni*. In addition, the continued expression of the flagella motility complex even into the stages of a mature biofilm was noted. While motility is important for the initial attachment of *Vibrio cholerae*, the expression of the flagella filament, *flaA* decreases in mature biofilms, as reviewed by Guttenplan and Kearns (2013). It is however unknown whether, after surface attachment, the flagellum remains functional or if it is lost and degraded or if it acts as a structural component of the biofilm (Teschler et al. 2015). Guttenplan and Kearns (2013) investigated the four model systems of *Bacillus*, *Vibrio*, *Pseudomonas*, and *Escherichia* and proposed that flagellar motility in the biofilm is regulated in two stages: firstly a fast inhibition stage at the level of flagellar function and, afterward, a slow inhibition at the level of flagellar gene transcription.

Nadell et al. (2015) investigated *V. cholerae* attached to surfaces and showed that the extracellular matrix prevented biofilm invasion, where non-motile cells were incapable of entering the existing biofilm and motile cells were only capable of colonizing and growing on the biofilm exterior.

3.8.2 *Motility in or Around the Biofilm*

Li et al. (2014) demonstrated that near-surface microbial swimming behavior differs in viscoelastic fluids, such as biofilms, compared to Newtonian fluids. Proteins and polysaccharides associated with the biofilm EPS matrix disperse into the surroundings, thereby imparting viscoelastic properties to the ambient fluid (Li et al. 2014). The viscoelastic environment allows puller swimmers to escape more readily from the biofilm environment, while the pusher swimmers appear to be perpetually trapped (Li et al. 2014). Disparate swimming strategies will interact differently with the viscoelastic environment, leading to increases or decreases in the swimming speed of both single and collective swimming bacteria. For example, helical bacteria demonstrate faster swimming speeds in a viscoelastic fluid compared to a Newtonian fluid with the same viscosity (Li et al. 2014).

Martinez et al. (2014) measured *E. coli* motility in polymer solutions and concluded that the flagella experienced a lower local viscosity than the cell itself. In this sense the flagella not only act as small rheometers able to detect local non-Newtonian behavior but also provide corridors of lower viscosity that will make it easier for another bacterium to follow in its wake (Martinez et al. 2014).

Planktonic bacteria in close proximity to a biofilm can tunnel deep into the biofilm by creating transient pores (Houry et al. 2012). These pores could benefit the biofilm by enhancing nutrient transport or harm it by allowing the penetration of antimicrobials (sometimes produced by the tunnelling bacteria themselves) leading to supplanting of the original biofilm occupants (Houry et al. 2012).

3.8.3 *Biofilm Detachment: Emigration*

The proposed advantages of detachment from biofilms are numerous and include (1) the ability to escape deteriorating environmental conditions once nutrients are depleted due to increased competition, or the presence of an antimicrobial agent, (2) the potential to reach and make use of new habitats, and (3) the opportunity to generate genetic variation (McDougald et al. 2012). To these advantages should be added the fact that detachment provides a mechanism to maintain metabolically active biofilms that are in equilibrium with the carrying capacity of the local environment, by balancing cell formation with accumulation. The continuous production and release of cells, both during biofilm development and after steady state is attained, has been proposed to be a mechanism for prokaryote proliferation and transmission (Bester et al. 2009). A prerequisite for the dissemination of microbes from biofilm-colonized habitats requires the capacity to detach and be actively propelled or passively transported to a different environment, whereafter reassociation with an existing biofilm or primary surface colonization can take place. Several biofilm detachment modes have been identified and described empirically, but a clear distinction between the different mechanisms and consensus regarding the underlying mechanisms and ecological purpose remains to be reached.

3.8.3.1 Modes of Biofilm Detachment

The mechanisms of biofilm detachment have been defined both in terms of the particle size of the detached biomass and the frequency of detachment (Bryers 1988). Grazing by higher organisms such as protozoa or nematodes may result in the physical disruption of the biofilm, and additional loss of attached biomass can be expected to occur as a result of abrasion by solid particulates. Both of these mechanisms may remove biomass ranging in size from small clumps to large aggregates containing numerous microbial cells and EPS. Substantial reductions in the amount of biomass may also occur at random intervals, due to the sloughing off of large aggregates from the biofilm. While the underlying mechanisms resulting in sloughing have not been identified, it is thought to occur when shear forces exerted by flowing liquid exceeds the cohesive strength of the biofilm matrix.

Biofilms can move over surfaces (Stoodley et al. 1999b) and through porous media under both laminar and turbulent flow conditions (Stoodley et al. 2005). Under turbulent flow, ripple-like structures form and move downstream (Stoodley et al. 1999b). Furthermore, the formation of filamentous biofilm streamers has been reported under turbulent as well as laminar flow conditions (Stoodley et al. 1998; Rusconi et al. 2010, 2011), with the latter being dependent on complex flow patterns created by varied channel geometries. An increase in turbulent flow velocity led to the breakage or detachment of some filaments (Stoodley et al. 1998), whereas spontaneous sloughing off of filaments occurred at unpredictable intervals and completely constricted laminar flow (Drescher et al. 2013). While it was thought that the development of filamentous streamers under laminar flow required the presence of unusual channel geometries such as corners, Parvinzadeh Gashti et al. (2015) observed the formation of streamers in straight microchannels. The accumulation of dense ridges of biofilm biomass was correlated to the abrupt partial detachment of one end of a ridge, leading to the formation of streamers. The authors determine a corresponding detachment rate of 0.15 events mm^{-2} h^{-1} for a single species *Pseudomonas* biofilm.

3.8.3.2 Erosion, Planktonic Cell Yield, and Seeding Dispersion

The detachment of single cells from the biofilm has similarly been attributed to various processes, including liquid-mediated erosion of cells from the biofilm surface, active growth, cell division and release of progeny cells by the biofilm, and most recently seeding dispersion. Despite evidence to the contrary, the detachment of single cells from biofilms continues to be ascribed primarily to liquid-mediated erosive action. Under conditions where substrate availability is governed by flow velocity, the observed detachment rates are often better correlated to substrate availability or biofilm growth rates, rather than shear removal forces (Peyton and Characklis 1992; Stewart 1993; Bester et al. 2009, 2013).

The rate at which single cells detached from various 120-h-old Pseudomonad biofilms, exposed to bulk liquid flow, was quantified at approximately 10^7 cells cm^{-2} h^{-1}, with an increase in the detachment rate evident as early as 6 h after the introduction of the cells into a sterile environment (Bester et al. 2009). The number of detached cells increased alongside biofilm development until both parameters reached a maximum after 96–120 h. The removal of the carbon source from 5-day-old Pseudomonad biofilms subjected to continuous flow resulted in a 40%, 94%, and 99% decrease in detachment rate after 1, 5, and 24 h, respectively. During the ensuing 8-day period of carbon starvation, the rate of cell detachment cells decreased by 1–2 orders of magnitude, compared to pre-starvation. A rapid increase in the number of detached cells was observed upon a reintroduction of carbon into the system, with a complete recovery of pre-starvation detachment rates within 24 h (Bester et al. 2011). The observed response in cell detachment was mirrored by the metabolic activity of the biofilm and clearly demonstrated the contribution of active biofilm carbon consumption and growth to the number of detached planktonic cells. To differentiate between shear-related erosion of attached biomass and active attached cell growth and release of progeny into the environment, the term "biofilm-derived planktonic cell yield" has been proposed (Bester et al. 2009).

Observation of the active dispersal of planktonic cells from within *P. aeruginosa* biofilm microcolonies (Sauer et al. 2002; Purevdorj-Gage et al. 2005) has been designated as the final stage of the biofilm development life-cycle model (Stoodley et al. 2002b). During dispersion, actively swimming planktonic cells are observed to exit from the interior of biofilm microcolonies, while a sessile, hollow structure remains behind (Purevdorj-Gage et al. 2005). Prior to cell dispersal taking place, an increase in cell lysis and death within the colonies have been reported. The latter is likely due to an accumulation of reactive oxygen and nitrogen species as well as bacteriophage-mediated cell lysis (Webb et al. 2003; Barraud et al. 2006). Interestingly, the dispersed population for both *P. aeruginosa* (Kirov et al. 2007) and *Pseudoalteromonas tunicata* (Mai-Prochnow et al. 2006) displayed significant variation in a number of phenotypic traits, such as colony morphology, motility, biofilm formation, metabolic activity, production of quorum sensing molecules, and virulence factors. The authors drew an analogy between this observed increase in phenotypic differentiation within a prokaryote and known eukaryotic dispersal strategies, which generate variation to enhance the probability of successful surface colonization under a variety of environmental conditions (Mai-Prochnow et al. 2006). While seeding dispersal has been documented for a number of bacterial species, not all strains within a species behave in this manner as evidenced by conflicting reports on the ability (or inability) of mucoid clinical isolates of *P. aeruginosa* to disperse (Purevdorj-Gage et al. 2005; Kirov et al. 2007). The relative contribution of each of these mechanisms to the extent of biofilm detachment will likely vary in space and time and depend on environmental conditions as well as biofilm-specific factors.

3.8.3.3 Rates and Extent of Biofilm Detachment

In addition to the abovementioned, others have attempted to quantify detachment rates from single species, defined consortia, and multispecies biofilms. In an elegant study by Stoodley et al. (2001), the rates at which aggregates detached from a defined four-species consortium biofilm and an undefined tap water biofilm, both cultivated under turbulent flow, were quantified. Detachment rates from the tap water biofilm were estimated at 3.7 detaching clusters mm^{-2} h^{-1} with an analysis interval of 1 h, up to 80.2 detaching clusters mm^{-2} h^{-1} when data was acquired at 20-min intervals. The majority of detached particles were single cells or small cell aggregates, whereas large clusters consisting of multiple cells detached less frequently but accounted for a significant fraction of the overall amount of biomass lost. The detachment rate from *P. aeruginosa* PAO1 biofilms has been estimated at 5.14×10^4 events min^{-1} mm^{-2}. Single cells accounted for a minimum of 70% of these events, whereas the remaining 30% corresponded to events consisting of 2 to >1000 cells, with aggregates consisting of >100 cells accounting for less than 0.28% of the total detachment events but containing up to 37% of the total number of detached cells for PAO1 (Wilson et al. 2004). A detachment rate of 1.8×10^7 CFU cm^{-2} h^{-1} quantified for 7-day-old *Staphylococcus aureus* biofilms. In contrast to *P. aeruginosa* biofilms, it was found that aggregates containing 11–100 cells detached at the greatest frequency (20% of all events) and contained the largest fraction of all detached biomass (54%) (Fux et al. 2004).

3.9 Conclusion

The recognition that microbes are predominantly associated with aggregates or biofilms in natural environments has naturally led to a shift in the perception of microbial processes, along with a concomitant increase in research efforts to address our lack of understanding. Extensive investigation into biofilm development by a selected number of "model" microorganisms has indeed greatly enhanced this understanding, and a notable outcome from these efforts was the establishment of the classical four- or five-stage model of biofilm development. This model, understandably, focuses on the attached, matrix-encased sessile cells and associated microenvironments, whereas planktonic cells are only seen to contribute to biofilm processes during initial attachment to an uncolonized surface and eventual detachment from a mature biofilm. This limitation has the potential to restrict our perspectives and understanding of surface-associated microbial communities.

Sessile cells acquire genetic and phenotypic characteristics during biofilm formation that distinguish them from planktonically derived cells. However, a number of authors have commented on the existence of a third phenotype (Bester et al. 2005). This phenotype is displayed by cells that detach from the biofilm and have been characterized with respect to differences in adhesion (Rollet et al. 2009), virulence (Uppuluri and Lopez-Ribot 2016), and antimicrobial susceptibility (Boles and Horswill 2008).

In addition, the biofilm or surface-associated region with greatly reduced flow rates may facilitate the persistence of an independently-replicating planktonic population alongside the attached biomass. Delineating the interactions between these attached and motile subpopulations, along with determining their relative contributions to global processes, would require concerted future research efforts.

It can be expected that the particular environment experienced by bacterial populations, whether conditions of plenty or scarcity, will drastically influence their behavior and productivity. In order to learn more about an individual cell in its habitat, it is critical to be aware of the characteristics of the environment in which it is studied, a good example being consideration that different conditions can turn cooperator into cannibal (González-Pastor et al. 2003). It can be argued that learning about an individual or species in its environment is only of interest to fundamental researchers. However, in the case where humans are dependent on the well-being versus demise of a microbial community, the importance of understanding the relationship that the community has with its environment becomes much more than fundamental curiosity, especially if it must be manipulated to achieve a desired outcome.

List of symbols with definitions and units

Mathematical symbol	Definition	Units
a	Acceleration	m s^{-2}
l	Characteristic length	m
ρ	Density	kg m^{-3}
d	Coasting distance	m
δ	Thickness of viscous sublayer	m
D	Diffusion coefficient	m^2 s^{-1}
F	Force	N or kg m s^{-2}
F_{inertia}	Force inertial	N or kg m s^{-2}
F_{viscous}	Force viscous	N or kg m s^{-2}
m	Mass	kg
ν	Kinematic viscosity	m^2 s^{-1}
Pe	Péclet number	Dimensionless
Re	Reynolds number	Dimensionless
t	Time	s
t_{a}	Time required for a solute to move a distance l under advective transport	s
t_{d}	Time required for a solute to move a distance l under diffusive transport	s
τ	Shear stress	Pa
τ_{w}	Shear stress at the wall	Pa
μ	Dynamic viscosity	Pa s
U	Velocity	m s^{-1}
V	Volume	m^3

(continued)

Mathematical symbol	Definition	Units
Shear stress, τ	The parallel component of the applied force, $F_{\parallel}$, divided by the initial area of the sample, A_0	$\tau = \frac{F_{\parallel}}{A_o}$
Shear rate, $\dot{\gamma}$	Various definitions dependent on physical configuration but for a parallel plate system defined as the velocity of the top plate relative to the stationary lower plate divided by the distance between the plates	$\dot{\gamma} = \frac{U}{l}$
Newtonian fluid	For a Newtonian fluid the shear stress, τ, is directly proportional to the shear rate, $\dot{\gamma}$, with the dynamic viscosity, μ, as the slope of the straight line	$\tau = \mu\dot{\gamma}$

Compliance with Ethical Standards

Conflict of Interest Otini Kroukamp declares that he has no conflict of interest. Elanna Bester declares that she has no conflict of interest. Gideon M. Wolfaardt declares that he has no conflict of interest.

Ethical Approval This article does not contain any studies with human participants or animals performed by any of the authors.

References

Agrawal SK (2012) Turbulent flows. Fluid mechanics and machinery. Tata McGraw Hill, New Delhi, pp 339–363

Alexander DE (2016) The biomechanics of solids and fluids: the physics of life. Eur J Phys 37 (5):053001

Allan JD (1995) Physical factors of importance to the biota. Stream ecology: structure and function of running waters. Kluwer Academic Publishers, Dordrecht, pp 45–81

Barai P, Kumar A, Mukherjee PP (2016) Modeling of mesoscale variability in biofilm shear behavior. PLoS One 11(11):e0165593

Barken KB, Pamp SJ, Yang L, Gjermansen M, Bertrand JJ, Klausen M et al (2008) Roles of type IV pili, flagellum-mediated motility and extracellular DNA in the formation of mature multicellular structures in *Pseudomonas aeruginosa* biofilms. Environ Microbiol 10(9):2331–2343

Barraud N, Hassett DJ, Hwang S-H, Rice SA, Kjelleberg S, Webb JS (2006) Involvement of nitric oxide in biofilm dispersal of *Pseudomonas aeruginosa*. J Bacteriol 188(21):7344–7353

Bearon R, Hazel A (2015) The trapping in high-shear regions of slender bacteria undergoing chemotaxis in a channel. J Fluid Mech 771:R3

Belas R (2014) Biofilms, flagella, and mechanosensing of surfaces by bacteria. Trends Microbiol 22 (9):517–527

Berg HC (2003) The rotary motor of bacterial flagella. Annu Rev Biochem 72:19–54

Berg HC, Brown DA (1972) Chemotaxis in *Escherichia coli* analysed by three-dimensional tracking. Nature 239(5374):500–504

Berke AP, Turner L, Berg HC, Lauga E (2008) Hydrodynamic attraction of swimming microorganisms by surfaces. Phys Rev Lett 101(3):038102

Bester E, Wolfaardt G, Joubert L, Garny K, Saftic S (2005) Planktonic-cell yield of a pseudomonad biofilm. Appl Environ Microbiol 71(12):7792–7798

Bester E, Edwards EA, Wolfaardt GM (2009) Planktonic cell yield is linked to biofilm development. Can J Microbiol 55(10):1195–1206

Bester E, Kroukamp O, Hausner M, Edwards E, Wolfaardt G (2011) Biofilm form and function: carbon availability affects biofilm architecture, metabolic activity and planktonic cell yield. J Appl Microbiol 110(2):387–398
Bester E, Wolfaardt GM, Aznaveh NB, Greener J (2013) Biofilms' role in planktonic cell proliferation. Int J Mol Sci 14(11):21965–21982
Billings N, Birjiniuk A, Samad TS, Doyle PS, Ribbeck K (2015) Material properties of biofilms—a review of methods for understanding permeability and mechanics. Rep Prog Phys 78(3):036601
Boles BR, Horswill AR (2008) *agr*-Mediated dispersal of *Staphylococcus aureus* biofilms. PLoS Pathog 4(4):e1000052
Brading MG, Jass J, Lappin-Scott HM (1995) Dynamics of bacterial biofilm formation. In: Lappin-Scott HM, Costerton JW (eds) Microbial biofilms. Cambridge University Press, Cambridge, pp 46–63
Brown E, Jaeger HM (2011) Material science. Through thick and thin. Science 333 (6047):1230–1231
Brumley D, Rusconi R, Son K, Stocker R (2015) Flagella, flexibility and flow: physical processes in microbial ecology. Eur Phys J Spec Top 224(17–18):3119–3140
Bryers J (1988) Modeling biofilm accumulation. In: Bazin M, Prosser J (eds) Physiological models in microbiology. CRC Press, Boca Raton, FL, pp 109–144
Cai Q, Li Z, Ouyang Q, Luo C, Gordon VD (2016) Singly flagellated *pseudomonas aeruginosa* chemotaxes efficiently by unbiased motor regulation. mBio 7(2):e00013
Cairns LS, Marlow VL, Bissett E, Ostrowski A, Stanley-Wall NR (2013) A mechanical signal transmitted by the flagellum controls signalling in *Bacillus subtilis*. Mol Microbiol 90(1):6–21
Caldwell D, Lawrence J (1988) Study of attached cells in continuous-flow slide culture. In: Wimpenny JWT (ed) CRC handbook of laboratory model systems for microbial ecosystems. CRC Press, Boca Raton, FL, pp 117–138
Conrad JC (2012) Physics of bacterial near-surface motility using flagella and type IV pili: implications for biofilm formation. Res Microbiol 163(9):619–629
De Beer D, Stoodley P, Lewandowski Z (1996) Liquid flow and mass transport in heterogeneous biofilms. Water Res 30(11):2761–2765
DiLuzio WR, Turner L, Mayer M, Garstecki P, Weibel DB, Berg HC et al (2005) *Escherichia coli* swim on the right-hand side. Nature 435(7046):1271–1274
Domka J, Lee J, Bansal T, Wood TK (2007) Temporal gene-expression in *Escherichia coli* K-12 biofilms. Environ Microbiol 9(2):332–346
Drescher K, Shen Y, Bassler BL, Stone HA (2013) Biofilm streamers cause catastrophic disruption of flow with consequences for environmental and medical systems. Proc Natl Acad Sci 110 (11):4345–4350
Dusenbery DB (2009) Living at micro scale: the unexpected physics of being small. Harvard University Press, Cambridge, MA
Ellison C, Brun YV (2015) Mechanosensing: a regulation sensation. Curr Biol 25(3):R113–R115
Fabbri S, Stoodley P (2016) Mechanical properties of biofilms. In: Flemming H-C, Neu TR, Wingender J (eds) The perfect slime: microbial extracellular polymeric substances (EPS). IWA Publishing, London, pp 153–178
Flemming H-C, Wingender J (2010) The biofilm matrix. Nat Rev Microbiol 8(9):623–633
Flemming H-C, Neu TR, Wozniak DJ (2007) The EPS matrix: the "house of biofilm cells". J Bacteriol 189(22):7945–7947
Fux C, Wilson S, Stoodley P (2004) Detachment characteristics and oxacillin resistance of *Staphyloccocus aureus* biofilm emboli in an in vitro catheter infection model. J Bacteriol 186 (14):4486–4491
Galajda P, Keymer J, Chaikin P, Austin R (2007) A wall of funnels concentrates swimming bacteria. J Bacteriol 189(23):8704–8707
Gauthier G, Gauthier M, Christen R (1995) Phylogenetic analysis of the genera *Alteromonas, Shewanella*, and *Moritella* using genes coding for small-subunit rRNA sequences and division

of the genus *Alteromonas* into two genera, *Alteromonas* (emended) and *Pseudoalteromonas* gen. nov., and proposal of twelve new species combinations. Int J Syst Bacteriol 45(4):755–761
Geesey G, Mutch R, Costerton J, Green R (1978) Sessile bacteria: an important component of the microbial population in small mountain streams. Limnol Oceanogr 23(6):1214–1223
Gilbert MJ (2006) Paper trails: port cities in the classical era of world history. World History Connected [Internet] 3(2). http://worldhistoryconnected.press.illinois.edu/3.2/gilbert.html
González-Pastor JE, Hobbs EC, Losick R (2003) Cannibalism by sporulating bacteria. Science 301 (5632):510–513
Goto T, Nakata K, Baba K, Nishimura M, Magariyama Y (2005) A fluid-dynamic interpretation of the asymmetric motion of singly flagellated bacteria swimming close to a boundary. Biophys J 89(6):3771–3779
Guasto JS, Rusconi R, Stocker R (2012) Fluid mechanics of planktonic microorganisms. Annu Rev Fluid Mech 44:373–400
Guttenplan SB, Kearns DB (2013) Regulation of flagellar motility during biofilm formation. FEMS Microbiol Rev 37(6):849–871
Harapanahalli AK, Younes JA, Allan E, van der Mei HC, Busscher HJ (2015) Chemical signals and mechanosensing in bacterial responses to their environment. PLoS Pathog 11(8):e1005057
Harwood CS, Fosnaugh K, Dispensa M (1989) Flagellation of *Pseudomonas putida* and analysis of its motile behavior. J Bacteriol 171(7):4063–4066
Hernandez-Ortiz JP, Stoltz CG, Graham MD (2005) Transport and collective dynamics in suspensions of confined swimming particles. Phys Rev Lett 95(20):204501
Hill J, Kalkanci O, McMurry JL, Koser H (2007) Hydrodynamic surface interactions enable *Escherichia coli* to seek efficient routes to swim upstream. Phys Rev Lett 98(6):068101
Houari A, Picard J, Habarou H, Galas L, Vaudry H, Heim V et al (2008) Rheology of biofilms formed at the surface of NF membranes in a drinking water production unit. Biofouling 24 (4):235–240
Houry A, Gohar M, Deschamps J, Tischenko E, Aymerich S, Gruss A et al (2012) Bacterial swimmers that infiltrate and take over the biofilm matrix. Proc Natl Acad Sci USA 109 (32):13088–13093
Hu J, Wysocki A, Winkler RG, Gompper G (2015) Physical sensing of surface properties by microswimmers – directing bacterial motion via wall slip. Sci Rep 5:9586
Ishikawa T (2009) Suspension biomechanics of swimming microbes. J R Soc Interface 6:815–834
Jarrell KF, McBride MJ (2008) The surprisingly diverse ways that prokaryotes move. Nat Rev Microbiol 6(6):466–476
Kalmokoff M, Lanthier P, Tremblay T-L, Foss M, Lau PC, Sanders G et al (2006) Proteomic analysis of *Campylobacter jejuni* 11168 biofilms reveals a role for the motility complex in biofilm formation. J Bacteriol 188(12):4312–4320
Karimi A, Karig D, Kumar A, Ardekani A (2015) Interplay of physical mechanisms and biofilm processes: review of microfluidic methods. Lab Chip 15(1):23–42
Kaya T, Koser H (2012) Direct upstream motility in *Escherichia coli*. Biophysical J 102 (7):1514–1523
Kirov SM, Webb JS, Che Y, Reid DW, Woo JK, Rice SA et al (2007) Biofilm differentiation and dispersal in mucoid *Pseudomonas aeruginosa* isolates from patients with cystic fibrosis. Microbiology 153(10):3264–3274
Kjelleberg S, Givskov M (2007) The biofilm mode of life. In: Kjelleberg S, Givskov M (eds) The biofilm mode of life: mechanisms and adaptations. Horizon Bioscience, Norfolk, pp 5–21
Klapper I, Rupp C, Cargo R, Purvedorj B, Stoodley P (2002) Viscoelastic fluid description of bacterial biofilm material properties. Biotechnol Bioeng 80(3):289–296
Korber DR, Lawrence JR, Sutton B, Caldwell DE (1989) Effect of laminar flow velocity on the kinetics of surface recolonization by Mot(+) and Mot(−) *Pseudomonas fluorescens*. Microb Ecol 18(1):1–19
Kumarasamy M, Maharaj PM (2015) The effect of biofilm growth on wall shear stress in drinking water PVC pipes. Pol J Environ Stud 24(6):2479–2483

Lambert C, Evans KJ, Till R, Hobley L, Capeness M, Rendulic S et al (2006) Characterizing the flagellar filament and the role of motility in bacterial prey-penetration by *Bdellovibrio bacteriovorus*. Mol Microbiol 60(2):274–286

Lauga E, Powers TR (2009) The hydrodynamics of swimming microorganisms. Rep Prog Phys 72 (9):096601

Lauga E, Brenner M, Stone H (2007) Microfluidics: the no-slip boundary condition. In: Tropea C, Yarin A, Foss J (eds) Springer handbook of experimental fluid mechanics. Springer, Berlin, pp 1219–1240

Lecuyer S, Rusconi R, Shen Y, Forsyth A, Vlamakis H, Kolter R et al (2011) Shear stress increases the residence time of adhesion of *Pseudomonas aeruginosa*. Biophys J 100(2):341–350

Leifson E, Cosenza B, Murchelano R, Cleverdon R (1964) Motile marine bacteria I. Techniques, ecology, and general characteristics. J Bacteriol 87(3):652–666

Lele PP, Hosu BG, Berg HC (2013) Dynamics of mechanosensing in the bacterial flagellar motor. Proc Natl Acad Sci 110(29):11839–11844

Lele PP, Roland T, Shrivastava A, Chen Y, Berg HC (2015) The flagellar motor of *Caulobacter crescentus* generates more torque when a cell swims backwards. Nat Phys 12:175–178

Lemon KP, Higgins DE, Kolter R (2007) Flagellar motility is critical for *Listeria monocytogenes* biofilm formation. J Bacteriol 189(12):4418–4424

Lewandowski Z, Beyenal H (2013) Introduction to biofilms and to biofilm research. In: Fundamentals of biofilm research, 2nd edn. CRC press, Boca Raton, FL, pp 1–66

Li G, Tam L-K, Tang JX (2008) Amplified effect of Brownian motion in bacterial near-surface swimming. Proc Natl Acad Sci 105(47):18355–18359

Li G, Bensson J, Nisimova L, Munger D, Mahautmr P, Tang JX et al (2011) Accumulation of swimming bacteria near a solid surface. Phys Rev E 84(4):041932

Li G-J, Karimi A, Ardekani A (2014) Effect of solid boundaries on swimming dynamics of microorganisms in a viscoelastic fluid. Rheol Acta 53(12):911–926

Loudon C (1999) How is fluid mechanics taught to biologists? In: Proceedings of the 3rd ASME/JSME joint fluids engineering conference, pp 18–23

Lushi E, Wioland H, Goldstein RE (2014) Fluid flows created by swimming bacteria drive self-organization in confined suspensions. Proc Natl Acad Sci 111(27):9733–9738

Macnab RM (2003) How bacteria assemble flagella. Annu Rev Microbiol 57(1):77–100

Mai-Prochnow A, Webb JS, Ferrari BC, Kjelleberg S (2006) Ecological advantages of autolysis during the development and dispersal of *Pseudoalteromonas tunicata* biofilms. Appl Environ Microbiol 72(8):5414–5420

Mampel J, Spirig T, Weber SS, Haagensen JA, Molin S, Hilbi H (2006) Planktonic replication is essential for biofilm formation by *Legionella pneumophila* in a complex medium under static and dynamic flow conditions. Appl Environ Microbiol 72(4):2885–2895

Martinez VA, Schwarz-Linek J, Reufer M, Wilson LG, Morozov AN, Poon WC (2014) Flagellated bacterial motility in polymer solutions. Proc Natl Acad Sci 111(50):17771–17776

Martins BMC, Locke JCW (2015 Apr) Microbial individuality: how single-cell heterogeneity enables population level strategies. Curr Opin Microbiol 24:104–112

McCarter LL (2005) Multiple modes of motility: a second flagellar system in *Escherichia coli*. J Bacteriol 187(4):1207–1209

McClaine JW, Ford RM (2002) Reversal of flagellar rotation is important in initial attachment of *Escherichia coli* to glass in a dynamic system with high-and low-ionic-strength buffers. Appl Environ Microbiol 68(3):1280–1289

McDougald D, Rice SA, Barraud N, Steinberg PD, Kjelleberg S (2012) Should we stay or should we go: mechanisms and ecological consequences for biofilm dispersal. Nat Rev Microbiol 10 (1):39–50

Merino S, Shaw JG, Tomás JM (2006) Bacterial lateral flagella: an inducible flagella system. FEMS Microbiol Lett 263(2):127–135

Mewis J (2012) Shear thickening effects in concentrated suspensions. In: Borsali R, Pecora R (eds) Structure and dynamics of polymer and colloidal systems. Kluwer Academic Publishers, Dordrecht, pp 333–362
Molaei M, Sheng J (2016) Succeed escape: flow shear promotes tumbling of *Escherichia coli* near a solid surface. Sci Rep 6:35290
Molaei M, Barry M, Stocker R, Sheng J (2014) Failed escape: solid surfaces prevent tumbling of *Escherichia coli*. Phys Rev Lett 113(6):068103
Moran J, McKean PG, Ginger ML (2014) Eukaryotic flagella: variations in form, function, and composition during evolution. BioScience 64(12):1103–1114
Morozov A, Marenduzzo D (2014) Enhanced diffusion of tracer particles in dilute bacterial suspensions. Soft Matter 10(16):2748–2758
Nadell CD, Drescher K, Wingreen NS, Bassler BL (2015) Extracellular matrix structure governs invasion resistance in bacterial biofilms. ISME J 9(8):1700–1709
Nealson KH (2003) Harnessing microbial appetites for remediation. Nat Biotechnol 21(3):243–244
Parvinzadeh Gashti M, Bellavance J, Kroukamp O, Wolfaardt G, Taghavi SM, Greener J (2015) Live-streaming: time-lapse video evidence of novel streamer formation mechanism and varying viscosity. Biomicrofluidics 9(4):041101
Patsios S, Goudoulas T, Kastrinakis E, Nychas S, Karabelas A (2015) A novel method for rheological characterization of biofouling layers developing in membrane bioreactors (MBR). J Membr Sci 482:13–24
Percival S, Knapp J, Wales D, Edyvean R (1999) The effect of turbulent flow and surface roughness on biofilm formation in drinking water. J Ind Microbiol Biotechnol 22(3):152–159
Perni S, Jordan SJ, Andrew PW, Shama G (2006) Biofilm development by *Listeria innocua* in turbulent flow regimes. Food Control 17(11):875–883
Persat A, Nadell CD, Kim MK, Ingremeau F, Siryaporn A, Drescher K et al (2015) The mechanical world of bacteria. Cell 161(5):988–997
Peterson BW, He Y, Ren Y, Zerdoum A, Libera MR, Sharma PK et al (2015) Viscoelasticity of biofilms and their recalcitrance to mechanical and chemical challenges. FEMS Microbiol Rev 39:234–245
Peyton BM, Characklis W (1992) Kinetics of biofilm detachment. Water Sci Technol 26 (9–11):1995–1998
Polin M, Tuval I, Drescher K, Gollub JP, Goldstein RE (2009) *Chlamydomonas* swims with two "gears" in a eukaryotic version of run-and-tumble locomotion. Science 325(5939):487–490
Purcell EM (1977) Life at low Reynolds number. Am J Phys 45(1):3–11
Purevdorj-Gage B, Costerton W, Stoodley P (2005) Phenotypic differentiation and seeding dispersal in non-mucoid and mucoid *Pseudomonas aeruginosa* biofilms. Microbiology 151 (5):1569–1576
Qian C, Wong CC, Swarup S, Chiam K-H (2013) Bacterial tethering analysis reveals a "run-reverse-turn" mechanism for *Pseudomonas* species motility. Appl Environ Microbiol 79 (15):4734–4743
Qin J, Li R, Raes J, Arumugam M, Burgdorf KS, Manichanh C et al (2010) A human gut microbial gene catalogue established by metagenomic sequencing. Nature 464(7285):59–65
Quinn GP, Keough MJ (2002) Experimental design and data analysis for biologists. Cambridge University Press, Cambridge
Rafaï S, Jibuti L, Peyla P (2010) Effective viscosity of microswimmer suspensions. Phys Rev Lett 104(9):098102
Rani SA, Pitts B, Beyenal H, Veluchamy RA, Lewandowski Z, Davison WM et al (2007) Spatial patterns of DNA replication, protein synthesis, and oxygen concentration within bacterial biofilms reveal diverse physiological states. J Bacteriol 189(11):4223–4233
Rao B (2010) Boundary layer and laminar flows. Fluid mechanics and machinery [Internet]. Tata McGraw Hill Education, New Delhi, pp 109–141

Rollet C, Gal L, Guzzo J (2009) Biofilm-detached cells, a transition from a sessile to a planktonic phenotype: a comparative study of adhesion and physiological characteristics in *Pseudomonas aeruginosa*. FEMS Microbiol Lett 290(2):135–142

Rothstein JP (2010) Slip on superhydrophobic surfaces. Annu Rev Fluid Mech 42:89–109

Rubin H, Atkinson J (2001) Surface water flows. Environmental fluid mechanics [Internet]. Marcel Dekker, New York, pp 277–359

Rusconi R, Stocker R (2015) Microbes in flow. Curr Opin Microbiol 25:1–8

Rusconi R, Lecuyer S, Guglielmini L, Stone HA (2010) Laminar flow around corners triggers the formation of biofilm streamers. J R Soc Interface 7(50):1293–1299

Rusconi R, Lecuyer S, Autrusson N, Guglielmini L, Stone HA (2011) Secondary flow as a mechanism for the formation of biofilm streamers. Biophysical J 100(6):1392–1399

Rusconi R, Guasto JS, Stocker R (2014) Bacterial transport suppressed by fluid shear. Nat Phys 10 (3):212–217

Sauer K, Camper AK, Ehrlich GD, Costerton JW, Davies DG (2002) *Pseudomonas aeruginosa* displays multiple phenotypes during development as a biofilm. J Bacteriol 184(4):1140–1154

Shen Y, Siryaporn A, Lecuyer S, Gitai Z, Stone HA (2012) Flow directs surface-attached bacteria to twitch upstream. Biophys J 103(1):146–151

Siryaporn A, Kuchma SL, O'Toole GA, Gitai Z (2014) Surface attachment induces *Pseudomonas aeruginosa* virulence. Proc Natl Acad Sci 111(47):16860–16865

Sokolov A, Aranson IS (2009) Reduction of viscosity in suspension of swimming bacteria. Phys Rev Lett 103(14):148101

Son K, Guasto JS, Stocker R (2013) Bacteria can exploit a flagellar buckling instability to change direction. Nat Phys 9(8):494–498

Stewart PS (1993) A model of biofilm detachment. Biotechnol Bioeng 41(1):111–117

Stewart PS (2003) Diffusion in biofilms. J Bacteriol 185(5):1485–1491

Stewart PS, Franklin MJ (2008) Physiological heterogeneity in biofilms. Nat Rev 6(3):199–210

Stocker R (2011) Reverse and flick: hybrid locomotion in bacteria. Proc Natl Acad Sci 108 (7):2635–2636

Stocker R, Seymour JR, Samadani A, Hunt DE, Polz MF (2008) Rapid chemotactic response enables marine bacteria to exploit ephemeral microscale nutrient patches. Proc Natl Acad Sci 105(11):4209–4214

Stoodley P, Lewandowski Z, Boyle JD, Lappin-Scott HM (1998) Oscillation characteristics of biofilm streamers in turbulent flowing water as related to drag and pressure drop. Biotechnol Bioeng 57(5):536–544

Stoodley P, Lewandowski Z, Boyle JD, Lappin-Scott HM (1999a) Structural deformation of bacterial biofilms caused by short-term fluctuations in fluid shear: an in situ investigation of biofilm rheology. Biotechnol Bioeng 65(1):83–92

Stoodley P, Lewandowski Z, Boyle JD, Lappin-Scott HM (1999b) The formation of migratory ripples in a mixed species bacterial biofilm growing in turbulent flow. Environ Microbiol 1 (5):447–455

Stoodley P, Wilson S, Hall-Stoodley L, Boyle JD, Lappin-Scott HM, Costerton J (2001) Growth and detachment of cell clusters from mature mixed-species biofilms. Appl Environ Microbiol 67 (12):5608–5613

Stoodley P, Cargo R, Rupp C, Wilson S, Klapper I (2002a) Biofilm material properties as related to shear-induced deformation and detachment phenomena. J Ind Microbiol Biotechnol 29 (6):361–367

Stoodley P, Sauer K, Davies D, Costerton JW (2002b) Biofilms as complex differentiated communities. Annu Rev Microbiol 56(1):187–209

Stoodley P, Dodds I, De Beer D, Scott HL, Boyle J (2005) Flowing biofilms as a transport mechanism for biomass through porous media under laminar and turbulent conditions in a laboratory reactor system. Biofouling 21(3–4):161–168

Sunthar P (2010) Polymer rheology. In: Krishnan JM, Deshpande AP, Kumar PBS (eds) Rheology of complex fluids. Springer, New York, pp 171–191

Tam D, Hosoi A (2011) Optimal feeding and swimming gaits of biflagellated organisms. Proc Natl Acad Sci 108(3):1001–1006

Taylor JR, Stocker R (2012) Trade-offs of chemotactic foraging in turbulent water. Science 338 (6107):675–679

Teodósio J, Simões M, Melo L, Mergulhão F (2011) Flow cell hydrodynamics and their effects on *E. coli* biofilm formation under different nutrient conditions and turbulent flow. Biofouling 27 (1):1–11

Teschler JK, Zamorano-Sánchez D, Utada AS, Warner CJ, Wong GC, Linington RG et al (2015) Living in the matrix: assembly and control of *Vibrio cholerae* biofilms. Nat Rev Microbiol 13 (5):255–268

Theves M, Taktikos J, Zaburdaev V, Stark H, Beta C (2013) A bacterial swimmer with two alternating speeds of propagation. Biophys J 105(8):1915–1924

Underhill PT, Hernandez-Ortiz JP, Graham MD (2008) Diffusion and spatial correlations in suspensions of swimming particles. Phys Rev Lett 100(24):248101

Uppuluri P, Lopez-Ribot JL (2016) Go forth and colonize: dispersal from clinically important microbial biofilms. PLoS Pathog 12(2):e1005397

Vigeant MA-S, Ford RM, Wagner M, Tamm LK (2002) Reversible and irreversible adhesion of motile *Escherichia coli* cells analyzed by total internal reflection aqueous fluorescence microscopy. Appl Environ Microbiol 68(6):2794–2801

Vogel S (1994) Life in moving fluids: the physical biology of flow. Princeton University Press, Princeton, NJ

Vogel S (2004) Living in a physical world. J Biosci 29(4):391–397

Waitukaitis SR, Jaeger HM (2012) Impact-activated solidification of dense suspensions via dynamic jamming fronts. Nature 487(7406):205–209

Wallace DF (2008) Weekend Journal; Essay: David Foster Wallace on life and work. Wall Street Journal [Internet]. 19 September; W.14. http://www.wsj.com/articles/SB122178211966454607

Watnick P, Kolter R (2000) Biofilm, city of microbes. J Bacteriol 182(10):2675–2679

Webb JS, Thompson LS, James S, Charlton T, Tolker-Nielsen T, Koch B et al (2003) Cell death in *Pseudomonas aeruginosa* biofilm development. J Bacteriol 185(15):4585–4592

West SA, Griffin AS, Gardner A (2007) Social semantics: altruism, cooperation, mutualism, strong reciprocity and group selection. J Evol Biol 20(2):415–432

White FM (2010) Viscous flow in ducts. Fluid mechanics [Internet], 7th edn. McGraw-Hill, New York, pp 347–456. https://docs.google.com/file/d/0B9JtpWUzKcwkNC11OC1mQ0NzZjg/edit?pli=1

Whiteley M, Bangera MG, Bumgarner RE, Parsek MR, Teitzel GM, Lory S et al (2001) Gene expression in *Pseudomonas aeruginosa* biofilms. Nature 413(6858):860–864

Wilking JN, Angelini TE, Seminara A, Brenner MP, Weitz DA (2011) Biofilms as complex fluids. MRS Bull 36(05):385–391

Wilson S, Hamilton MA, Hamilton GC, Schumann MR, Stoodley P (2004) Statistical quantification of detachment rates and size distributions of cell clumps from wild-type (PAO1) and cell signaling mutant (JP1) *Pseudomonas aeruginosa* biofilms. Appl Environ Microbiol 70 (10):5847–5852

Wolgemuth CW (2008) Collective swimming and the dynamics of bacterial turbulence. Biophysical J 95(4):1564–1574

Wood TK, González Barrios AF, Herzberg M, Lee J (2006) Motility influences biofilm architecture in *Escherichia coli*. Appl Microbiol Biotechnol 72(2):361–367

Chapter 4
Complex Structure but Simple Function in Microbial Mats from Antarctic Lakes

Ian Hawes, Dawn Sumner, and Anne D. Jungblut

Abstract Microbial mats growing under the permanent ice cover of Antarctic lakes occupy an exceptionally low-disturbance regime. Constant temperature, the absence of bioturbation or physical disturbance from wind action or ice formation allow mats to accumulate, as annual growth layers, over many decades or even centuries. In so doing they often assume decimetre scale, three-dimensional morphologies such as elaborate pinnacle structures and conical mounds. Here we combine existing and new information to describe microbial structures in three Antarctic lakes—simple prostrate mats in Lake Hoare, emergent cones in Lake Untersee and elaborate pinnacles in Lake Vanda. We attempt to determine whether structures emerge simply from uncoordinated organism-environment interactions or whether they represent an example of "emergent complexity", within which some degree of self-organisation occurs to confer a holistic functional advantage to component organisms. While some holistic advantages were evident from the structures—the increase in surface area allows greater biomass and overall productivity and nutrient exchange with overlying water—the structures could also be understood in terms of potential interactions between individuals, their orientation and their environment. The data lack strong evidence of coordinated behaviour directed towards holistic advantages to the structure, though hints of coordinated behaviour are present as non-random distributions of structural elements. The great size of microbial structures in Antarctic lakes, and their relatively simple community composition, makes them excellent models for more focused research on microbial cooperation.

I. Hawes (✉)
University of Waikato, Tauranga, New Zealand
e-mail: ian.hawes@waikato.ac.nz

D. Sumner
Department of Earth and Planetary Sciences, University of California, Davis, CA, USA

A. D. Jungblut
Life Sciences Department, Natural History Museum, London, UK

C. J. Hurst (ed.), *The Structure and Function of Aquatic Microbial Communities*,
Advances in Environmental Microbiology 7,
https://doi.org/10.1007/978-3-030-16775-2_4

4.1 Introduction

Photosynthetic microbial mats (biofilms) frequently show consistent physical patterns that hint at organisation. At one level, organisation is exemplified by metabolic zonation, which typically occurs on mm length scales (Franks and Stolz 2009). Such zonation results in organisms capable of oxygenic photosynthesis concentrated at the top of a mat, where light is most abundant, while deeper in the mats, lower irradiance and depleted oxygen enable the growth of organisms capable of a cascade of anaerobic metabolisms (Kühl and Fenchel 2000). Layering is often visible as colour changes, with an orange surface from abundant light-protective carotenoid pigments, transitioning to green light-harvesting phycobilin and chlorophyll pigments and pink layers corresponding to photosynthetic sulphur bacteria in anoxic zones. There are many examples of this metabolic "organisation" in the literature, and the close proximity of interacting metabolisms allows diffusion to be an effective coupling mechanism that provides key advantages to this mat structure (Seckbach and Oren 2010; Nadell et al. 2016). Viewed as an integrated unit, the microbial mat shows elegant metabolic coupling, facilitated by a one-dimensional, organised structure that allows a more efficient utilisation of resources than any one organism could do alone (Paerl and Pinckney 1996; Stahl 1995; Franks and Stolz 2009). Indeed, some argue that the organised complexity and physiological cooperativity of a microbial mat is analogous to that of eukaryotic tissues (Webb et al. 2003; Guerrero and Berlanga 2007; Stahl et al. 2018).

When microbial mat communities grow in low-disturbance environments, a higher level of organisation can occur, whereby elaborate three-dimensional structures emerge on scales much larger than those of either component organisms or metabolic zonation (Bosak et al. 2013). Such emergent structures are most frequently described from environments where conditions preclude metazoa large enough to physically disrupt developing microbial communities (Fenchel 1998). The most frequently described emergent microbial structures are from quiescent geothermal or hypersaline waters, where physical disruption is minimal and insects and other invertebrates are temperature- or salinity-excluded (Des Marais 1995; Oren 2010). There, a variety of macroscopic mat forms occur, overlain on metabolic zonation, which in quiescent locations often take the "ridge-pinnacle" form of regular-spaced cones of cm-scale size and spacing, connected by narrow ridges (Walter et al. 1976).

Microbial mats thus show emergent organisation on at least two levels: a three-dimensional arrangement of organisms within the mat that appears to produce an efficient metabolic unit and a three-dimensional organisation of the mat matrix to form spatially complex structures. In modern concepts of emergent organisation (Halley and Winkler 2008), the former would be termed a simple emergence, computable through a reductionist understanding of organism-environment interactions and close to an (dynamic) equilibrium (Corning 2002; de Wolf and Holvoet 2005). Whether the three-dimensional organisation is an example of complex emergence, within which some degree of self-organisation occurs to convey a holistic functional advantage, or whether it too is driven by simple organism-environment responses is not clear. At small scales, there is accumulating evidence for holistic, coordinated behaviour in biofilms (Stahl et al. 2018), perhaps controlled by cell to cell signalling and "programmed death" that would allow a mechanism for complex emergence of structure, though to date we are aware of no definitive examples from nature where coordinated behaviour for holistic benefit is shown (Battin et al. 2007). One of the few recent studies to suggest a holistic function for macroscale ridge-pinnacle structure suggests that a mechanism exists to space pinnacles to optimise overall access to water column resources by minimising inter-pinnacle competition (Petroff et al. 2010). This degree of self-organisation requires information transfer amongst pinnacles to achieve the overall optimum condition, which the authors propose via resource depletion "shells" around structures. Other authors argue for mechanisms of complex structure emergence that require much less self-organisation, suggesting that cones initiate as surface irregularities that are consequences of un-oriented movement (Shepard and Sumner 2010) that then propagate vertically either through phototaxis (Stahl 1995) or through differential growth of tips due to enhanced access to limiting resources (Tice et al. 2011). The question of whether 3-D microbial mat structures derive from internal or external drivers, or a mix of the two, extends beyond semantics; it relates to whether ridge-pinnacle structure is the inevitable outcome of individual behaviour in response to environmental drivers and has no holistic function or whether individual behaviour within microbial mats has, over time, been structured to facilitate enhanced overall function through a self-organised structure.

In this contribution, we examine the extent to which microbial mat structures are organised and consider whether this organisation is essentially mechanistic, with organisms responding to environmental stimuli or whether there is evidence that organised structures contribute to holistic functions that enhance overall performance. As case studies, we focus on similarities and differences between mats in three perennially ice-covered Antarctic lakes (Fig. 4.1). Antarctic lakes are particularly favourable locations for the development of complex microbial structures and contain some of the best developed modern examples (Hawes et al. 2011, 2013; Mackey et al. 2015; Sumner et al. 2016). They share with geothermal and hypersaline waters the lack of bioturbation from macro-invertebrates, due in this case to the combination of extreme environmental conditions and geographic isolation (see discussion in Karanovic et al. 2014), but furthermore, the ice cover ensures that lake water is exceptionally still, over long time periods, with no wind-induced turbulence (Wharton et al. 1993; Andersen et al. 2011; Hawes et al. 2013) that could generate sufficient shear stress to disturb mat communities (Tice et al. 2011). Consequently, large-scale structures, overlain on an internal mat zonation, reach decimetre proportions in these lakes (Fig. 4.2; Andersen et al. 2011; Hawes et al. 2011, 2013). Microbial mats with obvious internal one-dimensional structure are also common in shallow Antarctic habitats (Howard-Williams et al. 1989; Vincent 2000; Jungblut et al. 2012), but their prolonged winter freezing and summer water movement appear to prevent the development of large-scale three-dimensional organisation, and thus our focus remains on lake mats developing in perennially liquid, under-ice habitats.

Fig. 4.1 The location in Antarctica of the three lakes primarily referred to in this chapter. Public domain image produced by the US National Aeronautics and Space Administration

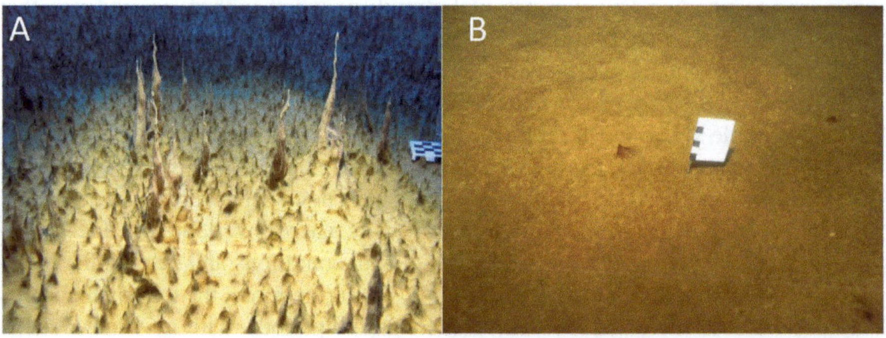

Fig. 4.2 Examples of contrasting microbial mats in perennially ice-covered Antarctic lakes. (**a**) Complex emergent structures at 22 m depth in Lake Vanda, Antarctica. (**b**) Microbial mat with little emergent complexity at 11 m depth in Lake Hoare. The checkers are 1 cm in both images, which are by courtesy of Tyler Mackey

We focus on three lakes that allow comparison of three distinctly different macroscale mat topographies: Lake Hoare where prostrate forms prevail (Hawes et al. 2014, 2016); Lake Untersee where a unique conical growth form is widespread (Andersen et al. 2011); and Lake Vanda, where dense fields of tall cuspate pinnacles dominate (Fig. 4.2; Hawes et al. 2013; Sumner et al. 2016). Although early literature refers to Antarctic lake mats rather generically as "modern stromatolites" (e.g. Parker et al. 1981), for the most part, this is misleading. Though laminated, they are mostly unlithified (Mackey et al. 2015), being entirely built from alternating annual laminations of sediment and microbially derived organic material (Hawes et al. 2001; Sutherland and Hawes 2009). Thus, the potential structuring effects of spatially constrained mineral precipitation (Petroff et al. 2013) are not in play, and Antarctic lakes instead present some of the best developed microbe-soft sediment accretion structures on modern Earth. Lake Vanda has the added advantage that it has undergone recent increases in lake level, with newly inundated areas of lake bed providing dated underwater landscapes allowing an examination of the evolution of structural complexity over time (Hawes et al. 2013; Sumner et al. 2016).

4.2 Lake Environments

4.2.1 Lake Hoare

Lake Hoare (77.63°S, 162.88°E) is a closed-basin lake near the eastern end of Taylor Valley in southern Victoria Land, Antarctica. The lake is 4.2 km long and 1.0 km wide and has maximum and mean depths of 34 and 14 m (Spigel and Priscu 1998). It is dammed to the north-east by the Canada Glacier, which provides an inflow of glacial meltwater (Wharton et al. 1992). Other sources of inflow come from Andersen Creek entering the north-east corner of Lake Hoare and drainage from Lake Chad in the south-east. No outflows from Lake Hoare exist, and water loss is restricted to sublimation of ice and evaporation of meltwater during summer (Doran et al. 1994). The ice cover of Lake Hoare is perennial, except for small areas at the lake margins, which melt during summer. The ice cover was 3.5 m thick in 1983 (Wharton et al. 1992), but the thickness has increased to ~5 m since that time (Doran et al. 2002). There are 3 months of complete darkness during winter and 3 months of continuous light during summer (Dana et al. 1998). Howard-Williams et al. (1998) summarised the optical properties of the ice cover and the water column. Net transmission of photosynthetically active radiation (PAR) in the lake centre ranged from <1% to 3%, with a spectral transmission peak at wavelengths between 450 and 550 nm. Vertical extinction coefficients for downwelling PAR within the water column from beneath the ice to a depth of 33 m were typically 0.12–0.22 m^{-1} (Howard-Williams et al. 1998), resulting in a 1% irradiance depth of between just below the ice cover and 9 m.

Lake Hoare shows weak density stratification in the upper waters, and although there is controversy over the extent of mixing of surface water (Spigel and Priscu 1998; Tyler et al. 1998), the density gradient is continuous to at least 13 m depth. There is a pronounced inflexion in the density-depth profile at 13–15 m from the surface (the depth of the lake varies temporally), which divides the lake into upper and lower compartments. The upper compartment is characterised by low concentrations of dissolved nutrients, particularly nitrate, which is probably limiting planktonic production (Lizotte and Priscu 1992). Upper waters also contain lower bicarbonate, with a higher pH (up to 8.6, Cathey et al. 1981) than the lower compartment (pH 7.9). The lake is anoxic below 25–26 m.

Benthic microbial mats, comprised primarily of cyanobacteria, diatoms and other bacteria, with various microbial eukaryotes at low abundance, line the lake from the edge into the anoxic zone at 25 m (Wharton et al. 1983; Hawes and Schwarz 1999). Mats in Lake Hoare are often prostrate, with little surface elaboration, though in some shallow areas rounded pinnacles and lift-off structures can form (Wharton et al. 1983). The matrices of the mats are dominated by narrow, filamentous cyanobacteria, attributable on morphological and molecular bases to the genera *Leptolyngbya* and *Pseudanabaena* with occasional *Phormidium* (Sutherland and Hawes 2009; Zhang et al. 2015). Diatoms comprise 10–15% of the mat community, though this increases at depth. Sixteen taxa of diatoms have been recognised, with *Psammothidium chlidanos*, *Diadesmis contenta* var. *parallela* and *Navicula gregaria* the most frequent (Sutherland and Hawes 2009). These mats are adapted to low irradiance through efficient light harvesting and utilisation, and compensation points of <1 μmol photons m^{-2} s^{-1} allow net production to occur to >20 m depth (Hawes and Schwarz 1999, 2001; Vopel and Hawes 2006).

4.2.2 Lake Untersee

Lake Untersee (Fig. 4.1) is located at 71.33°S 13.75° E in the Otto-von-Gruber-Gebirge (Gruber Mountains) of central Dronning Maud Land. The lake is 563 m above sea level, with an area of 11.4 km^2, and is the largest surface lake in East Antarctica. Lake Untersee has two subbasins; the largest, 160 m deep, lies adjacent to the Anuchin Glacier and is separated by a sill at 50 m depth from a smaller, 100-m-deep basin in the southwest corner (Wand et al. 1997). The deep basin is close to homothermal and well oxygenated to the bottom. In contrast the shallow basin is density stratified below the sill depth (c. 50 m) and anoxic at its base. Profiles of conductivity, temperature and dissolved oxygen indicate that the deep basin and upper part of the anoxic basin are part of the same body of well-mixed water (Kaup et al. 1988; Haendel et al. 1995; Wand et al. 1997, 2006; Andersen et al. 2011). The unusually high pH of this mixed water mass (pH 10.4) has been attributed to weathering of the predominant anorthosite rock in the lake's catchment in the absence of an effective connection to the atmosphere (Kaup et al. 1988; Andersen et al. 2011). The lake ice transmits approximately 5% of incident irradiance and vertical extinction coefficient for scalar PAR of 0.033 m^{-1}, (Andersen et al. 2011) resulting in a 1% surface irradiance depth of ~50 m.

Benthic microbial mats occur to at least 100 m in Lake Untersee and comprise mostly oscillatorian cyanobacteria primarily as *Phormidium autumnale* and *Leptolyngbya* spp., with some unicellular *Chamaesiphon* and nitrogen-fixing *Nostoc*, with few chlorophytes and almost no diatoms (Andersen et al. 2011; authors' unpublished molecular data). These microbial mats form two distinct macroscale morphologies, pinnacles and large cones, with the latter, which can be up to 50 cm tall, being most widespread. Cones are built from sub-mm scale laminations of fine glacially derived sediments and organic material, and radiocarbon dating suggests that they accumulate over centuries (Andersen et al. 2011). As in lakes Vanda and Hoare, these cyanobacteria are rich in phycoerythrin, and mats have an overall purple-pink appearance.

4.2.3 Lake Vanda

Lake Vanda (Fig. 4.1) lies in the Wright Valley, one of the McMurdo Dry Valleys of southern Victoria Land (77.50°S, 161.67°E). It occupies a closed basin with a perennial ice cover 3.5–4.0 m thick; though for several weeks each summer, the ice around parts of the lake shore melts to produce a discontinuous, open-water moat. In recent years inflow has exceeded outflow, and the lake level has risen by 12 m between 1960 and 2012 (Hawes et al. 2013). The endorheic nature of the lake and historical changes in water balance have left Lake Vanda with an unusual physical structure, with an inverse temperature gradient stabilised by increasing solute content with depth. In December 2013, temperature and conductivity were constant from 4 m to 23 m depth at ~4 °C and ~560 μS cm^{-1}, increasing abruptly as a pycnocline from 23 to 28 m depth, then stabilising at 6 °C and 1020 μS cm^{-1} between 28 and 45 m. These two upper cells are individually mixed by thermohaline convection (Spigel and Priscu 1998), while below 45 m, there is a continuous gradual increase in temperature and conductivity. The ice cover transmits 15–20% of incident photosynthetically active radiation (PAR), biased to wavelengths below 550 nm (Hawes and Schwarz 2001). The lake water is extraordinarily clear, with a vertical extinction coefficient for downwelling scalar PAR of 0.06 m^{-1} (Howard-Williams et al. 1998), resulting in a 1% surface irradiance depth of 48–63 m. Low concentrations of dissolved reactive phosphorus (DRP) appear to limit phytoplankton growth (Vincent and Vincent 1982).

In Lake Vanda, benthic microbial mats are widespread, attain high biomass, form elaborate pinnacles and extend to at least 50 m depth (Figs. 4.2 and 4.3; Wharton 1994; Hawes and Schwarz 2001; Hawes et al. 2013). These mats comprise mostly cyanobacteria, particularly species of *Phormidium* and *Leptolyngbya* (Sumner et al. 2016), with pennate diatoms (species of *Navicula*, *Nitzschia*, *Caloneis* and *Stauroneis*) throughout and occasional strands of moss below the 25 m pycnocline (Love et al. 1983). Lake Vanda mats appear to have a higher cyanobacterial diversity than Lake Hoare, with up to 12 compared to 8 operational taxonomic units (OTUs) based on clone library analysis (Zhang et al. 2015). Vanda mats are annually laminated, with the upper 2–4 laminae constituted an orange-brown zone, rich in myxoxanthophyll and dominated by intertwined *Leptolyngbya* trichomes. Below the upper zone, green-/pink-pigmented subsurface zones are present, containing up to six phycobilin-rich laminae, also colonised by *Leptolyngbya*, *Oscillatoria* and

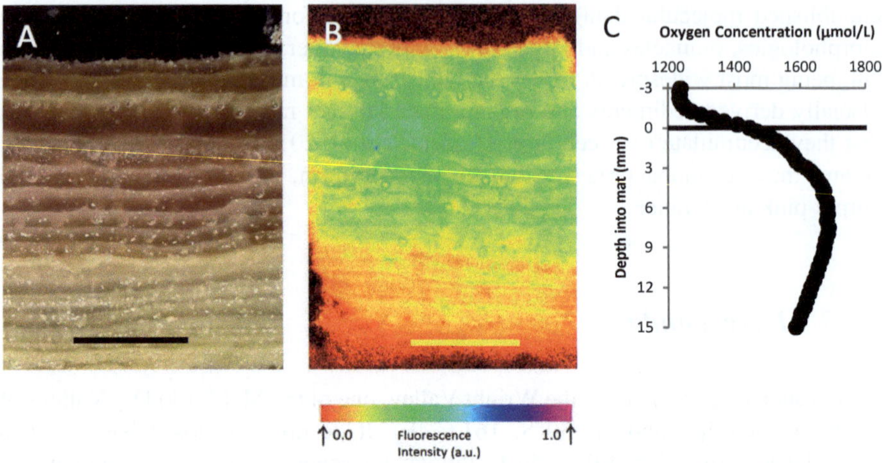

Fig. 4.3 A vertical section of microbial mat from 10 m depth in Lake Hoare, 2010. (**a**) A photograph showing mm-scale, annual laminations and the zonation from pink-brown at the surface, though pink, to colourless. (**b**) A false-colour composite fluorescence image (blue excitation, red emission) of a similar section. False colour scale is in arbitrary units of chlorophyll fluorescence. The scale bars in (**a**) and (**b**) represent 10 mm. (**c**) An in situ dissolved oxygen profile through a mat at the same depth from which the sections were obtained, shown on a similar vertical scale

Phormidium morphotypes (Hawes et al. 2013, 2016; Sumner et al. 2016). Recent, well-constrained, increases in lake level have allowed mats from different depths to be examined as a proxy for rate of development (Hawes et al. 2013). At depths inundated for <30 years, mats were accreting, adding one lamina (0.3 mm) and ~0.2 μg cm^{-2} of chlorophyll-a per year. At depths that have been submerged for >40 years, mat chlorophyll-a appeared to have reached an equilibrium and did not increase over time. The older mats formed tall pinnacles that were only beginning to emerge in shallower mats.

4.3 One-Dimensional Structures

Lake Hoare microbial mats frequently lack well-developed, complex three-dimensional structures (Fig. 4.2), though one-dimensional organisation is well developed. The properties of 1-D structure in these mats take two forms; lamination and zonation (Hawes et al. 2014). Lamination, composed of alternating sediment-rich and sediment-poor bands, is a product of annual accumulation (Hawes et al. 2001). Sediments enter the lake in summer via streams that typically flow for 1–2 months, and settle as a distinct optically dense band over autumn-winter when darkness prevents photosynthetic growth. The resumption of growth the following spring, when light returns but inflows are not running, forms a hyaline band (organic accumulation without sediment) overlying the relatively opaque (sediment without growth) band formed in winter (Fig. 4.3a). Hawes et al. (2001) used confocal laser scanning microscopy to show that trichomes are relatively sparse and oriented mostly vertically in the hyaline bands, including at the surface in summer, but tend to be more abundant, intertwined and oriented parallel to the mat surface in the opaque bands. Such laminations through trapping and binding of episodic depositions of sediment are reported from microbial mat communities in other sedimentary environments such as marine stromatolites (e.g. Reid et al. 2000).

Zonation is overlain on this annual banding pattern and in Antarctic lakes is recognisable by pigmentation bands that comprise multiple laminae. The orange-brown surface zone contains a higher proportion of carotenoids and live diatoms than the lower zone, where carotenoids are rare and cyanobacterial phycobilin pigments are responsible for pink coloration (Hawes et al. 2016). However, cyanobacteria dominate biovolume throughout the pigmented zone, with a similar morphotypic composition, dominated by *Leptolyngbya* in both zones, with *Pseudanabaena* present in both but most abundant in the upper pink-brown zone (Sutherland and Hawes 2009; Hawes et al. 2016).

In addition, Lake Hoare mats have two unusual features that are not common in temperate microbial mats: (1) the >18 mm thickness of the pigmented zone, which occupies 9 years of vertical accumulation (Fig. 4.3a, b) and (2) the persistence of dissolved oxygen concentrations exceeding that of the ambient water column to more than 15 mm into the mat (Fig. 4.3c). These are despite the mat receiving <1% of lake-incident irradiance (Vopel and Hawes 2006; Hawes et al. 2014). Fluorescence imaging (Fig. 4.3b, see Vopel and Hawes 2006 for methods) shows that chlorophyll-a is present throughout the pigmented zone of the mat, only reaching the limits of detection at 15 mm or more below the mat surface. The bulge in oxygen concentration associated with the pigmented zones is indicative of a broad photosynthetic zone, but also that oxic conditions extend well below the region containing pigments associated with oxygenic photosynthesis. This is different to what is normally seen in temperate mats, and even in shallow water (seasonally frozen) mats in Antarctica (Hawes et al. 1999), where oxygenic photosynthesising organisms are usually concentrated in the upper 1–2 mm, and a sharp oxycline is present at the base of the oxygenic, pigment-containing zone. In temperate system, the oxycline within the mat can be associated with an increase in sulphide and a shift in dominant metabolism from oxygen- to sulphur-based and moves vertically in response to diel changes in irradiance (e.g. Denis et al. 2012). Indeed, in temperate mats, this redox change is the basis of the 1-D metabolic structure that is often interpreted as having the functional role of facilitating nutrient cycling by positioning all components of nutrient cycling pathways within diffusion distance of each other (e.g. Paerl and Pinckney 1996).

The unusual oxygenation features of Lake Hoare mats may in part reflect the 24-h daily irradiance that they receive, albeit at low photon fluxes. Estimates of photosynthesis in Lake Hoare show that photic zones of microbial mats export oxygen to the water column and to underlying mats during summer and that oxygen consumption in mats below the photic zone quickly falls to close to zero despite an abundance of organic carbon (Hawes et al. 2014). Vopel and Hawes (2006) demonstrated that the rate of photosynthesis is linearly related to irradiance and argued for a compensation photon flux as low as 0.1 μmol m^{-2} s^{-1}. Hawes et al. (2014) went on to confirm persistent light-limitation of photosynthesis but also showed that net flux of oxygen to the water column continued through the 24-h light cycle. However, the export of dissolved oxygen at low irradiances around midnight was due to ongoing effusion of dissolved oxygen produced during brighter parts of the day: under the night-time photon flux of ~1 μmol photon m^{-2} s^{-1}. Gross photosynthesis was close to zero, and a net consumption of oxygen was actually occurring when flux was integrated through the mat, implying that ~1 μmol photon m^{-2} s^{-1} may be closer to the effective community compensation point than 0.1 μmol photon m^{-2} s^{-1}.

The ability of the Lake Hoare microbial mat to be photosynthetic at low irradiance is enhanced by efficient absorption and utilisation of light. Hawes and Schwarz (2001) showed how the spectral absorption characteristics of ice (which absorbs red light strongly) results in light that penetrates to 10 m depth in Lake Hoare having a dominant wavelength of close to 500 nm and an overall blue-green hue. They showed how phycoerythrin allowed benthic mats to capture effectively these green wavebands and absorb up to 50% of light incident to the mat using these photosynthetic pigments. A variety of methodologies have shown how the cyanobacteria function at close to the maximum quantum yield of oxygenic photosynthesis, fixing the equivalent of 1 mol carbon per 12 mol photons absorbed (Hawes and Schwarz 1999, 2001; Vopel and Hawes 2006; Hawes et al. 2014). Low respiration rate is, however, also essential to both low-compensation irradiance and deep-penetrating oxygen (Hawes and Schwarz 1999; Hawes et al. 2014). The low respiration rate, despite an apparent abundance of oxygen and organic carbon in the mat matrix (Sutherland and Hawes 2009), is unexpected. It has been suggested that low respiration rate is due to the recalcitrant nature of the cyanobacterial sheaths and extracellular polymeric material that is suspected to form much of the organic laminae (Sutherland and Hawes 2009). What happens to dissolved oxygen concentration in mat communities over the 4 months of winter darkness is not yet known.

While we can hypothesise that the unusual oxygen profiles in Lake Hoare mats reflect the absence of release of labile organics from the photosynthetic mat components to the underlying mat, it is clear that the 1-D structure of these Antarctic lake mats is not based around redox gradients that facilitate coupling of microbial processes. The absence of an oxycline within the mats implies the absence of processes that link aerobic and anaerobic metabolisms to the carbon acquisition processes at diffusion-efficient distance scales—at least during the summer months. Nutrient accumulation does occur in these mats, with concentrations of all inorganic- and organic-dissolved forms of N and P increasing severalfold from the water column to the mat (Quesada et al. 2008), but the absence of abrupt redox gradients means that benefits from closely linked, diverse metabolisms are unlikely, and the enigma of apparently abundant carbon resources in oxygen-rich, aphotic lower zones of microbial mats remains incompletely resolved.

So does the one-dimensional arrangement of cyanobacteria in the mat have another function? It seems unlikely that the structure of the mat, with pigments arranged within a sequence of laminae representing up to 10 years of growth (Fig. 4.3), has the function of enhancing the ability to harvesting light. The most effective arrangement of pigments to intercept light would be to concentrate them in a narrow layer at the mat surface, as this would minimise attenuation due to non-photosynthetic mat components (Kühl and Fenchel 2000). Thus, one might expect these potentially motile organisms to migrate vertically to the upper, better illuminated part of the mat. At present the reason for the absence of migration is not clear, but we have previously argued (Hawes et al. 2014) that the only available cue for migration is light and that the rapid scattering of light within mats (Kühl and Fenchel 2000) and the low component of the red wavelengths that appear to be important in phototaxis (Ng et al. 2003) prevent coordinated vertical migration from occurring. The alternate hypothesis then is that pigments in deeper, older laminae are persistent legacies of previous growth and that the 1-D structure seen in these mats—annual lamination and some zonation of pigments and dominant morphotypes—has no holistic function; rather it is simply a consequence of how these mats grow and age.

4.4 Three-Dimensional Structure

4.4.1 Lake Untersee

Above we argue that the 1-D structure that is evident in vertical sections of microbial mat from Lake Hoare, at 10 m depth, is largely a consequence of the way that the mat grows and confers no advantage in the acquisition of the key-limiting resource at that location. In the second example considered here, Lake Untersee, Dronning Maud Land, the mats have a 1-D laminated structure that superficially resembles that of Lake Hoare, but this 1-D structure is folded into the emergence of conical macro-scale structures (Andersen et al. 2011). Here we will consider how these structures may form and how this formation may create a holistic advantage for the capture of otherwise limiting resources.

Lake Untersee actually contains two types of emergent structure, pinnacles and cones (Andersen et al. 2011; Fig. 4.4). Surveys using drop cameras and divers have shown that the cones are the most frequent structures seen in the lake and that these are present from less than 15 m depth (the minimum depth surveyed) to in excess of 100 m, equivalent to optical depths of ~3–0.25% of surface incident irradiance (author's unpublished data). This encompasses the optical depth of the mats discussed above from Lake Hoare (~1% surface light). Cones themselves are all sizes, but often large, up to 50 cm tall, and rise steeply (average slope 60°) from the surrounding sediment with a spacing length of metres (Andersen et al. 2011). A characteristic feature is a fibrous cluster of a broad-trichome cyanobacterium (identified by sequence analysis of 16S rRNA gene as *Phormidium autumnale*; author's unpublished data) on the apex of the cones, particularly evident when viewed from above (Fig. 4.4b). In fact, 16S high-throughput sequencing showed that *Phormidium*

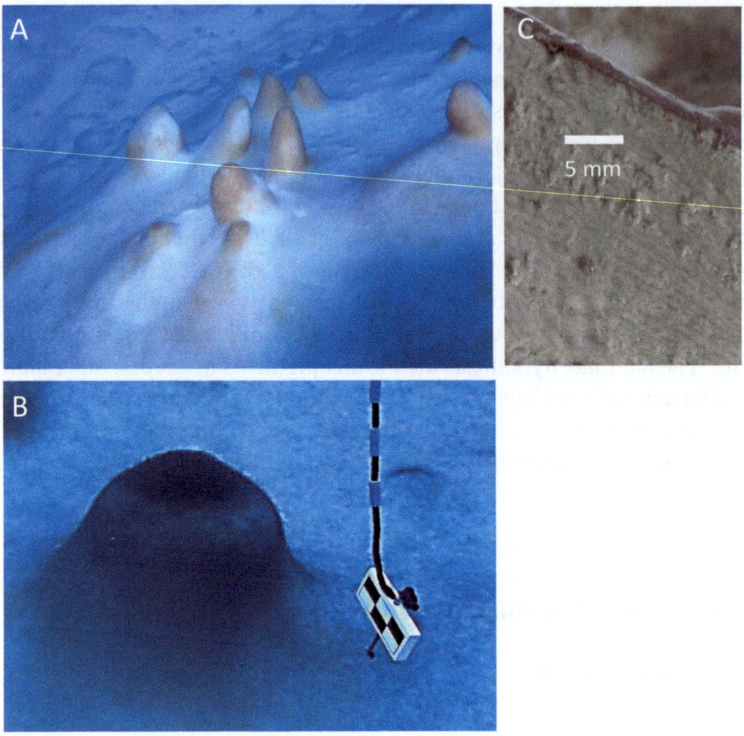

Fig. 4.4 (**a**) A field of conical mounds at 20 m depth in Lake Untersee. For comparison with other lakes, 20 m physical depth in Lake Untersee is the depth to which 2–3% of incident irradiance penetrates. (**b**) A close-up of a small conical mound, with a nascent structure in the upper right. The scale bar shows cm. (**c**) Cross section of the cone imaged in (**b**) after return to the surface, section close to "shoulder" of the cone. All images are taken with no artificial lighting

dominates the cyanobacterial assemblages on the sides and top of the cone from where the fibrous cluster of a broad-trichome cyanobacterium protrudes (authors' unpublished data). This is quantifiable as an elevated concentration of photosynthetic pigments at the cone apex (Table 4.1).

Vertical sections of cones reveal that the cone structure is built as a modification of the simple laminar accrual process described in Lake Hoare, but the internal structure is quantitatively different to that seen in Lake Hoare (Fig. 4.4c). In particular, zonation is restricted to a single pink-pigmented, upper zone less than 1 mm thick, overlying an unpigmented organosedimentary zone, and the scale of lamination in Lake Untersee is much finer, sub-mm scale laminated clay-organic sediment. Although arranged differently, the absolute abundance of pigments per unit area was similar to lakes Vanda and Hoare (Table 4.1). Potential benefits of the conical geometry include an increase in mat surface area for a given area of lake bed. Given that the concentration of pigments on the cone walls are similar to those on the adjacent flat areas (Table 4.1), the increase in surface area of cones relative to flats

Table 4.1 Comparison of pigment concentrations in Lake Untersee at 25 m depth, Lake Hoare at 10 m and Lake Vanda at 20 and 30 m

Lake	Location	Chlorophyll (μg cm^{-2})	Phycoerythrin (μg cm^{-2})
Untersee	Top of cone	24.2 ± 12.1	38.0 ± 14.6
	Side of cone	11.9 ± 2.3	8.1 ± 4.4
	Flat mat	9.5 ± 3.2	5.6 ± 3.5
Hoare	Flat mat	14.5 ± 0.4	36.0 ± 5.3
Vanda	Flat mat (20 m)	11.9 ± 1.2	24.7 ± 1.5
	Flat mat (30 m)	15.1 ± 0.5	46.6 ± 0.5
	Pinnacle (20 m)	17.8 ± 2.8	30.5 ± 3.4
	Pinnacle (30 m)	33.9 ± 9.1	103.0 ± 32.1

$N = 3$ (Untersee—from Andersen et al. 2011) or $N = 5$ (Hoare and Vanda—from Hawes and Schwarz 2001, and unpublished)

will inevitably increase the overall biomass per unit lake bed. A regular-truncated cone that is 20 cm tall, with a 60° wall slope and a flat apex 6 cm across would have more than twice the surface area of an equivalent flat disc (996 vs. 380 cm^2). Allowing for enhanced pigment concentration at the apex, the overall average chlorophyll-a concentration per unit bed area of a cone would be approximately 35.5 μg cm^{-2} compared to 9.5 μg cm^{-2} for flat mat.

The very thin laminations in Lake Untersee compared to Lake Hoare suggest a lower rate of sediment and organic carbon accrual, even though the irradiance in Lake Untersee is higher than in Lake Hoare. New estimates (author's unpublished data) of sedimentation in Lake Untersee, based on four replicate passive traps at 20 m depth, deployed for 2 years, yield an accumulation rate of 6 g m^{-2}, which compares to 41 g m^{-2} for Lake Hoare (Wharton et al. 1989). New in situ measurements of oxygen concentration profiles at 20 m depth and a photon flux of 30 μmol m^{-2} s^{-1} in Lake Untersee (Fig. 4.5), obtained as described in Vopel and Hawes (2006), show that the zone of net oxygenic photosynthesis is restricted in Lake Untersee to ~1 mm, much narrower than Lake Hoare at 10 m depth (Fig. 4.5). However, the flux of oxygen to the overlying water calculated from this profile was 190 μmol m^{-2} h^{-1} or approximately half of that in Lake Hoare at 10 m depth with an irradiance of 3 μmol photons m^{-2} s^{-1} (Vopel and Hawes 2006). Thus Lake Untersee mats show lower gross photosynthesis and photosynthetic efficiency than Lake Hoare mats. As in Lake Hoare, there is no evidence for a sharp oxycline below the zone of oxygenic photosynthesis; instead there is a steady decline deep into the mat, repeating the apparent paradox of coincident oxygen and organic carbon in the aphotic mat seen in Lake Hoare.

Two likely reasons for the lower accumulation of extracellular organic carbon in Lake Untersee compared to Lake Hoare are geochemical and taxonomic differences. Lake Untersee has an unusual water chemistry, with a pH of >10.4 generated by weathering of anorthosite rock floor under an ice cover that effectively eliminates atmospheric contact (Kaup et al. 1988; Andersen et al. 2011). High pH persists some distance into the sediment, with a maximum coincident with the photosynthesis-induced dissolved oxygen maximum at the mat surface (Fig. 4.5). Recently, using

Fig. 4.5 Profiles of dissolved oxygen and pH through the surface layer of a cone from Lake Untersee. Profile obtained using a manual micromanipulator, and freshly calibrated Unisense Ltd. (http://www.unisense.com/) microelectrodes connected to a Unisense UW-M underwater picoammeter

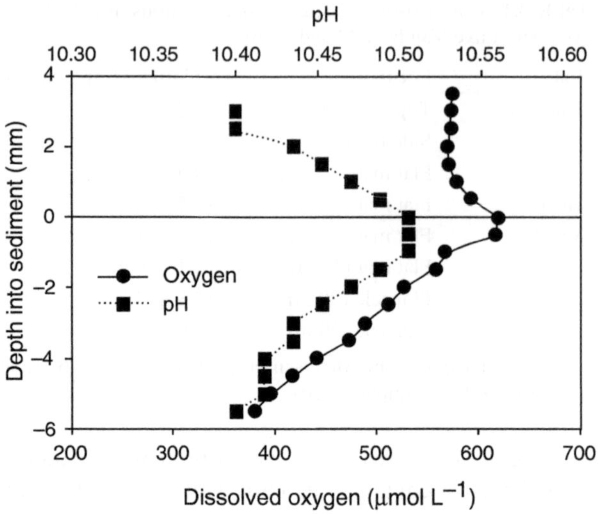

techniques described by Hawes et al. (2011), we measured concentrations of dissolved inorganic carbon (DIC) in water from Lake Untersee at 12 μmol L^{-1}, amongst the lower values recorded in natural waters. At pH 10.4, less than 50% of DIC would be in a photosynthetically available form, e.g. as bicarbonate. Thus the difference between the rate of accumulation of lamina thickness between Lake Untersee and Lake Hoare may be due to limitation by both sediment deposition and carbon accrual rates.

A second substantial difference between the lakes is the dominance of *Phormidium* taxa in the phototrophic community in Lake Untersee, whereas in Lake Hoare mats are dominated by a mix of *Leptolyngbya* and diatoms. *Phormidium* tends to be a highly motile group, occurs at high pH in Antarctic lakes (Mackey et al. 2015) indicating an affinity for bicarbonate and is characterised as a "surface-smoothing" taxon (Mackey et al. 2015). In contrast *Leptolyngbya* is less motile, but is associated with topographical enhancement through vertically oriented growth (Reyes et al. 2013; Mackey et al. 2015). The surface topography, with filaments of *Phormidium autumnale* projecting from the top of each mound in a loose, woolly surficial layer, may represent a behavioural response to optimise carbon accrual at low DIC availability; the ability of cyanobacteria to move along light and DIC gradients is well known (see review by Stahl 1995). At low irradiance a tendency for trichomes to accumulate at the surface of topographic highs, through orientation towards the light, and then to extend vertically through a pH and DIC gradient, is consistent with known behaviours. Enhanced accumulation of sediment, on the flattened apices of cones is also likely due to the enhanced trapping ability of the "woolly cap" of *P. autumnale* trichomes and may also contribute to the differential accumulation of mass on the apex of the cone compare to the cone sides and surrounding lake floor.

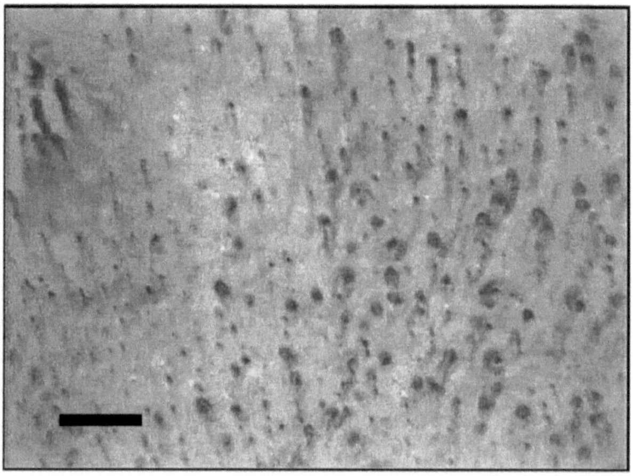

Fig. 4.6 Vertical image of the floor of Lake Untersee at 45 m depth obtained with a low-resolution SeaView (http://www.seaview.com/) drop camera lowered through a hole in the ice. The dark spots are the apices of mounds, and the streaks trailing from the apices are trails of cyanobacteria. The bar represents ~2 m

Once a biomass maximum is established on a cone apex, the morphology is likely to become a self-reinforcing feature. Bosak et al. (2012) showed that the tips of pinnacles grow faster than the surrounding bases because of higher density of active cells and, like many other authors, suggested that this reflects enhanced access to water column nutrients for vertical elements projecting though the hydraulic laminar sub-layer. This advantage is most effective in situations where near-laminar flow occurs at the lake bed. Unlike many ice-covered lakes, most of the water column of Lake Untersee is mixed at least some of the time (Wand et al. 1997). Mixing in the lake is indicated by homogenous conductivity profiles with depth through the >140 m water column and is thought to be driven by a mix of solar heating and the generation of cold, dilute and buoyant water through subsurface melting of the Anuchin Glacier that forms the northern boundary of the lake (Wand et al. 1997; Andersen et al. 2011). Thus water movement, albeit slight, occurs in the lake, and sufficient water movement may occur to result in significant near-bed velocity gradients. Evidence in favour of slow flow at the lake bed that interacts with cone apices is provided by the apparent unidirectional drift of trichomes from apical concentrations of *P. autumnale* evident in vertical images of the lake floor taken with a drop camera (Fig. 4.6). Divers operating in the lake have never detected any evidence of water movement, even when disturbing bottom sediments during sample collection, and we suspect that at most flows are slow and laminar rather than turbulent. However, it is possible that the presence of tall, smooth cones within moving water may enhance nutrient supply to taller "canopy" elements of benthic communities over what would accrue on flats as they would be placed in higher velocity parts of the laminar flow gradient (Ghisalberti et al. 2014).

Nepf (2012) reviewed the hydrodynamics of vegetated channels, providing an understanding of the role of canopy density on near-bed turbulence regime. She noted that sparse canopies, as in Lake Untersee, resemble a simple boundary layer regime with a velocity gradient predicted by distance from the channel bed, enhancing the probability that the cone apices would experience higher velocity environments than cone bases. While the conical structures allow an increase in surface area for uptake of nutrients, a possible specific functional advantage of the conical structures in Lake Untersee may thus emerge as placing the most actively growing part of the microbial population in the location where they are most hydrodynamically favoured to exploit otherwise limiting nutrients, in this case perhaps DIC.

The growth of large conical structures in Lake Untersee may be the consequence of the responses of the constituent organisms growing under unusual field conditions. Cones may naturally develop on local topographic irregularities, which propagate through enhanced growth and sediment accumulation on the apex. This model implies no complex, holistic organisation of structure. However, one observation challenges this simple interpretation. The spacing of cones shows marginally significant regularity, based on analysis of six down-looking images of cones, using nearest neighbour analysis (Clarke and Evans 1954). Cone spacing was on scales of decimetres or more, yet the R_n statistic (n-sample nearest neighbour risk) of the nearest neighbour analysis for the image in Fig. 4.6 was 1.29 (where 1 is random and >1.2 indicates a significant element of regularity—analysis was at $N > 200$ and $p > 0.05$).

Overall the R_n for the six analysed images was 1.36 ± 0.06 (average $\pm$ s.d.). While this statistic suggests a low level of ordering, any emergence of regular spacing may imply information transfer between cones, and how this may occur is not clear. The potential for competitive interaction for water column resources to result in regular spacing in pinnacle mats has been argued for on the basis of elegant experimentation and modelling by Petroff et al. (2010). Those arguments suggest that under still conditions, spacing on cm distance scales will result from pulsed activity on a 24 periodicity (i.e. photosynthesis). Even in Antarctic lakes, with 24 h of daylight, sufficient day-night variation in irradiance exists to drive a diel photosynthetic pattern (Hawes et al. 2014). Thus in still water, this mechanism might explain a cm scale of spacing of pinnacles (Andersen et al. 2011) but not a greater than dm-scale spacing of large cones, a point noted by Petroff et al. (2010). However, as discussed above, laminar flow of water is suspected in Lake Untersee, and Petroff et al. (2010) went on to argue that larger-scale spacing patterns may be propagated by the competitive interaction mechanism in directional flowing water. They suggest that laminar flow would result in longitudinal regularity, or ridges, rather than the distribution of cones seen in Lake Untersee.

As with Lake Hoare, the emergent structures in Lake Untersee, here a mixture of one-dimensional laminations elaborated to large conical forms, appear to largely result from uncoordinated responses of organisms to growth conditions. However, the enigmatic tendency for cones to be spaced in a non-random pattern over dm distances is unexplained. A mechanism whereby communication between cones could occur over such distances is not clear, and given the expectation of minimal competitive interaction at such scales, neither is any holistic advantage that such communication may confer.

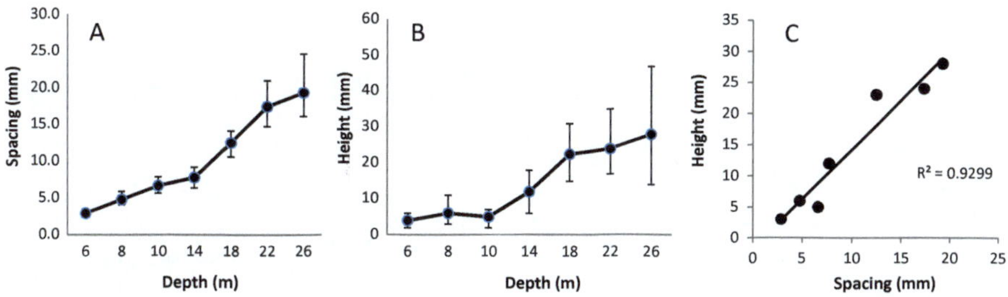

Fig. 4.7 (**a**) Spacing and (**b**) height of pinnacles along a depth gradient in Lake Vanda. (**c**) The relationship between spacing and height at each sample depth. Spacing and height are not normally distributed, and the median and quartiles are shown. Modified after Hawes et al. (2013)

4.4.2 Lake Vanda

If there is a "model" emergent structure for microbial mats that has dominated recent literature, it is that of cm-scale spaced, cuspate pinnacles, associated with narrow-trichome cyanobacterial taxa (Petroff et al. 2010; Tice et al. 2011; Bosak et al. 2012). This morphology and taxonomic dominance is common in Lake Vanda (Hawes et al. 2013; Zhang et al. 2015; Sumner et al. 2016). For logistic reasons, published information is largely limited to the upper part of the lake, from 6 to 26 m depth where a time series of mat age that has resulted from a gradual increase in lake level allows insights into the development of these microbial pinnacles (Hawes et al. 2013; Sumner et al. 2016). The number of annual growth laminations and accumulation of biomass along the depth gradient in the upper water column of Lake Vanda was consistent with an accumulation of mat material over the duration of inundation calculated from rising water level (Hawes et al. 2013). The gradual increase in height and spacing of pinnacles with depth (Fig. 4.7) was therefore consistent with temporal development. Recently, we deployed drop cameras across Lake Vanda and found that pinnacles extend to depths of at least 55 m (Fig. 4.8), though most pinnacles at this depth appear to have collapsed, perhaps a result of declining irradiance over time as the lake level has risen. The potential for insights into the mechanisms of pinnacle growth from Lake Vanda are high because pinnacles can be ordered by age, are particularly large and are readily measured. Thus, we focus on both the organisation of fields of pinnacles and possible functions of the pinnacle structure itself.

4.4.2.1 Size and Spacing

Distributions of pinnacle height at any given depth are skewed by the presence of exceptionally large pinnacles at all depths (Fig. 4.9). There was a close, positive correlation between median size at each depth and median spacing (Fig. 4.7c, $n = 7$, $r^2 = 0.96$, $p < 0.0005$). Nearest neighbour analysis of the Vanda pinnacles showed, as in Lake Untersee, a slight tendency away from random, towards even spacing. For

Fig. 4.8 The appearance of pinnacle mats from a range of depths in Lake Vanda. The physical depth and optical depth, as % incident irradiance at noon, are indicated. Where visible, the red laser dots are 3 cm apart

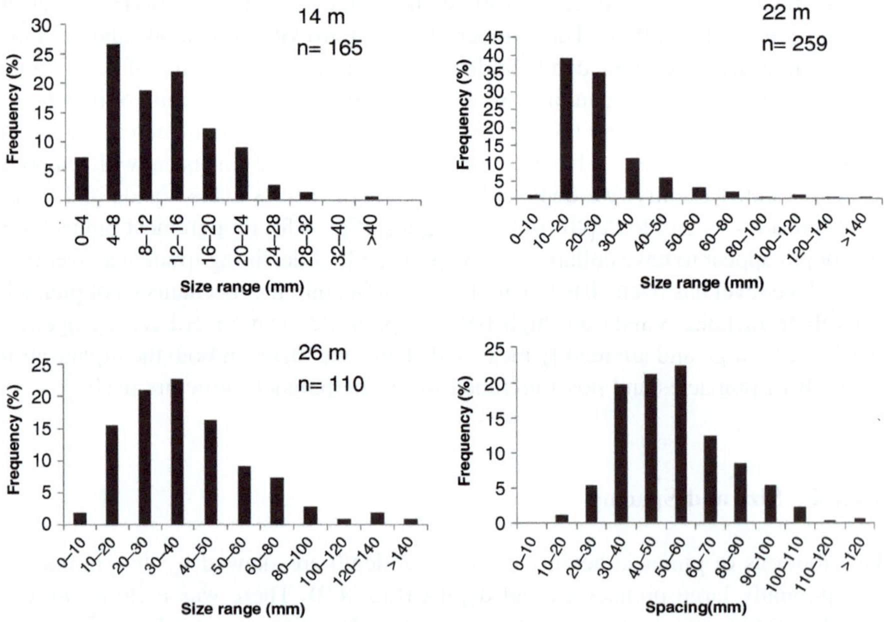

Fig. 4.9 Log-normal distributions of pinnacle height at three depths in Lake Vanda and of spacing distance at 21 m depth

Fig. 4.10 Pinnacle mats at 22 m water depth in Lake Vanda. At the centre, a cluster of large pinnacles shows a tendency to merge at the base but for the pinnacles themselves to diverge

three quadrats taken at 21 m depth, R_n was calculated from 80 measures of nearest neighbour difference as 1.25 ± 0.01, while for three quadrats at 8 m depth R_n was 1.25 ± 0.02 ($N = 3$, mean $\pm$ s.d.), suggesting a weakly significant effect ($p > 0.05$). When pinnacle spacing was examined more closely, it was apparent that deviation from randomness was largely driven by truncation of the distribution at the closely spaced end. This was evident for both the larger, well-spaced pinnacles at 22 m depth (Fig. 4.9) and also the small, closely packed pinnacles at 8 m (8 m data not shown).

Two simple ways to explain this observation are that (1) at short separation, it is not possible to detect accurately small spacing between pinnacles, due to their finite basal diameter, or (2) interactions lead either to divergence or merging of pinnacles at short range. Figure 4.10 shows how closely associated tall pinnacles can tend to both merge at the base and lean away from each other, suggesting that both mechanisms may be in play at least some of the time. Divergence of closely associated pinnacles is consistent with the concept of competition for nutrients from the water column, or perhaps to minimise mutual shading. In contrast, the correlation between spacing and size is consistent with the basal merging of pinnacles to produce fewer, larger individuals as the microbial mat ages. The distribution of pinnacles in Lake Vanda provides only partial support for the view of emergence of structural complexity driven by inter-pinnacle competition for water-column resources. Rather the simplest explanation is of random initiation (see Sumner et al. 2016), merging as basal area increases, with direct inter-pinnacle interactions only being evident at close range. This type of distribution pattern can be fully accommodated by simple interactions between organisms and environment rather than needing to invoke a more holistic mechanism for pinnacle growth.

4.4.2.2 Structure

While data from Lake Vanda suggest that pinnacle fields tend to be structured by simple mechanisms that occur at the individual pinnacle level, the prevalence of the pinnacle structure across a range of optical depths spanning at least 1–15% incident irradiance implies that it has some functional advantage. It is noteworthy to compare the pinnacles of microbial mats in Lake Vanda at 1.5% surface irradiance (45 m depth, Fig. 4.8), with the much flatter structure of Lake Hoare at similar optical depth (Fig. 4.2a). It appears that, based on molecular taxonomy, the two lakes share dominant cyanobacteria (Zhang et al. 2015), and thus behavioural differences, as proposed for Lake Untersee, are less likely. What are the functional advantages that the formation of pinnacles confer on the microbial communities in Lake Vanda, and why is this advantage enhanced for a small subset of the pinnacle population resulting in the frequent occurrence of a few "outsize" pinnacles?

The evolution of pinnacle morphology and zonation in Lake Vanda is described by Sumner et al. (2016). In brief, all pinnacles and flat mats share a laminated and zoned structure; the outer 3–4 laminae are orange-brown in colour, with green and purple subsurface laminae. The transition from the surface orange-brown zone to the green and purple zones is abrupt and only occasionally crosses lamina boundaries (Fig. 4.11). The outer laminae were typically less than 0.5 mm thick each (except at pinnacle tips), whereas the thicknesses of green/purple subsurface laminae ranged from less than 1 to greater than 3 mm thick. The shapes of the green/purple subsurface laminae towards the tips of pinnacles differed from those of the orange-brown laminae; the tops of these laminae were conical to geniculate (Fig. 4.11), rather than cuspate. Thus, the overall thickness of a single green/purple lamina was more uniform across the pinnacle than any orange-brown laminae were, and no green/purple laminae mimicked the cuspate morphology of pinnacle tips. Sumner et al. (2016) noted that at young stages, pinnacles only comprised the outer orange-brown zone, with the green and purple zones only extending into the pinnacles at later stages of development. However, once the pinnacles contain green and purple zones, the interior laminae grow and thicken, which causes the lateral expansion of pinnacles and results in a columnar pinnacle morphology (Fig. 4.12; Sumner et al. 2016). 16S rDNA clone library techniques show the orange-brown laminae in the pinnacles contain almost exclusively *Leptolyngbya* ribotypes (80% *L. antarctica*), while the green and pink zones contain 40–50% *L. antarctica* and the remainder larger ribotypes, primarily *Phormidium* and *Tychonema* (Sumner et al. 2016).

As indicated in Fig. 4.12, tall, columnar pinnacles take considerable time to emerge, and it is important to recall that pinnacle growth is a function of the differential between growth of the apex and the surrounding base. Hawes et al. (2013) calculated that during the first decades of growth, prostrate mat accumulated at approximately 0.33 mm year^{-1}, while pinnacle apices extended at 0.25 mm

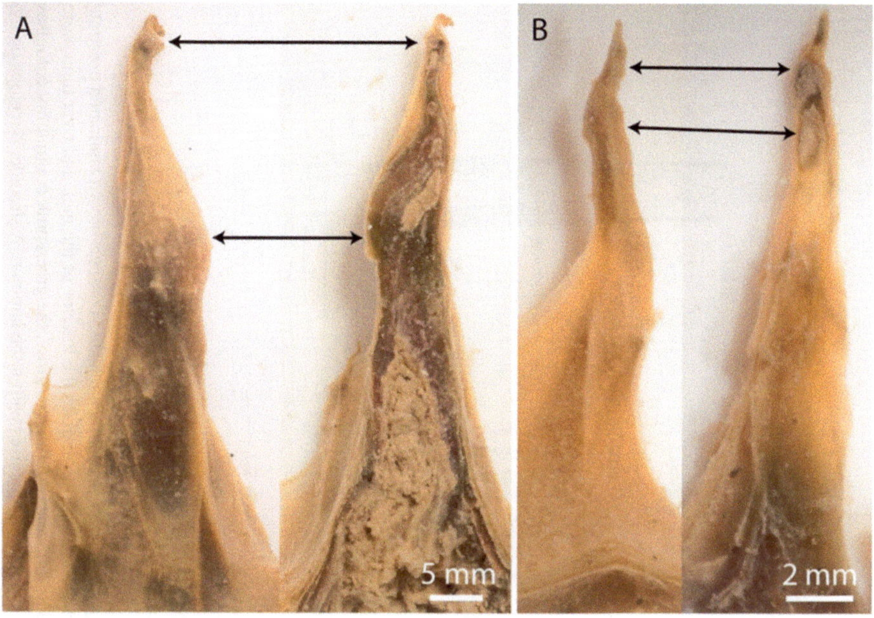

Fig. 4.11 Pinnacles from 19 m depth in Lake Vanda. (**a**) A pinnacle developing "columnar" attributes with the exterior dominated by orange-brown laminae. The vertical section on the right shows the thick purple and green subsurface zones that cause the columnar shape. (**b**) A smaller, cuspate pinnacle that has similar exterior orange-brown laminae. The vertical section on the right illustrates the near absence of purple and green subsurface zones except near the base and near sediment at the crest

year^{-1}, a net growth rate of 0.48 mm year^{-1} for pinnacles. Thus the differential between the rate of growth of pinnacles and the base mat is slight—on average. However, some pinnacles increased in height by more than 2 mm year^{-1} over a decade (Sumner et al. 2016). Thus, the large size of pinnacles in Lake Vanda attests to both their age and variations in their growth rate. Hawes et al. (2013) calculated that the pinnacles at 20 m depth may be the result of 60–80 years of undisturbed growth.

Positive differential growth relative to background mat requires that the pinnacle morphology must provide a growth advantage to the apex. This may relate to interactions between flow of water around the pinnacle fields and nutrient supply, as discussed above for Lake Untersee. As in Untersee, parts of Lake Vanda are convectively mixed, and currents of up to 1 cm s^{-1} have been reported in the lake (Ragotzkie and Likens 1964). While DIC is not depleted in Lake Vanda as in Lake Untersee, dissolved reactive phosphorus is thought to be limiting (Vincent and Vincent 1982). Ghisalberti et al. (2014), like others, argue that tall organisms—or in this case pinnacles—benefit by extracting more nutrients from water columns due to the effect of the near-bed velocity gradient.

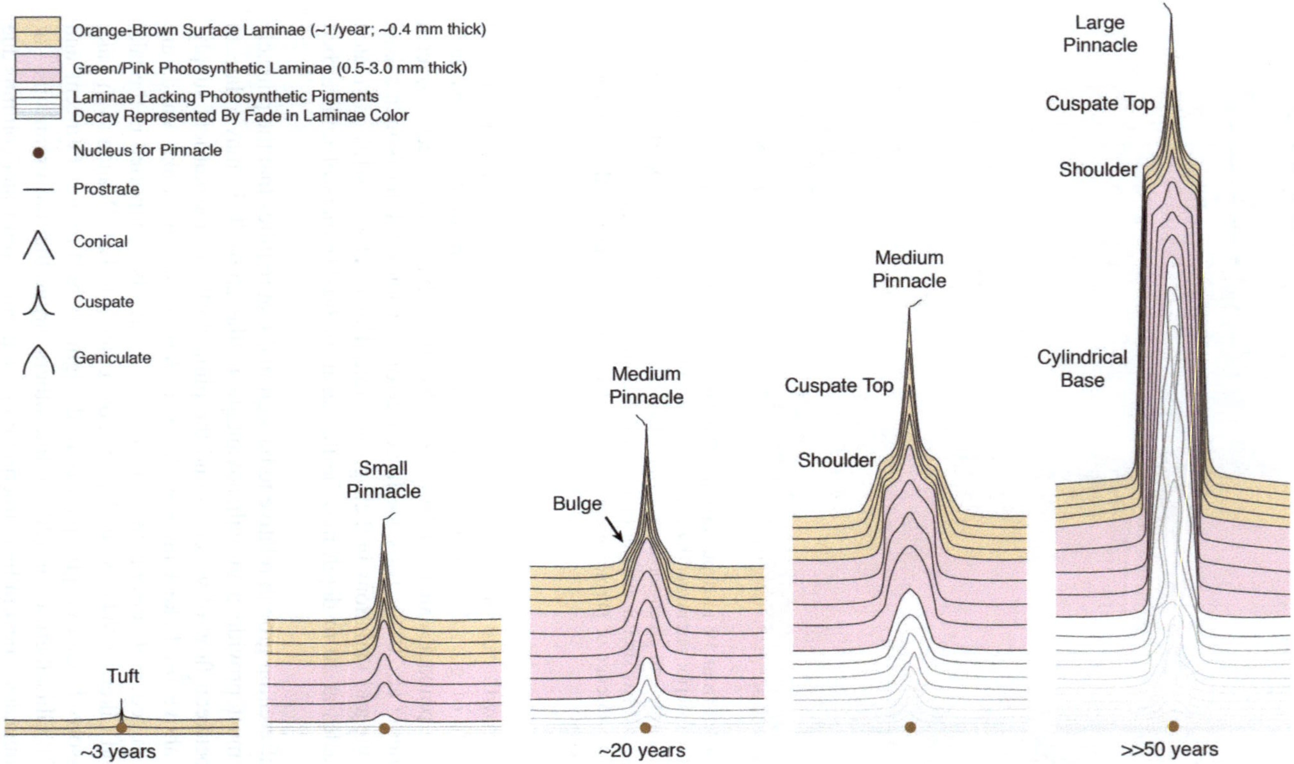

Fig. 4.12 Sketch of pinnacle geometry and growth history. After several years of prostrate growth, microbial tufts nucleate in orange-brown laminated prostrate mats and grow into small cuspate pinnacles. Once thick enough, a green/pink cyanobacterial community colonises the interior of the mat, and over time more deeply buried layers begin to degrade. Apical growth exceeds the flat surrounding mat, and as pinnacles continue to grow, the green/pink community colonises the interiors of pinnacles and begins to swell. Formerly cuspate laminae become geniculate as biomass increases within the laminae. A shoulder develops on the pinnacle and propagates upward through time. Eventually, the cores of pinnacles begin to degrade. Modified after Sumner et al. (2016)

Table 4.2 Organic N:P molar ratios for pigmented layers of microbial mat taken from pinnacles and flat mats at two depths in Lake Vanda

Depth (m)	Flat	Pinnacle
16	38.4 ± 7.0	34.3 ± 5.1
18	44.4 ± 10.5	42.0 ± 8.4

For flat mats $N = 8$, for pinnacles $N = 16$; mean ± s.d.

Nepf (2012) points out that the movement of water within a canopy decreases proportionally to the density of the canopy elements, enhancing this effect. Canopies significantly affect turbulence, and when the product of volume-specific frontal area of canopy elements and their height exceeds 0.23, the canopy slows water sufficient to generate a shear layer at the top of the canopy (Nepf 2012). The product of volume-specific frontal area and height in Lake Vanda pinnacle fields at 20 m is 0.3—exceeding the critical value of Nepf (2012). Thus, a shear layer is likely near the tops of pinnacles. If limiting nutrients are being derived from the water column, the density of the pinnacle field in Lake Vanda is sufficient to affect water turbulence, leading to enhanced nutrient supply and thus growth to pinnacle apices. Indeed, the presence of a velocity gradient displaced from the lake bed to the level of the median pinnacle height may also provide a growth advantage to isolated extra-tall pinnacles and thus reinforce the differential between median and quartile sizes.

While this model is attractive, there is little evidence to suggest that enhanced nutrient availability is effective. The N:P ratios of pigmented outer layers taken from pinnacles and from adjacent flat mats are both high, consistent with P limitation, but there are no significant differences that would suggest an enhanced nutrient supply, relative to growth, for pinnacles over flat mats (Table 4.2). Conditions in Lake Vanda, where currents, though measured, are slight (no visible movement of disturbed sediment has ever been seen by divers operating in Lake Vanda), may not be well suited to hydraulic enhancement of growth of pinnacles. Indeed, it has been shown that nutrient concentrations in microbial mat interstitial waters can exceed that in overlying water, suggesting that internal recycling is a significant potential source of nutrients (Quesada et al. 2008) and this may be expected to be most effective over short-length scales rather than in elongated pinnacles.

An alternative growth model is differential photosynthesis rates as observed by Bosak et al. (2012). To determine whether net photosynthetic capacity per unit area increased towards the apices of the pinnacles, we collected three pinnacles and returned them to a lakeside laboratory and maintained them at ambient photon flux (40 μmol m^{-2}) and temperature (4 °C) in static lake water for 3 h to reach equilibrium with respect to oxygen concentration profile. Oxygen profiles were measured with a Unisense microelectrode system and the gradient of dissolved oxygen at equilibrium was measured at approximately 1 cm intervals along the length of each pinnacle, and this used to calculate the diffusive flux of oxygen out of the microbial mat using Fick's law (methods fully described in Vopel and Hawes 2006). No consistent evidence emerged for any enhanced net rate of photosynthesis along the length of the pinnacles (Fig. 4.13). Similarly, Sumner et al. (2016) did not

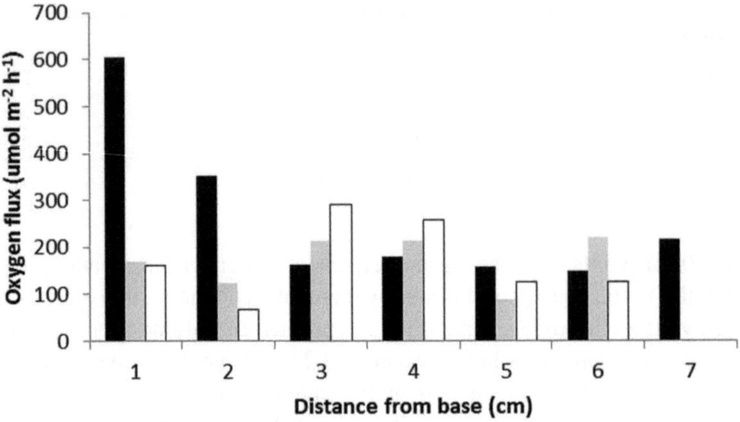

Fig. 4.13 Oxygen efflux rates at simulated ambient light and temperature conditions at approximately 1 cm intervals along three representative pinnacles from Lake Vanda

observe increased photosynthetic potential near pinnacle tops using PAM fluorescence techniques.

While photosynthesis potential per unit area is not enhanced towards pinnacle apices, there is no doubt that surface area is. As calculated for Lake Untersee, the increase in surface area conveyed by the pinnacle morphology results in an increase in pigment biomass per unit of lake bottom (Table 4.1) and, if sufficient irradiance is received, will result in a greater overall accumulation of carbon per unit lake bottom area than for flat mats. The simplest interpretation of cumulative data on pinnacles in Lake Vanda is that the functional benefit of forming elaborate pinnacle structures is to enhance surface area. The pinnacles are physiologically and compositionally simply flat mats that have been stretched and with a more vertical orientation. Photosynthetic tissue is thus spread over a larger area than in flat oriented mats, resulting in the same number of thinner laminae each accumulating less falling sediment per unit area, due to steeper slope. Reduced sediment load minimises competition for light absorption between sediment and photosynthetic pigments, high surface area allows ready access to water column nutrients, and thin laminae allow ready access to nutrients regenerated in decomposing material below. Within these stretched outer pinnacle laminae, trichomes of *Leptolyngbya* are oriented parallel to the axis of the pinnacle (authors unpublished data and Sumner et al. 2016), often intertwined to form the network of meshlike skeletons that if growing by extension will naturally result in enhanced apical growth. We argue that the small growth advantage that pinnacles have over flat mats relates to this process of increased surface area and oriented extension.

The surface area mechanisms will only be effective if the steep sides of pinnacles receive adequate illumination. The angular distribution of light in Lake Vanda at 10 m depth is moderately diffuse due to the scattering nature of the lake ice cover (Sumner et al. 2016). Radiance is maximal at across a 60° arc about the vertical, and substantial light is still present at angles close to 60° from zenith. The light reaching the steeply sloping faces of pinnacles is only slightly less than that reaching horizontal surfaces, particularly as the thinning of pigmented laminae will increase the penetration of light through the pinnacle walls (c.f. Hawes and Schwarz 2001). As in Lake Hoare and Lake Untersee, the pinnacles in Lake Vanda appear to be predictable and to emerge on the basis of simple organism response to environmental variables, over long periods of stable conditions.

4.5 Conclusions

In this brief review of the function of structure in microbial mats in Antarctic lakes, we have focused on features of these unusual systems that may provide insights into general features of microbial mats. Structures formed in Antarctic lakes can be particularly well developed, due to the extreme lack of disturbance that allows them to accumulate over very long time periods (decades perhaps to centuries). Structure is evident on both 1-D and 3-D bases. 1-D structure comprises two elements, an annual banding due to temporal separation of sedimentation and carbon accrual and a pigment zonation superficially similar to the metabolic zonation seen in some more typical mats, though lacking any linkage to oxygen and sulphide gradients. Instead the pigment zonation can be reconciled to the persistence of organisms, particularly cyanobacteria, in buried laminae for many years, and penetration of just enough light to allow these to survive and even grow in some instances. There is no clear indication in data available to date that this zonation provides any holistic advantage to the mat components through, for example, cascades of nutrient cycling elements. Further investigations are needed to better understand the bacterial and microbial eukaryote assemblages as well as functional potential of the communities, using in-depth environmental genomics analyses.

Antarctic lakes provide excellent opportunities to increase understanding of the emergence of complex 3-D structures in microbial mats, because they display a wide range of such structure most of which are non-lithifying and which attain substantial sizes across environmental gradients. Limited data available to date confirm findings from elsewhere that the interaction of dominant organisms present and growth conditions are important in determining the outcome of 3-D emergence. We note that elaboration occurs slowly, resulting from small differences in growth rate in various parts of a microbial mat community. For example the spectacular pinnacles in Lake Vanda, which reach over 10 cm high, accrue at annual rates on the order of mm or less, but the rate of emergence of complex features in other less stressful locations may scale with growth rate. Thus rather small differentials in growth may be responsible for emergence of structures. Functional benefits of these structures may relate to enhanced resource availability through increased surface area, and possibly interactions of structures with water movement to enhance delivery of resources.

While we were able to show how the increase in surface area of elaborated structures over flat ones allows greater biomass and overall productivity on a lake bed area basis, the structures that formed could be easily understood in terms of potential interactions between individuals, their orientation and their environment. The data lack strong evidence of coordinated behaviour, directed towards holistic advantages to the structure. The possibility of some level of coordinated behaviour is, however, suggested by the spacing of the structural elements. Initiation of structures in the lakes considered here was close to random, and the growth of pinnacles in Lake Vanda appears to result in a gradual increase in spacing, following a log-normal distribution, due to merging of near neighbours. However, though a hint of regular spacing in both Lake Vanda and Lake Untersee does argue for some inter-communication between pinnacles, more focused research will be required to investigate this intriguing possibility. The large size of structures in Antarctic lakes make them particularly suitable to investigation of any role of cell to cell signalling over short or long spatial scales in structural emergence that occur on small spatial scales in biofilms (Battin et al. 2007; Decho et al. 2010).

Acknowledgments The information presented in this contribution is derived from field and laboratory work that would have been impossible without the support of many people and organisations. Logistic support was provided by Antarctica New Zealand, the US Antarctic Program and Antarctic Logistics Centre International. We thank all of our Antarctic field colleagues, in particular Drs Dale Andersen and Tyler Mackey, without whom the underwater research would not have been possible. We also thank the reviewers and editors for helpful comments and hard work to see the volume to production.

Compliance with Ethical Standards

Funding This study was funded by NASA Astrobiology: Exobiology and Evolutionary Biology (NNX08AO19G and NN13AE77A), the New Zealand Ministry of Business Innovation and Employment (CO1605 and UOWX1401), the National Science Foundation (MCM-LTER grant number 1115245) and the Tawani Foundation.

Conflict of Interest Ian Hawes declares that he has no conflict of interest. Dawn Sumner declares that she has no conflict of interest. Anne D. Jungblut declares that she has no conflict of interest.

Ethical Approval This article does not contain any studies with human participants or animals performed by any of the authors.

References

Andersen DT, Sumner DY, Hawes I, Webster-Brown J, McKay CP (2011) Discovery of large conical stromatolites in Lake Untersee, Antarctica. Geobiology 9:280–293

Battin TJ, Sloan WT, Kjellberg S, Damis H, Head IM, Curtis TO, Eberl L (2007) Microbial landscapes: new paths to biofilm research. Nat Rev Microbiol 5:76–81

Bosak T, Liang B, Wu T-D, Templer SP, Evans A, Vali H, Guerquin-Kern J-L, Klepac-Ceraj V, Sim MS, Mui J (2012) Cyanobacterial diversity and activity in modern conical microbialites. Geobiology 10:384–401

Bosak T, Knoll AH, Petroff AP (2013) The meaning of stromatolites. Annu Rev Earth Planet Sci 41:21–44

Cathey DD, Parker BC, Simmons GM, Yongue WH, Van Brunt MR (1981) The microfauna od algal mats and artificial substrates in Southern Victoria Land lakes of Antarctica. Hydrobiologia 85:3–15
Clark PJ, Evans FC (1954) Distance to nearest neighbour as a measure of spatial relationships in populations. Ecology 35:445–453
Corning PA (2002) The re-emergence of "emergence": a venerable concept in search of a theory. Complexity 7:18–30
Dana GL, Wharton RR, Dubayah R (1998) Solar radiation in the McMurdo Dry Valleys, Antarctica. In: Priscu JC (ed) Ecosystem dynamics in a polar desert: The McMurdo Dry Valleys, Antarctica. American Geophysical Union, Washington, DC, pp 38–64
De Wolf T, Holvoet T (2005) Emergence versus self-organisation: different concepts but promising when combined. In: Brueckner S, Di Marzo Serugendo G, Karegeoros A, Nagpal R (eds) Engineering self organising systems: methodologies and applications. lecture notes in computer science. Springer, Berlin, pp 1–15
Decho AW, Norman RS, Visscher PT (2010) Quorum sensing in natural environments: emerging views from microbial mats. Trends Microbiol 18:73–80
Denis L, Gevaert F, Spilmont N (2012) Microphytobenthic production estimated by in situ oxygen microprofiling: short-term dynamics and carbon budget implications. J Soils Sediments 12:1517–1529
Des Marais DJ (1995) The biogeochemistry of hypersaline microbial mats. Adv Microb Ecol 14:251–274
Doran PT, Wharton RA, Lyons WB (1994) Paleolimnology of the McMurdo Dry Valleys, Antarctica. J Paleoliomnol 10:85–114
Doran PT, Priscu JC, Lyons WB, Welsh JE, Fountain AG, McKnight DM, Moorhead DL, Virginia RA, Wall DH, Clow GD, Fritsen CH, McKay CP, Parsons AN (2002) Antarctic climate cooling and terrestrial ecosystem response. Nature 415:517–520
Fenchel T (1998) Formation of laminated cyanobacterial mats in the absence of benthic fauna. Aquat Microb Ecol 14:235–240
Franks J, Stolz JF (2009) Flat laminated microbial mat communities. Earth Sci Rev 96:163–172
Ghisalberti M, Gold DA, Laflamme M, Clapham ME, Narbonne GM, Summons RE, Johnston DT, Jacobs DK (2014) Canopy flow analysis reveals the advantage of size in the oldest communities of multicellular eukaryotes. Curr Biol 24:305–309
Guerrero R, Berlanga M (2007) The hidden side of the prokaryote cell: rediscovering the microbial world. Int Microbiol 10:157–168
Haendel D, Kaup E, Loopman A, Wand U (1995) Physical and hydrochemical properties of water bodies. In: Bormann O, Fritzsche D (eds) The Schirmacher Oasis, Queen Maud Land, East Antarctica, and its surroundings. PGM Ergh 289. Perthes, Gotha, pp 259–319
Halley JD, Winkler DA (2008) Classification of emergence and its relation to self-organization. Complexity 13:10–15
Hawes I, Schwarz A-M (1999) Photosynthesis in an extreme shade environment: benthic microbial mats from Lake Hoare, a permanently ice-covered Antarctic lake. J Phycol 35:448–459
Hawes I, Schwarz A-M (2001) Absorption and utilization of irradiance by cyanobacterial mats in two ice-covered Antarctic lakes with contrasting light climates. J Phycol 37:5–15
Hawes I, Schwarz A-M, Smith R, Howard-Williams C (1999) Environmental conditions during freezing, and response of microbial mats in ponds of the McMurdo Ice Shelf, Antarctica. Antarct Sci 11:198–208
Hawes I, Moorhead D, Sutherland D, Schmeling J, Schwarz A-M (2001) Benthic primary production in two perennially ice-covered Antarctic lakes: patterns of biomass accumulation with a model of community metabolism. Antarct Sci 13:18–27
Hawes I, Sumner DY, Andersen DT, Mackey TJ (2011) Legacies of recent environmental change in the benthic communities of Lake Joyce, a perennially ice covered, Antarctic lake. Geobiology 9:394–410

Hawes I, Sumner D, Andersen D, Jungblut A, Mackey T (2013) Timescales of growth response of microbial mats to environmental change in an ice-covered Antarctic lake. Biology 2:151–176
Hawes I, Giles H, Doran PT (2014) Estimating photosynthetic activity in microbial mats in an ice-covered Antarctic lake using automated oxygen microelectrode profiling and variable chlorophyll fluorescence. Limnol Oceanogr 59:674–688
Hawes I, Jungblut AD, Obryk MK, Doran PT (2016) Growth dynamics of laminated microbial mats in response to variable irradiance in an Antarctic lake. Freshw Biol 61:396–410
Howard-Williams C, Pridmore RD, Downes MT, Vincent WF (1989) Microbial biomass, photosynthesis and chlorophyll a related pigments in the ponds of the McMurdo Ice Shelf, Antarctica. Antarct Sci 1:125–131
Howard-Williams C, Schwarz A-M, Hawes I (1998) Optical properties of the McMurdo Dry Valley Lakes, Antarctica. In: Priscu JC (ed) Ecosystem dynamics in a Polar Desert: the McMurdo Dry Valleys, Antarctica. American Geophysical Union, Washington, DC, pp 189–205
Jungblut AD, Wood S, Hawes I, Webster-Brown J, Harris C (2012) The Pyramid Trough Wetland: environmental and biological diversity in a newly created Antarctic protected area. FEMS Microb Ecol 82:356–366
Karanovic T, Gibson J, Hawes I, Andersen D, Stevens M (2014) Three new species of Diacyclops (Copepods: Cyclopoida) from continental Antarctica. Antarct Sci 26:250–260
Kaup E, Loopman A, Klokov V, Simonov I, Haendel D (1988) Limnological investigations in the Untersee Oasis during the summer season 1983/84. In: Martin J (ed) Limnological studies in Queen Maud Land. Academy of Sciences, Tallinn, Estonia, pp 43–56
Kühl M, Fenchel T (2000) Bio-optical characteristics and the vertical distribution of photosynthetic pigments and photosynthesis in an artificial Cyanobacterial mat. Microb Ecol 40:94–103
Lizotte MP, Priscu JC (1992) Photosynthesis-irradiance relationships in phytoplankton from the physically stable water column of a perennially ice-covered lake (Lake Bonney, Antarctica). J Phycol 28:179–185
Love FG, Simmons GM, Parker BC, Wharton RA, Seaburg KG (1983) Modern conophyton-like microbial mats discovered in Lake Vanda, Antarctica. Geomicrobiol J 3:33–48
Mackey TJ, Sumner DY, Hawes I, Jungblut A, Andersen DT (2015) Branched columnar stromatolites in Lake Joyce. Antarct Geobiol 13:373–390
Nadell D, Drescher K, Foster KR (2016) Spatial structure, cooperation and competition in biofilms. Nat Rev Microbiol 14:589–600
Nepf HM (2012) Hydrodynamics of vegetated channels. J Hydraul Res 50:262–279
Ng WO, Grossman AR, Bhaya D (2003) Multiple light inputs control phototaxis in *Synechocystis* sp. strain PCC6803. J Bacteriol 185:1599–1607
Oren A (2010) Mats of filamentous and unicellular cyanobacteria in hypersaline environments. In: Seckbach J, Oren A (eds) Microbial mats: modern and ancient microorganisms in stratified systems. Springer, Dordrecht, pp 387–400
Paerl HW, Pinckney JL (1996) A mini-review of microbial consortia: their roles in aquatic production and biogeochemical cycling. Microb Ecol 31:225–247
Parker BC, Simmons GM, Love FG, Wharton RA, Seaburg KG (1981) Modern stromatolites in Antarctic Dry Valley lakes. Bioscience 31:656–661
Petroff AP, Sim MS, Maslov A, Krupenin M, Rothman DH, Bosak T (2010) Biophysical basis for the geometry of conical stromatolites. Proc Natl Acad Sci 107:9956–9961
Petroff AP, Beukes NJ, Rothamn DH, Bosak T (2013) Biofilm growth and fossil form. Phys Rev. https://doi.org/10.1103/PhysRevX.3.041012
Quesada A, Fernández-Valiente E, Hawes I, Howard-Williams C (2008) Benthic primary production in polar lakes and rivers. In: Vincent WF, Laybourn-Parry J (eds) Polar lakes and rivers–Arctic and Antarctic aquatic ecosystems. Oxford University Press, Oxford, pp 179–196
Ragotzkie RA, Likens GE (1964) The heat balance of two Antarctic lakes. Limnol Oceanogr 9:412–425
Reid RP, Visscher PT, Decho AW, Stolz JF, Beboutk BM, Dupraz C, Macintyre IG, Paerl HW, Pinckney JL, Prufert-Bebout L, Steppe TF, DesMarais DJ (2000) The role of microbes in

accretion, lamination and early lithification of modern marine stromatolites. Nature 406:989–992

Reyes K, Gonzalez NI, Stewart J, Ospino F, Nguyen D, Cho DT, Ghahremani N, Spear JR, Johnson HA (2013) Surface orientation affects the direction of cone growth by *Leptolyngbya* sp. strain C1, a likely architect of coniform structures Octopus Spring (Yellowstone National Park). Appl Environ Microbiol 79:1302–1308

Seckbach J, Oren A (2010) Microbial mats: modern and ancient microorganisms in stratified systems. Cellular origin, life in extreme habitats and Astrobiology 14. Springer

Shepard RN, Sumner DY (2010) Undirected motility of filamentous cyanobacteria produces reticulate mats. Geobiology 8:179–190

Spigel RH, Priscu JC (1998) Physical limnology of the mcmurdo dry valley lakes. In: Priscu JC (ed) Ecosystem dynamics in a Polar desert: the mcmurdo dry valleys, Antarctica. AGU, Washington, DC, pp 153–189

Stahl LJ (1995) Physiological ecology of cyanobacteria in microbial mats and other communities. New Phytol 131:1–32

Stahl LJ, Bolhuis H, Cretoiu MM (2018) Phototrophic marine benthic microbiomes: the ecophysiology of these biological entities. Environ Microbiol. https://doi.org/10.1111/1462-2920.1449

Sumner DY, Jungblut A-D, Hawes I, Andersen DT, Mackey TJ, Wall K (2016) Growth of elaborate microbial pinnacles in Lake Vanda, Antarctica. Geobiology 14(6):559–574. https://doi.org/10.1111/gbi.12188

Sutherland DL, Hawes I (2009) Annual growth layers as proxies of past growth conditions for benthic microbial mats in a perennially ice-covered Antarctic lake. FEMS Microb Ecol 67:279–292

Tice MM, Thornton DCO, Pope MC, Olszewski TD, Gong J (2011) Archean microbial mat communities. Annu Rev Earth Planet Sci 39:297–319

Tyler SW, Cook PG, Butt AZ, Thomas JM, Doran PT, Lyons WB (1998) Evidence of deep circulation in two deep perennially ice-covered Antarctic Lakes. Limnol Oceanogr 43:625–635

Vincent WF (2000) Cyanobacterial dominance in polar regions. In: Whitton BD, Potts M (eds) The ecology of cyanobacteria. Kluwer Academic Publishers, Dordrecht, pp 321–340

Vincent WF, Vincent CL (1982) Factors controlling phytoplankton production in Lake Vanda (77°S). Can J Fish Aquat Sci 39:1602–1609

Vopel K, Hawes I (2006) Photosynthetic performance of benthic microbial mats in Lake Hoare, Antarctica. Limnol Oceanogr 51:1801–1812

Walter MR, Bauld J, Brock TD (1976) Distribution and structure of recent stromatolitic algal mats. In: Walter MR (ed) Stromatolites. Elsevier, Amsterdam, pp 273–310

Wand U, Schwarz G, Brüggemann E, Aüuer K (1997) Evidence for physical and chemical stratification in Lake Untersee (central Dronning Maud Land, East Antarctica). Antarct Sci 9:43–45

Wand U, Samarkin V, Nitzsche H, Hubberten H (2006) Biogeochemistry of methane in the permanently ice-covered Lake UNtersee, central Dronning Maud Land, East Antarctica. Limnol Oceanogr 51:1180–1194

Webb JS, Givskov M, Kjelleberg S (2003) Bacterial biofilms: prokaryotic adventures in multicellularity. Curr Opin Microbiol 6:578–585

Wharton RA Jr, Parker BC, Simmons GM Jr (1983) Distribution, species composition and morphology of algal mats in Antarctic dry valley lakes. Phycologia 22:355–365

Wharton RA Jr, McKay CP, Clow GD, Andersen DT (1993) Perennial ice covers and their influence on Antarctic lake ecosystems. In: Green WJ, Friedmann EI (eds) Physical and biogeochemical processes in Antarctic lakes. Antarctic research series. American Geophysical Union, Washington, DC, pp 53–70

Wharton RA Jr (1994) Stromatolitic mats in Antarctic lakes. In: Bertrand-Sarfati J, Monty C (eds) Phanerozoic stromatolites II. Kluwer, Dordrecht, pp 53–70

Wharton RA Jr, Simmons GM, McKay CP (1989) Perennially ice-covered Lake Hoare, Antarctica: physical environment, biology and sedimentation. Hydrobiologia 172:305–320

Wharton RA Jr, McKay CP, Clow GD, Andersen DT, Simmons GM, Love FG (1992) Changes in ice cover thickness and lake level of Lake Hoare, Antarctica: implications for local climate change. J Geophys Res 97:3503–3513

Zhang L, Jungblut AD, Hawes I, Andersen DT, Sumner DY, Mackey TJ (2015) Cyanobacterial diversity in benthic mats of the McMurdo Dry Valley Lakes, Antarctica. Polar Biol 38:1097–1110

Chapter 5
Fungal Decomposers in Freshwater Environments

Vladislav Gulis, Rong Su, and Kevin A. Kuehn

Abstract Streams, rivers, and freshwater marshes often depend on plant litter as a source of carbon, nutrients, and energy that drive ecosystem processes. Decomposition of this organic matter, such as leaves, wood, or emergent macrophytes, is mediated mostly by fungi, whereas the role of bacteria is minor. Fungal colonization leads to enzymatic breakdown of major plant polymers and fungal biomass accrual (often around 10% of total detrital dry mass), which makes decaying plant material more palatable to detritivorous invertebrates. Representatives of almost all major groups of fungi can be isolated from decaying plant litter collected in freshwater ecosystems or detected using molecular techniques; however, ascomycetes, including their asexual stages (e.g., aquatic hyphomycetes in streams), predominate. In recent years, utilization of radioisotopic approaches (e.g., acetate incorporation into ergosterol) to estimate fungal growth rates and production has facilitated the construction of partial carbon budgets for decaying plant litter that illustrate the importance of fungal decomposers in both lotic and lentic systems. For example, some estimates suggest that 23–60% of leaf litter carbon loss in streams can be explained by fungal assimilation (production plus respiration), which does not include fungal-mediated losses as fine particulate or dissolved organic carbon. Estimates of fungal contribution to plant carbon loss can be even higher (47–65%) in standing-dead emergent macrophyte systems in wetlands. The effects of environmental variables on fungal activity and plant litter decomposition in freshwaters, including inorganic nutrient availability and eutrophication, have also received considerable attention in the recent years. Molecular approaches are now becoming increasingly important in both streams and wetlands to assess the effects of environmental variables on litter-associated fungal assemblages. However, there are considerable differences in

V. Gulis (✉)
Department of Biology, Coastal Carolina University, Conway, SC, USA
e-mail: vgulis@coastal.edu

R. Su · K. A. Kuehn
Department of Biological Sciences, The University of Southern Mississippi, Hattiesburg, MS, USA

C. J. Hurst (ed.), *The Structure and Function of Aquatic Microbial Communities*, Advances in Environmental Microbiology 7,
https://doi.org/10.1007/978-3-030-16775-2_5

fungal dynamics and assemblages between streams and freshwater wetlands, which are discussed here in detail.

5.1 Introduction

Streams and rivers as well as inland ecosystems with standing waters (wetlands, ponds, littoral zones of lakes) receive considerable input of allochthonous terrestrially derived organic carbon (C) in the form of leaf litter and wood from riparian trees (Webster and Meyer 1997; Benstead et al. 2009) as well as dead autochthonous organic matter originating from submerged and emergent macrophytes (Mitsch and Gosselink 2007). Increasing lines of evidence indicate that fungi are key players in the decomposition of this plant litter under aerobic conditions, as indicated by the high rates of fungal secondary production observed in plant litter and the concomitant accrual of fungal biomass (Gulis and Suberkropp 2003c; Suberkropp et al. 2010; Kuehn et al. 2011; Kuehn 2016).

However, historically, comprehensive assessments of fungal-mediated decay processes have been hindered by the inability to quantify fungal biomass and rates of fungal production in field-collected plant material. The major impediment to such data has been the pervasive, intimate association of fungi with plant litter they colonize, which conceals their filamentous somatic body (hyphae) and makes estimation of their biomass and growth rates extremely difficult. However, compelling evidence has accumulated on the usefulness of ergosterol, a membrane lipid found in higher fungi (Weete et al. 2010), as a biomarker for quantifying fungal biomass in decaying plant matter and soils (Newell et al. 1988; Gessner and Chauvet 1993; Gessner 2005; Gulis and Bärlocher 2017). Additional techniques have also been developed for measuring instantaneous growth rates of fungi via incorporation of [1-^{14}C]-acetate into ergosterol, which allows for the estimation of fungal growth and production rates (Newell and Fallon 1991; Suberkropp and Gessner 2005; Gulis and Bärlocher 2017). Both of these methods have been increasingly used within a variety of aquatic and terrestrial ecosystems, where they have yielded important information concerning the role and contribution of fungi to ecosystem carbon and nutrient cycling pathways (Gessner et al. 2007; Rousk and Baath 2007; Gulis et al. 2008; Suberkropp et al. 2010; Kuehn et al. 2011; Clemmensen et al. 2013; Wallander et al. 2013; Kuehn 2016).

Despite the critical roles that fungi play in carbon, nutrient, and energy flow in freshwater ecosystems, there are considerable differences in fungal communities, biomass and production dynamics, and litter processing between lotic (i.e., flowing) and lentic (i.e., standing water) ecosystems. This can be explained by differences in environmental conditions and microhabitats where fungi-driven decomposition of plant litter typically takes place. In streams, aquatic fungi are mostly associated with benthic coarse particulate organic matter. In wetlands dominated by emergent macrophytes, fungi start decomposing senescent plant litter in the standing-dead position before it ultimately collapses to the benthic sediments where decomposition continues. Thus, the dynamics of fungal decomposers in streams and wetlands are fundamentally different and need to be discussed separately. Fungi are also common within the pelagic zone of lake ecosystems (Wurzbacher et al. 2010, 2014); however, data concerning their functional role in ecosystem processes is limited in comparison to streams and wetlands. As a consequence, they are not discussed here.

5.2 Fungi in Running Waters

5.2.1 Aquatic Hyphomycetes

Representative of almost all major groups of fungi (*Chytridiomycota*, *Zygomycota*, *Ascomycota*, and *Basidiomycota*) can be isolated from submerged plant litter collected in streams and rivers or detected using molecular techniques (Tsui and Hyde 2003; Nikolcheva and Bärlocher 2004; Shearer et al. 2007). However, some of the species reported from running waters are transient organisms, occurring as dormant propagules or displaying little metabolic activity and, hence, contributing little to ecosystem processes. Aquatic hyphomycetes are arguably the best-known group of fungi associated with decaying leaf litter in streams where they complete their life cycle under submerged or amphibious conditions. They are an ecological group of ~300–320 species, with phylogenetic affinities to ascomycetes or basidiomycetes (Webster 1992; Descals 2005; Shearer et al. 2007; Duarte et al. 2013). Molecular analyses of environmental samples indicate that ascomycetes dominate plant litter-associated fungal communities in streams in terms of both the number of operational taxonomic units (OTUs) and relative abundance (Nikolcheva and Bärlocher 2004; Seena et al. 2008). The majority of aquatic hyphomycetes that were cultured and subjected to phylogenetic analyses (based mostly on 28S and ITS rDNA sequences) have been placed in the class *Leotiomycetes*, with fewer species associated with *Dothideomycetes* (Baschien et al. 2013; Duarte et al. 2013). However, sexual stages are known for less than 10% of the species (Shearer et al. 2007); instead, aquatic hyphomycetes reproduce mainly via asexual spores (i.e., conidia) which often have unique morphologies (Fig. 5.1). Due to the characteristic morphology of conidia in many species, it is often possible to identify taxa by their individual spores within environmental samples (Gulis et al. 2005). The shape of the spores is an adaptation to lotic environment, allowing conidia to attach firmly to their substrate as well as facilitating dispersal and concentration of conidia in the foam at the water surface (Webster and Descals 1981; Descals 2005). As autumn-shed leaves or twigs enter the water, they are quickly colonized by conidia from the neuston or by conidia drifting in the water column.

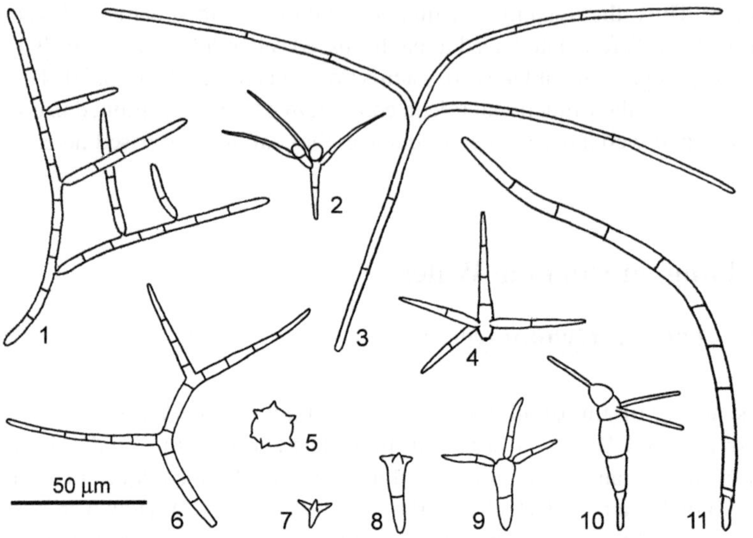

Fig. 5.1 Conidia of aquatic hyphomycetes demonstrating the variety of shapes and sizes. (1) *Varicosporium elodeae*. (2) *Tetracladium marchalianum*. (3) *Tetrachaetum elegans*. (4) *Triscelophorus* sp. (5) *Goniopila monticola* (or *Margaritispora aquatica*). (6) *Tricladium kelleri*. (7) *Heliscella stellata*. (8) *Heliscus lugdunensis*. (9) *Heliscina antennata*. (10) *Culicidospora gravida*. (11) *Anguillospora longissima*

5.2.2 Litter-Associated Fungal Biomass, Production, and Sporulation

Allochthonous organic matter in the form of leaves, twigs, branches, dissolved organic matter (DOM), etc. fuels forested stream ecosystems, with some estimates suggesting up to 99% contribution to stream carbon and energy budgets (Fisher and Likens 1973; Webster and Meyer 1997). Annual streambed litter inputs are typically on the order of 500 g dry mass m^{-2} $year^{-1}$, but may exceed 1000 g m^{-2} $year^{-1}$ in small forested headwater streams (Webster and Meyer 1997; Benstead et al. 2009). Decomposition of this plant litter is a key ecosystem process, which is driven mostly by fungi. Multiple concurrent estimates of fungal and bacterial biomass associated with decomposing leaf litter in streams suggest that fungi contribute 88–99.9% of total microbial (fungi plus bacteria) biomass (Baldy et al. 1995, 2007; Baldy and Gessner 1997; Hieber and Gessner 2002; Gulis and Suberkropp 2003c; Duarte et al. 2010; Suberkropp et al. 2010; Tant et al. 2013). A similar pattern of fungal dominance has also been reported for submerged wood across streams in the USA, while bacterial participation is more important on fine particulate organic matter (FPOM) (Findlay et al. 2002a; Tant et al. 2013).

Leaf litter entering streams typically has low levels of fungal colonization by epiphytic fungi and early terrestrial saprotrophs. Shortly after submergence, spores of aquatic fungi attach and germinate usually within several hours. Molecular evidence suggests that fungal diversity may be somewhat higher during early stages of colonization than at later decomposition stages due to the presence of terrestrial fungi and spores of some aquatic fungi that may be unable to establish on a particular type of substrate (Nikolcheva et al. 2003). Once established, hyphae of aquatic fungi pervasively extend within the leaf matrix, and the increase in fungal biomass associated with plant litter and their rates of growth and production can be detected using ergosterol-based methods (see above). In general, the dynamics of fungal biomass and parameters of fungal activity, such as sporulation, growth rate, and production, depend on the type of plant litter (e.g., leaves vs. wood), the intrinsic litter quality (e.g., nutrient and lignin content) (Gessner and Chauvet 1994; Ferreira et al. 2006b; Gulis et al. 2008), and environmental variables including water chemistry (see Sect. 5.2.4). Fungal biomass often increases sharply at early stages of decomposition, reaches a maximum, and then either levels off or decreases slightly during later stages of leaf decay (Fig. 5.2, see also Ferreira et al. 2006a; Gulis et al. 2006). Peak litter-associated fungal biomass is typically attained within 2–10 weeks after litter submergence depending on litter type and environmental conditions and can reach as high as 23–25% of total detrital dry mass (Table 5.1). However, typical peak values for decomposing leaves of deciduous trees are lower, on the order of 8–12%. Fungal biomass associated with randomly collected leaf litter from streams during various seasons (i.e., a composite of litter of different species at different stages of decomposition) averages around 4–6% of litter dry mass (Methvin and Suberkropp 2003; Carter and Suberkropp 2004; Suberkropp et al. 2010).

Estimating fungal growth rates and production may give a better understanding of the importance of fungal decomposers in facilitating carbon losses from decaying plant litter, since estimates of fungal biomass alone do not provide any insights into carbon losses as conidia, mycelial fragments, through respiration or due to invertebrate feeding and hyphal death. Fungal growth and production rates, as determined from ^{14}C-acetate incorporation into ergosterol, usually peak early after litter submergence when fungal biomass is still relatively low (Suberkropp and Weyers 1996; Gessner and Chauvet 1997). Fungal growth rate estimates from decaying leaf litter range from <0.01 to 0.64 day^{-1} (Methvin and Suberkropp 2003; Carter and Suberkropp 2004; Suberkropp et al. 2010; Gulis et al. unpublished), which are generally lower than growth rates of bacteria. However, due to much higher litter-associated fungal biomass (typically >90% of microbial biomass), fungal production is often much higher (1–627×) than bacterial production in all studies where both groups were followed simultaneously (Weyers and Suberkropp 1996; Baldy et al. 2002; Pascoal and Cassio 2004; Pascoal et al. 2005; Suberkropp et al. 2010).

After litter submergence, sporulation rates of aquatic hyphomycetes typically peak earlier than fungal biomass (Fig. 5.2), often within 1–2 weeks, suggesting that aquatic hyphomycetes tend to invest considerable resources into reproduction early. Some estimates show that these fungi can invest up to 46–80% of their production into conidia (Suberkropp 1991) and convert up to 7% of initial plant litter carbon into spores (Suberkropp 1991; Hieber and Gessner 2002; Ferreira et al. 2006b). Since large amounts of spores are released from decomposing litter, the concentration of conidia in stream water can reach up to 25,000 spores L^{-1} in some

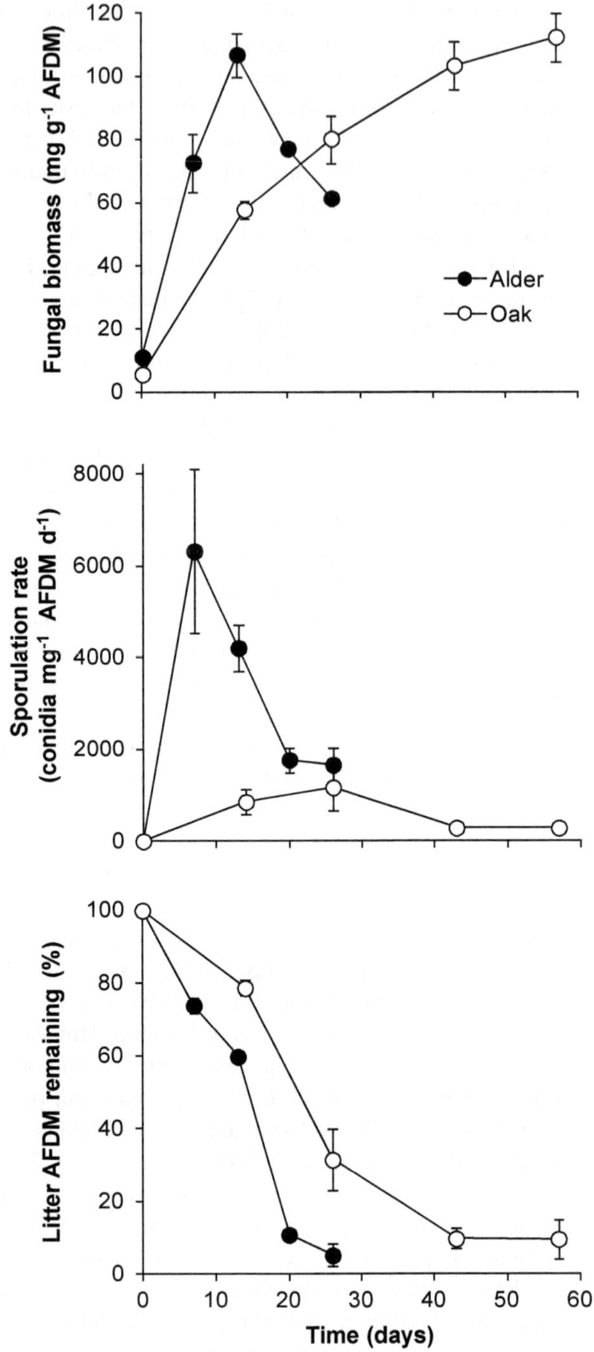

Fig. 5.2 Dynamics of fungal biomass (top panel), sporulation rates of aquatic hyphomycetes, and litter mass loss (bottom panel) associated with two types of decomposing leaves in a Portuguese stream. All data are on the basis of litter ash-free dry mass (AFDM). Symbols indicate means ±1 SE ($n = 4–8$). Note the relatively fast litter breakdown and fungal dynamics due to high water temperature and relatively high dissolved nutrient concentrations. Data from Gulis et al. (2006) and unpublished data from the same authors

Table 5.1 Estimates of fungal biomass associated with decomposing plant litter in streams

Fungal biomass (mg g^{-1} detrital mass)[a]	No. of streams	Litter type[b]	References
61–158	2	LB (7)	Suberkropp et al. (1993), Gessner and Chauvet (1994)
78–226	4	LB (1)	Methvin and Suberkropp (2003), Carter and Suberkropp (2004)
34–73	6	RCL-A	Methvin and Suberkropp (2003), Carter and Suberkropp (2004), Suberkropp et al. (2010)[c]
17–104	2	RCL-M	Suberkropp et al. (2010)[c]
1–249	10	WV	Simon and Benfield (2001), Stelzer et al. (2003), Gulis et al. (2004, unpublished), Ferreira et al. (2006b)
2–25	4	WS	Diez et al. (2002), Spanhoff and Gessner (2004)
5–56	9	RCWS	Gulis et al. (2008, unpublished)

[a]Ergosterol concentrations were converted to fungal biomass assuming an average concentration of 5.5 mg g^{-1} dry mass (Gessner and Chauvet 1993) unless more specific data was available
[b]LB = leaves in litter bags with the number of leaf types in parentheses (maximum fungal biomass from decomposition experiments is given); RCL = randomly collected naturally occurring leaf litter (RCL-A = annual means, RCL-M = monthly means); WV = wood veneers and WS = wood sticks (range of fungal biomass from decomposition experiments); RCWS = randomly collected naturally occurring wood sticks 7–40 mm diam
[c]Data from 6 years of monthly sampling in two streams

streams during the autumn-winter season when in-stream leaf litter standing stocks are at their highest. These spores can be easily captured on membrane filters, identified, and counted, an approach often employed in ecological studies of aquatic hyphomycetes (e.g., Suberkropp 1991; Gulis and Suberkropp 2004; Bärlocher 2005).

5.2.3 *Contribution to Plant Litter Decomposition*

Fungal decomposer communities in streams are well equipped to break down all major plant polymers. Enzymatic capabilities of many aquatic hyphomycetes have been assessed in pure cultures (see review by Shearer 1992). Enzymes that hydrolyze cellulose and hemicelluloses are common, while ligninolytic capabilities of many aquatic hyphomycetes may be limited. Nevertheless, some aquatic hyphomycetes have been reported to degrade lignin-like substrates (Zare-Maivan and Shearer 1988; Abdullah and Taj-Aldeen 1989), while other freshwater fungi are known to solubilize lignin from wood (Bucher et al. 2004). Overall, there is little specialization among species indicating that aquatic hyphomycetes are a generalist group; however, some substrate preferences and distinct communities associated with leaves vs. wood have been reported (Gulis 2001; Ferreira et al. 2006b). In field studies, the activity of several lignocellulolytic enzymes was positively related with litter-associated fungal biomass (Tank et al. 1998; Gulis et al. unpublished), suggesting that the enzymes were likely derived from fungi rather than other microorganisms. The activity of pectin-degrading enzymes that correlated with higher fungal activity has been shown to be crucial for litter mass loss, since it facilitates the maceration of leaf litter by releasing whole plant cells (Jenkins and Suberkropp 1995).

The different fates of plant litter C and the proportion of C channeled through fungi as litter decomposes can be estimated by constructing partial litter decomposition budgets. In general, a fraction of initial litter C is converted into microbial (including fungal) production, while some C is lost as CO_2 due to microbial respiration. Unfortunately, litter-associated respiration measured in the field cannot be easily partitioned among microbial groups colonizing plant litter (i.e., fungal vs. bacterial). However, experiments in laboratory microcosms simulating stream conditions (Suberkropp 1991) with plant litter colonized by pure cultures of aquatic hyphomycetes show remarkably similar decomposition patterns to those found in streams (Gulis and Suberkropp 2003a, b; Ferreira and Chauvet 2011). In these studies, fungal assimilation (i.e., production + respiration) accounted for 23–56% of plant carbon loss depending on temperature and dissolved nutrient availability. However, these values are likely underestimates of fungal contribution, since cumulative fungal production was estimated from accumulated biomass and C losses as conidia (i.e., not via [^{14}C]-acetate-to-ergosterol approach), which would not include hyphal losses with FPOM. In addition, fungal-generated FPOM and DOM losses were also not taken into account. In streams, when litter-associated fungal production is accurately measured, it is possible to calculate fungal respiration and assimilation by using estimates of fungal growth efficiency obtained in laboratory microcosms (24–60%). Such estimates resulted in fungal assimilation accounting for 29–39% of leaf litter C losses, while bacteria contributed only 4–13% (Pascoal and Cassio 2004). The low contribution of bacteria is especially surprising since the experiment was conducted in a polluted river where bacteria may also metabolize DOM from the water column. The values for fungal contribution to litter decomposition given above are clearly underestimates since they also did not account for fungi-mediated FPOM and DOM losses from leaf litter, which may be comparable to respiratory losses (Gessner et al. 2007). When these losses were estimated, fungi accounted for up to 98% of the microbial contribution to leaf mass loss (Baldy et al. 2007).

Several studies have noted strong significant relationships between litter-associated fungal biomass, sporulation rate of aquatic hyphomycetes, percentage of initial litter C converted into conidia, and rates of litter decomposition (Gessner and Chauvet 1994; Gessner et al. 2007; Gulis et al. 2009). Recently, we found that fungal parameters (biomass or production) measured at early stages of decomposition (day 28) can be used to accurately predict final decomposition rates of leaf litter and wood after 4–5 months (Fig. 5.3; Gulis et al. unpublished). Cumulative spore production by the end of the experiment also showed a strong relationship with litter decomposition rates (Fig. 5.3), suggesting that fungi are major decomposers of plant litter in streams.

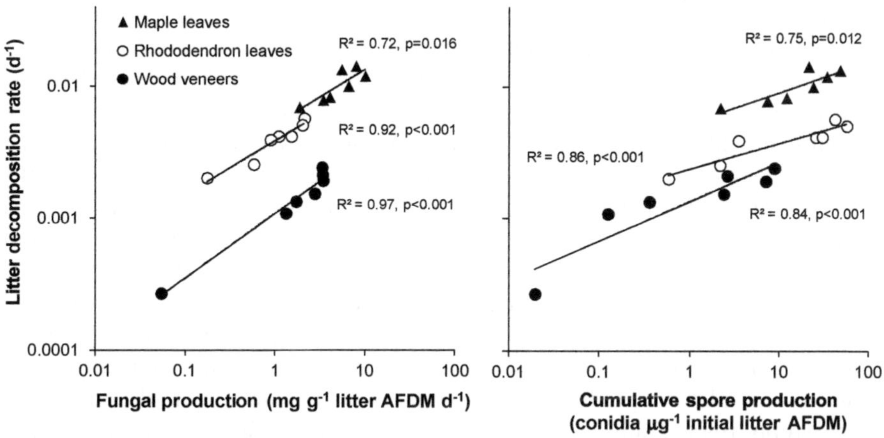

Fig. 5.3 Relationships between final plant litter decomposition rates and fungal production at early stages of breakdown (day 28) (left panel) and cumulative spore production of aquatic hyphomycetes from plant litter by the last day of decomposition experiment (day 111 for maple leaf litter, day 144 for rhododendron leaves and wood veneers) (right panel). Data points are means from seven dissolved inorganic nutrient treatments set up in replicated streamside channels. R^2 and p values of linear regressions based on log-log transformed data are shown. Gulis et al. unpublished

5.2.4 Effects of Environmental Factors, Including Dissolved Nutrients, on Fungal Activity and Organic Carbon Processing

A variety of parameters can potentially affect litter-associated microbial activity and decomposition rates. Intrinsic parameters of the organic substrate include the type of plant litter (e.g., leaves of deciduous trees vs. grasses vs. wood) and litter quality, such as lignin content (or carbon quality) and the concentrations of nitrogen (N) and phosphorus (P), i.e., litter C:N and C:P ratios. In general, fungal activity and litter decomposition are negatively affected by high lignin content and high initial C: nutrient stoichiometric ratios of the substrate as was summarized earlier (e.g., Stelzer et al. 2003; Ferreira et al. 2006b; Gessner et al. 2007). The major external factors exerting control over fungal activity and breakdown rates are water temperature and water chemistry. The effects of stream pollution on aquatic fungi including nutrient enrichment, acidification, effects of heavy metals, organic xenobiotics, nanoparticles, and thermal pollution have been recently summarized by Ferreira et al. (2014, 2015). Here we address the effects of dissolved nutrients as this has received considerable attention in the recent years, since in most streams affected by human activities, the magnitude of their effects on fungal processes and litter decomposition is greater than that of other environmental parameters.

Pollution of streams and rivers often leads to elevated concentrations of dissolved inorganic nutrients. Fungi colonizing plant litter can obtain N and P from both the substrate they grow on and the water column (Suberkropp 1995; Cheever et al. 2013).

Mining N and P from organic substrates requires considerable energy and resource expenditures to produce extracellular enzymes to attack recalcitrant organic molecules in order to cleave amino or phosphate groups (Sinsabaugh et al. 2014). Thus, from an energetic perspective, fungi should preferentially use dissolved inorganic nutrients. In addition, plant litter C:N and C:P ratios are considerably higher than C:nutrient ratios of microbial, including fungal, biomass (Danger and Chauvet 2013; Grimmett et al. 2013; Gulis et al. 2017). The mismatch between the resource stoichiometric ratios and microbial biomass elemental ratios or nutrient limitation of fungal activity can be alleviated by external nutrient supply. Indeed, experiments in laboratory microcosms simulating stream conditions and whole-stream nutrient addition experiments have shown that elevated dissolved nutrient concentrations stimulate fungal activity (growth and sporulation rates, maximum fungal biomass attained, cumulative production, respiration) and plant litter decomposition (Suberkropp 1998; Gulis and Suberkropp 2003a, b, c; Suberkropp et al. 2010; Ferreira and Chauvet 2011). In some studies, the relationships between nutrient concentrations and fungal parameters or decomposition rates followed asymptotic saturation-type models (e.g., Michaelis-Menten) where large responses were triggered by relatively small increases in nutrients (Rosemond et al. 2002; Ferreira et al. 2006b). Stimulation of fungal activity by inorganic nutrients is generally more pronounced for low-quality (low N and P, high lignin) substrates, such as wood (Stelzer et al. 2003; Gulis et al. 2004, 2008; Ferreira et al. 2006b), pointing to more severe nutrient limitation of fungi on these substrates. Correlative studies based on natural or anthropogenic gradients of dissolved nutrients show mixed results, since in many cases nutrient enrichment was accompanied by the presence of other pollutants (pesticides, heavy metals, etc.) with corresponding negative effects on fungal activity. However, a recent pan-European study of 100 streams (Woodward et al. 2012) and a meta-analysis by Ferreira et al. (2015) suggest that stimulation of microbially driven decomposition by dissolved nutrients is widespread.

Most studies that dealt with the effects of nutrients on aquatic fungi have involved enrichments by *both* N and P; however, very little is known about how N and P concentrations and varying N:P *ratios* affect fungal activity. The situation is complicated by the ability of fungi to obtain these nutrients from both the substrate and the water. Recent studies suggest that fungal activity and litter decomposition may be stimulated to a greater extent by dissolved N rather than P availability (Fig. 5.4; see also Cheever et al. 2012, 2013). This phenomenon may be related to fungal biomass stoichiometry and elemental homeostasis. Until recently, little was known about C:N and C:P ratios of mycelium in aquatic fungi, since it is impossible to separate it from the substrate for direct measurements. Indirect estimates of fungal elemental stoichiometry can be obtained from field studies by plotting increases in litter-associated fungal carbon vs. increases in detrital N or P (Fig. 5.5), assuming that changes in nutrient content of detritus are due to fungal immobilization of inorganic N and P. Such estimates are generally in agreement with data from laboratory experiments in liquid cultures (Danger and Chauvet 2013; Grimmett et al. 2013). Our recent experiments performed in defined media and, more

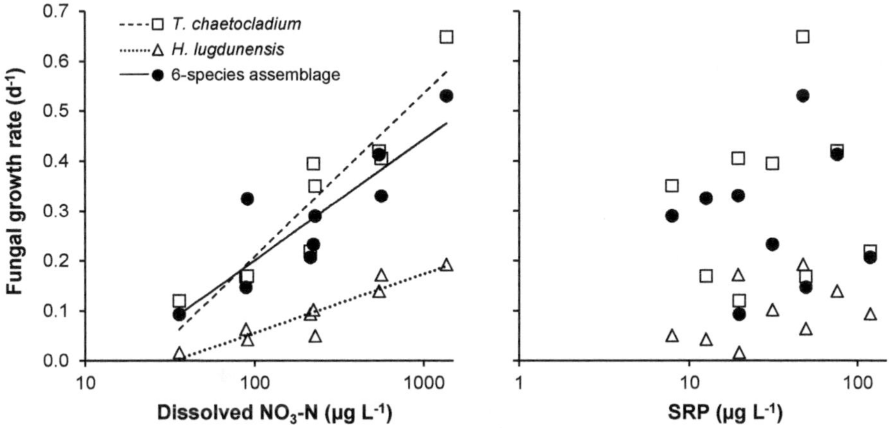

Fig. 5.4 Effects of nitrate-N and soluble reactive phosphorus (SRP) concentrations on growth rates of two aquatic hyphomycetes and six-species aquatic fungal assemblage associated with submerged decomposing maple leaf litter in experimental microcosms simulating stream conditions. Data points are means for 15-day-old samples from nine dissolved inorganic nutrient treatments. Fungal growth rates were determined based on ^{14}C-acetate-to-ergosterol approach (Gulis and Bärlocher 2017). For the left panel, the effects of NO_3–N on fungal growth rate were significant (TC $R^2 = 0.88$, $p < 0.001$; HL $R^2 = 0.88$, $p < 0.001$; assemblage $R^2 = 0.76$, $p = 0.002$). Data are from Gulis et al. (2017)

importantly, with fungi growing on plant litter in laboratory microcosms (Gulis et al. 2017) suggest that aquatic hyphomycetes are homeostatic with respect to their C:N ratio (i.e., fungal C:N does not change with varying dissolved N availability), but are not homeostatic with respect to their C:P ratio (Fig. 5.6). This indicates that fungi may be able to take up and store P in excess when it is available. Overall, fungal biomass accrual and concomitant immobilization of dissolved nutrients during litter decomposition have important implications for higher trophic levels in stream ecosystems (see below).

5.2.5 *Fungal Importance at the Ecosystem Scale*

The majority of studies that followed the dynamics of fungal biomass, production, and sporulation have dealt with decomposition of a single type and cohort of plant litter, typically enclosed in litter bags and placed in a stream. However, streams naturally contain a high diversity of litter types at different stages of decay due to the lateral litter input ("blow-in") occurring throughout the year. Thus, periodic measurements of standing stock of plant litter on areal basis (per m^2 of stream bottom) combined with periodic estimates of fungal parameters per g dry mass of naturally occurring plant litter would provide insights into fungal importance at the ecosystem scale. Estimates of leaf litter-associated fungal biomass on areal basis range from <1

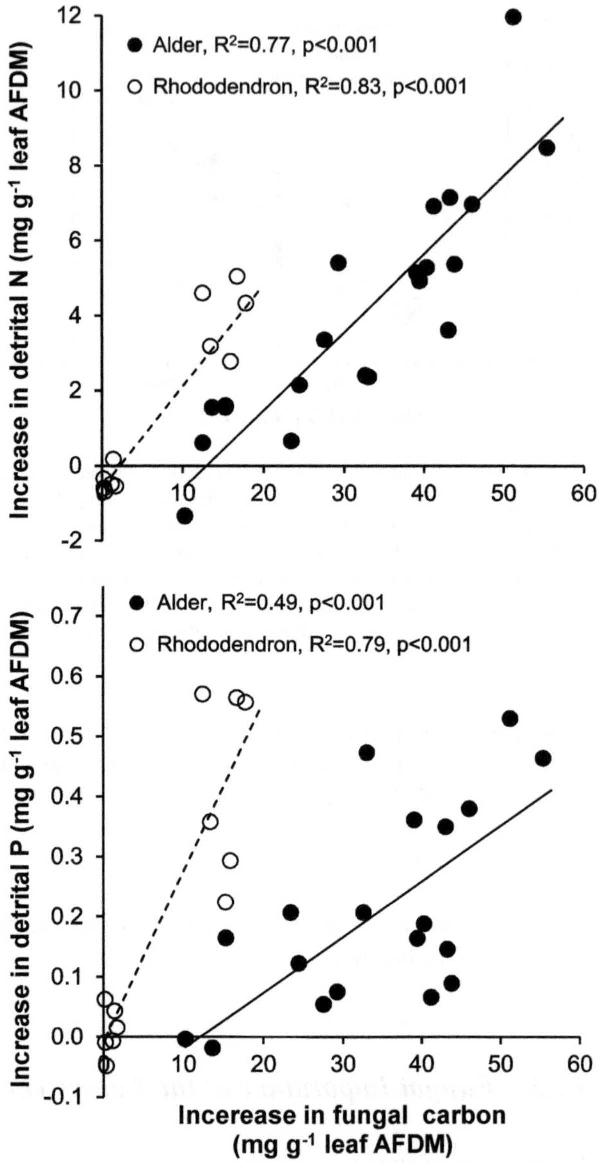

Fig. 5.5 Relationships between increases in fungal C and litter N content (top panel) and fungal C and litter P content (bottom panel) of rhododendron leaf litter at Coweeta Long Term Ecological Research site in North Carolina and alder leaf litter in a Portuguese stream. Slopes of regressions were used to calculate fungal elemental ratios (molar), C:N = 4.3, C:P = 91, and N:P = 21 for rhododendron, and 6, 278 and 50 for alder litter, respectively. Data are from Gulis et al. (2006), Ferreira et al. (2006b) and Tant et al. (2013)

to 23 g C m^{-2} and are highly seasonal in temperate streams due to seasonal patterns of deciduous leaf input and, hence, in-stream litter standing stocks that peak in autumn (Suberkropp 1997; Methvin and Suberkropp 2003; Carter and Suberkropp 2004; Suberkropp et al. 2010). Mean annual fungal biomass estimates in these studies ranged from 1 to 9 g C m^{-2}. Fungal biomass associated with small wood (7–40 mm diam) was estimated to average 4–7 g C m^{-2} in two streams with little seasonal variation (Gulis et al. 2008).

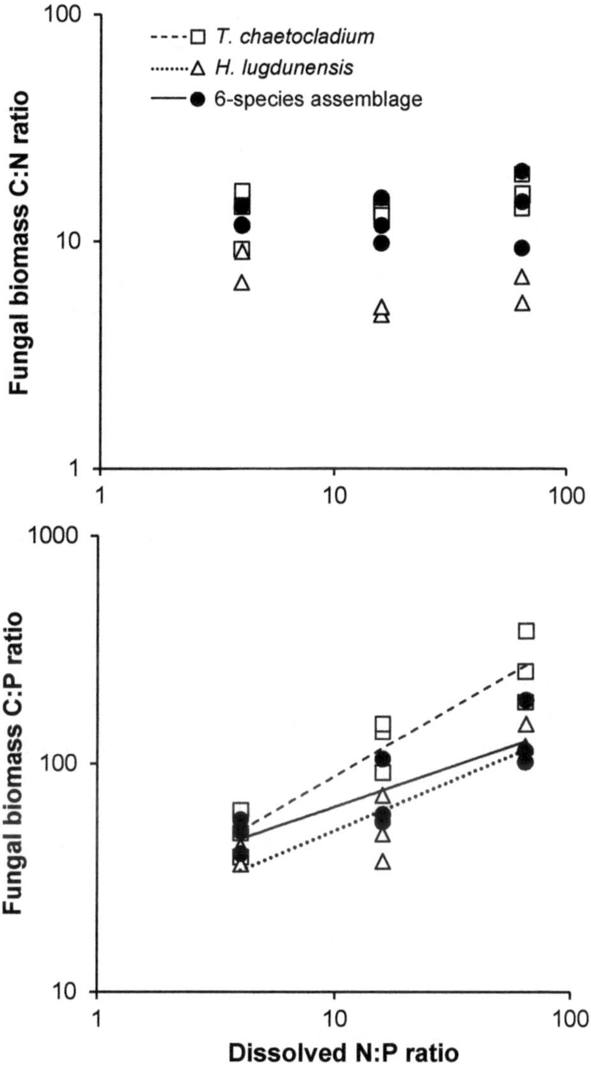

Fig. 5.6 Relationships between dissolved inorganic N:P ratios (molar) and C:N or C:P ratios of fungal biomass. Data are from experiments in laboratory microcosms simulating stream conditions with fungi (two monocultures and a six-species fungal assemblage) grown on maple leaf litter (Gulis et al. 2017). $NaNO_3$ and KH_2PO_4 were used as dissolved N and P sources in microcosm solutions. Data points are means of three replicates. For the bottom panel, the slopes of regression of log-log transformed data are different from zero suggesting that fungi are not homeostatic with respect to their C:P elemental ratio (TC $R^2 = 0.89$, $p < 0.001$; HL $R^2 = 0.76$, $p = 0.011$; assemblage $R^2 = 0.72$, $p = 0.004$)

Areal estimates of annual fungal production associated with leaf litter in five streams ranged from 8 to 23 g C m^{-2} $year^{-1}$ (Suberkropp 1997; Methvin and Suberkropp 2003; Carter and Suberkropp 2004). Taking into account annual leaf litter input into these streams (~190–246 g C m^{-2} $year^{-1}$) and fungal growth efficiency (33%, Suberkropp 1991; Gulis and Suberkropp 2003b), annual fungal assimilation (production + respiration) accounted for 10–29% of annual litter input. However, these streams exhibited relatively low litter retention, i.e., considerable amount of leaf litter tends to be flushed from these streams during winter storm events. In contrast, in two small highly retentive streams with high litter inputs that maintained mean annual litter standing stocks up to 450 g C m^{-2}, annual fungal

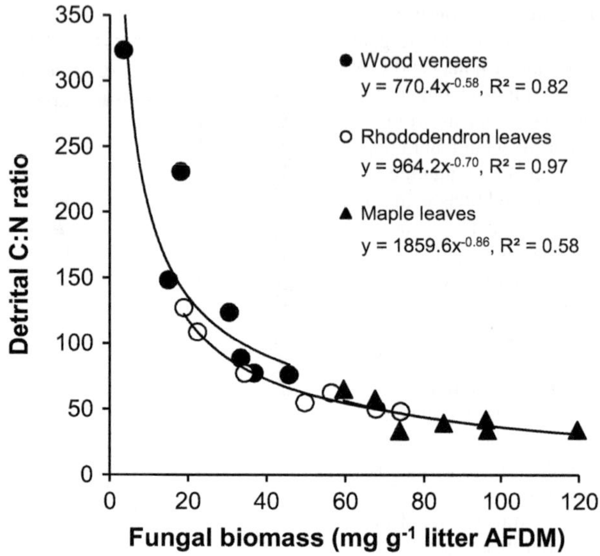

Fig. 5.7 Relationship between litter-associated fungal biomass and detrital C:N ratios (molar) at later stages of decomposition (day 77) for three substrate types. Data (Gulis et al. unpublished) are fitted separately using a power function; R^2 are shown for each substrate. When data for all substrates are combined, linear regression based on log-log transformed data gives $R^2 = 0.91, p = 9.0 \times 10^{-12}$, and $n = 21$

production reached 49–290 g C m^{-2} year^{-1} with estimated fungal assimilation of 35% to >100% of the annual leaf litter input (Suberkropp et al. 2010). Corresponding annual bacterial production from leaf litter in these streams ranged from 7 to 13 g C m^{-2} year^{-1}, which is more than an order of magnitude lower than annual fungal production. Estimates of annual fungal production associated with small wood in these same two highly retentive streams were 13–17 g C m^{-2} year^{-1}, which translated into a fungal assimilation of 45–57% of annual wood inputs to these streams (Gulis et al. 2008, unpublished). These findings further emphasize the key role that fungi play in carbon and energy flow from submerged plant litter in stream ecosystems.

Increases in fungal biomass as litter decomposes (see above) and concomitant fungal uptake and immobilization of dissolved N and P lead to fungi exerting strong control over the C:N and C:P stoichiometry of decomposing plant litter (Fig. 5.7; see also Tant et al. 2013; Manning et al. 2015, 2016). These increases in N and P content of plant litter together with the digestion of recalcitrant plant polymers by extracellular fungal enzymes improve the resource quality and palatability of plant litter to stream invertebrates (Suberkropp 1992), which lead to increased secondary production of some metazoan groups (Cross et al. 2006). N- and P-rich fungal biomass have been shown to directly contribute to diets of detritivorous invertebrates (Cross et al. 2007), thus facilitating the transfer of C and nutrients to higher trophic levels in aquatic ecosystems. In some cases, incorporation rates of fungal C accounted for up to 100% of daily growth rates of aquatic invertebrate larvae (Chung and Suberkropp 2009).

5.3 Fungi Associated with Emergent Macrophytes in Freshwater Wetlands

5.3.1 Characteristics of Freshwater Marshes and Emergent Plant Decay

Freshwater marshes are widely regarded as important transition zones that occur at the interface of terrestrial and aquatic habitats. A common feature of these freshwater ecotones is the presence of emergent macrophytes, such as *Phragmites*, *Typha*, and *Juncus*. These wetland-adapted plants exhibit high aboveground biomass production frequently exceeding 1 kg C m^{-2} $year^{-1}$. As a result, emergent marsh plants constitute an important reservoir of carbon and nutrients and are often the principal carbon and nutrient pool depicted in most wetland elemental budgets (Mitsch and Gosselink 2007; Reddy and Delaune 2008).

Most of the plant biomass produced in marshes is not consumed by herbivores but instead enters the detrital pool where microbial decomposers (bacteria and fungi) and detritus-feeding consumers (macroinvertebrates) participate in its breakdown and mineralization. Despite the well-known importance of plant detritus in these ecosystems (Moore et al. 2004; Hagen et al. 2012), *natural* plant decomposition in freshwater marshes and the role of litter-associated fungal decomposers have rarely been investigated. Most studies examining emergent plant decomposition have focused on microbial decay processes occurring either at or within the wetland surface sediments (e.g., Polunin 1984; Webster and Benfield 1986). This approach continues today (e.g., Rothman and Bouchard 2007; Fennessy et al. 2008; Gingerich and Anderson 2011) and is likely grounded in the false perception among wetland researchers that emergent plant decomposition occurs solely at or within the marsh sediments, where it is driven by aerobic and anaerobic bacterial communities (Gutknecht et al. 2006; Kayranli et al. 2010; Morrissey et al. 2014). As a result, our understanding of fungal importance in wetland biogeochemical cycles has not been fully realized and today remains notably absent in conceptual or quantitative models describing wetland carbon and nutrient cycles (Batzer and Sharitz 2006; Mitsch and Gosselink 2007; Reddy and Delaune 2008). Such a paradigm contrasts sharply with our knowledge of litter decomposition in streams (see above), where fungi are widely recognized as a key microbial group that drives litter decomposition and facilitates carbon and nutrient transfer to invertebrate consumers (Gessner et al. 2007; Gulis et al. 2009; Findlay 2010).

Fundamental details to recognize when examining emergent plant decomposition are both the spatial and temporal conditions under which plant material naturally decomposes. In most emergent macrophytes, abscission and collapse of plant matter to the sediments or overlying surface waters do not occur immediately following plant senescence and death. A significant portion of the plant detrital mass remains in a standing-dead position (Asaeda et al. 2002; Christensen et al. 2009) where it can undergo considerable initial microbial transformation and decay before it collapses to the water or sediments (Gessner 2001; Kuehn et al. 2011; Su et al. 2015). As a result, emergent plant decay is a complex sequential process that involves two distinct spatial phases separated in time: an initial decay phase under aerial standing-dead conditions followed by a second decay phase under submerged or surface sediment conditions after the eventual collapse of standing litter. When plant litter decomposition studies have closely simulated these natural decay conditions, then fungi have been found to be important drivers of emergent macrophyte decomposition (Gessner et al. 2007; Kuehn 2008, 2016).

5.3.2 Fungal Diversity in Freshwater Marshes

Mycologists have known for over a century (Saccardo 1898) that filamentous fungi pervasively colonize and reproduce on and within both standing and collapsed litter of emergent macrophytes (Kuehn 2016). To date, much of our knowledge of fungal diversity associated with plant litter comes from traditional studies where fungal reproductive structures (e.g., ascomata) were microscopically observed and identified either directly from field-collected material or after culturing fungi in the laboratory. To date, very few published studies have employed molecular-based techniques to characterize litter-associated fungal communities in freshwater marshes (e.g., Buesing et al. 2009).

Unlike stream systems where aquatic hyphomycetes typically dominate on submerged leaf litter (see above), fungal communities inhabiting emergent plant litter are often quite diverse. For example, an earlier compilation of literature by Gessner and Van Ryckegem (2003) found that over 600 species of fungi had been recorded from the common reed (*Phragmites australis*) alone, with members of the *Ascomycota* being the most common. Many studies have documented distinct temporal changes in fungal taxa during litter decomposition (Pugh and Mulder 1971; Apinis and Taligoola 1974; Van Ryckegem and Verbeken 2005b, c; Van Ryckegem et al. 2007). For example, Pugh and Mulder (1971) observed that prevalent fungi during the standing decay phase of *Typha latifolia* consisted of common phylloplane taxa, such as *Aureobasidium*, *Cladosporium*, and *Epicoccum*. These initial fungal colonizers were followed by an increased occurrence of ascomycetes (e.g., *Leptosphaeria* spp.). Predatory nematode-trapping fungi, *Arthrobotrys* and *Dactylaria*, were the most common taxa observed following the collapse of standing litter to the marsh benthic sediments.

In addition to temporal changes during emergent litter decay, fungal taxa may also vary in their spatial colonization of plant litter. Several researchers have reported that certain fungal taxa may preferentially occupy specific plant tissues, such as leaves, leaf sheaths, or the nodes and internodes of culms (Pugh and Mulder 1971; Apinis et al. 1975; Poon and Hyde 1998; Van Ryckegem and Verbeken 2005a; Van Ryckegem et al. 2007), which is likely a result of changes in decay conditions (aerial vs. submerged) and differences in the intrinsic quality of plant litter, such as concentrations of recalcitrant compounds (lignocellulose) or nutrients (N and P). For example, Van Ryckegem and Verbeken (2005a) examined the fungal communities associated with standing-dead culms of *P. australis* in four tidal marshes that varied in salinity (freshwater to mesohaline). A distinct vertical distribution of fungal taxa was observed along the *P. australis* shoot axis, which was influenced by flooding height, frequency of inundation, and salinity.

5.3.3 Fungal Contribution to Litter Decomposition: Biomass and Rates of Production

Despite the abundant evidence indicating fungal colonization of decaying plant litter, our quantitative understanding of fungal decomposers in wetlands has lagged behind the body of data on other microbial groups, such as bacteria, due primarily to methodological limitations (see Sect. 5.1 above). However, application of ergosterol-based methods in freshwater marshes has produced compelling evidence that fungi are important participants in plant litter decomposition within these systems, particularly during the initial standing decay phase. Significant accumulation of fungal biomass has been reported in standing emergent plant litter, with peak values often accounting for as much as 5–10% of the total detrital dry mass (Newell et al. 1995; Bärlocher and Biddiscombe 1996; Kuehn and Suberkropp 1998b; Kuehn et al. 1999, 2011; Gessner 2001; Findlay et al. 2002b; Newell 2003; Welsch and Yavitt 2003; Ohsowski 2008; Su 2014; Su et al. 2015; Kuehn 2016). Recently, Kuehn et al. (2011) and Su et al. (2015) estimated the contribution of fungal decomposers during standing litter decay of *Typha angustifolia* and *Typha domingensis* leaves in a temperate (Michigan) and subtropical (Alabama) freshwater marsh, respectively. During senescence and early standing litter decay, significant increases in both ergosterol and chitin (glucosamine) were observed in *Typha* leaves (Fig. 5.8), which revealed the rapid colonization of *Typha* litter by fungal decomposers. Increases in both ergosterol and glucosamine concentrations correlated with losses in leaf carbon, providing strong evidence that increased growth of fungal decomposers coincides with accelerated losses in *Typha* detrital mass (see also Gessner 2001). Analogous to spatial patterns in fungal diversity (above), differences in fungal biomass have also been noted among specific plant tissues. For example, Kuehn et al. (1999) quantified fungal biomass concentrations in *Erianthus giganteus* during shoot senescence and early standing decomposition. Significantly higher fungal biomass was observed in leaf versus culm litter, which was consistent with the more recalcitrant nature and lower N and P concentrations of culm litter.

In addition to accumulating large quantities of biomass, fungal communities inhabiting standing litter also exhibit high rates of secondary production. Using the ^{14}C-acetate-to-ergosterol incorporation method, Kuehn et al. (2011) reported fungal biomass production rates ranging between 0.2 and 3.3 mg fungal C g^{-1} detrital C day^{-1} in standing *T. angustifolia* litter in a north temperate freshwater marsh (Fig. 5.9). Significant differences in fungal production rates were observed during the study period, with peak daily production occurring just after leaf

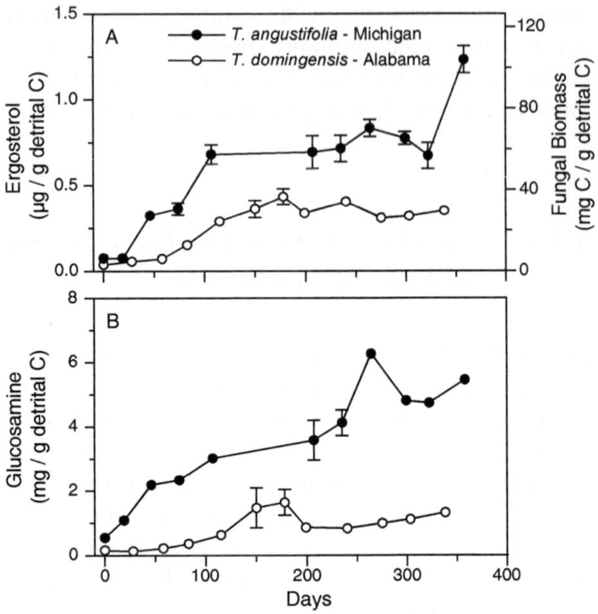

Fig. 5.8 Dynamics of fungal biomass (**a**) and glucosamine (**b**) concentrations in *Typha angustifolia* (Michigan) and *Typha domingensis* (Alabama) leaves during standing litter decomposition. Symbols indicate means ±1 SE ($n = 6$). Data from Kuehn et al. (2011) and Su et al. (2015)

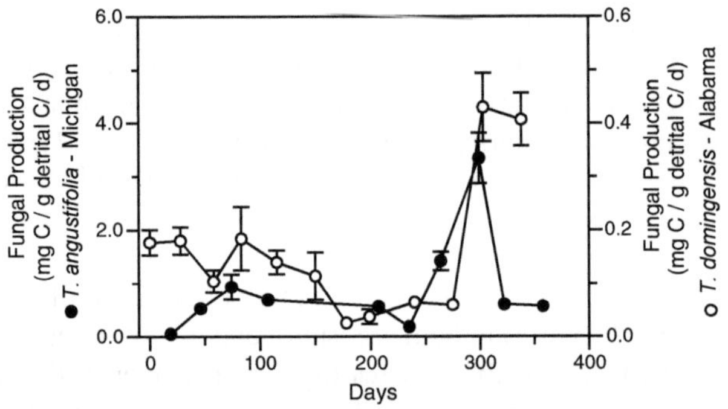

Fig. 5.9 Rates of fungal production associated with *T. angustifolia* (Michigan) and *T. domingensis* (Alabama) leaves during standing litter decomposition. Symbols indicate means ±1 SE ($n = 6$). Data from Kuehn et al. (2011) and Su et al. (2015)

senescence in late fall (74 days, November) and again during the summer season (300 days, July). Similar production patterns were also reported by Su et al. (2015) for fungal communities inhabiting *T. domingensis* in a subtropical freshwater marsh, with peak production rates being observed in early summer (~300 days, June). However, in contrast to Kuehn et al. (2011), fungal production associated with standing *T. domingensis* leaves was roughly an order of magnitude lower than production rates observed in *T. angustifolia* (Fig. 5.9), which was consistent with the lower accumulation of fungal biomass in standing *T. domingensis* leaf litter (Fig. 5.8).

When integrated over the entire study period, estimated cumulative production of fungi associated with standing *T. domingensis* and *T. angustifolia* leaf litter totaled 39 and 123 mg fungal C per g initial detrital C, respectively, indicating that a large

Table 5.2 Partial litter decay budgets, associated microbial parameters, and estimated contribution of fungal decomposers during standing-dead leaf litter decay of *T. angustifolia* and *T. domingensis*

	Michigan	Alabama
Parameter	*T. angustifolia*	*T. domingensis*
Total leaf mass loss (mg C g^{-1} initial leaf C)[a]	556 ± 60	371 ± 39
Cumulative fungal production (mg C g^{-1} initial leaf C)[b]	123 ± 10	39 ± 4
Cumulative respiration (mg C g^{-1} initial leaf C)[b]	–	135 ± 23
Mean fungal biomass (mg C g^{-1} initial leaf C)[a]	33 ± 11	17 ± 1
Fungal production to biomass (P:B) ratio	3.5 ± 1.3	2.3 ± 0.1
Turnover time of fungal biomass (day)	99 ± 34	145 ± 7
Fungal yield coefficient (%)[c]	22 ± 3	11 ± 1
Fungal contribution to overall leaf C loss (%)[d]	69 ± 5[e]	47 ± 5

Data from Kuehn et al. (2011) and Su (2014)
[a]Values for total leaf mass loss and fungal biomass are means ±1 SD
[b]Values for cumulative fungal production and respiration are means ±1 SD, as estimated using Monte Carlo simulation analysis
[c]Fungal yield coefficient (%) = total net fungal production/total leaf mass loss ×100
[d]The contribution of fungi to overall carbon loss from standing-dead leaf litter was determined by dividing total fungal assimilation (cumulative production + respiration) by the total leaf C mass loss. Assumes that respiratory activity is entirely due to fungal organisms
[e]Assumes fungal growth efficiency of 32% (Gulis and Suberkropp 2003b)

portion of the observed leaf carbon lost was converted into fungal biomass (see fungal yield, Table 5.2). Furthermore, estimates of cumulative fungal production associated with *Typha* litter were significantly related with cumulative losses in leaf litter carbon (Fig. 5.10), providing additional evidence that fungi are likely the primary drivers of emergent plant decomposition during the standing decay phase. If respiratory activities in standing litter are taken into account, which are likely from fungi (see below), then total fungal assimilation (production + respiration) could account for even a greater proportion of leaf carbon losses (Table 5.2).

After standing litter collapses, commencement of the benthic decomposition phase is often accompanied by a notable shift in the environmental conditions (e.g., increased water availability, exogenous nutrients, decreased temperature fluctuations), which may lead to shifts in litter-associated microbial communities and concomitant changes in the biomass and production of both fungi and bacteria. Earlier, Kuehn et al. (2000) observed a rapid decrease in litter-associated ATP concentrations, fungal biomass (ergosterol), and production rates of fungi following the transition of standing *Juncus effusus* litter to a submerged environment. This initial decrease was followed by an increase in fungal biomass and production rates during later stages of submerged litter decomposition, reflecting a possible shift in fungal community to taxa better adapted to an aquatic or semiaquatic existence (see fungal diversity above). Similar initial decreases in fungal biomass and activity have also been reported for *P. australis* litter following its transition from an aerial to a submerged environment (Tanaka 1991; Kominkova et al. 2000; Van Ryckegem et al. 2007).

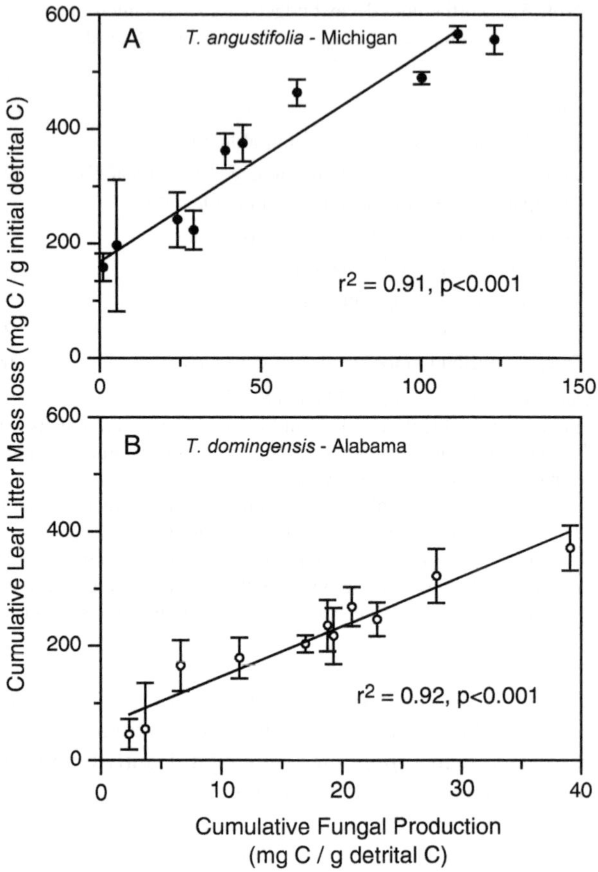

Fig. 5.10 Relationship between cumulative fungal production and cumulative litter carbon loss of (**a**) *T. angustifolia* (Michigan) and (**b**) *T. domingensis* (Alabama) leaves during standing litter decomposition phase. Symbols indicate means ±1 SE ($n = 6$). Data from Kuehn et al. (2011) and Su et al. (2015)

Despite litter submergence and initial declines in fungal biomass, fungi continue to be a quantitatively important microbial group on and within decaying plant litter. Simultaneous estimates of fungal and bacterial biomass associated with submerged plant litter reveal that fungal decomposers often account for >90% of the total microbial biomass (Findlay et al. 1990, 2002b; Newell et al. 1995; Sinsabaugh and Findlay 1995; Kominkova et al. 2000; Kuehn et al. 2000, 2014; Su et al. 2007). Furthermore, several studies have reported that rates of fungal production are often comparable to or greatly exceed corresponding rates of bacterial production (Newell et al. 1995; Kuehn et al. 2000, 2014; Findlay et al. 2002a, b; Su et al. 2007, but see Buesing and Gessner 2006). For example, Su et al. (2007) simultaneously quantified bacterial and fungal biomass and production and extracellular degradative enzyme activities during the decomposition of collapsed benthic leaf litter of *T. angustifolia* in two hydrologically different (i.e., permanently inundated vs. exposed) Lake Erie coastal marshes. Microbial biomass and production associated with decaying *Typha* litter were dominated by fungi at both sites, accounting for greater than 94% and 99% of the total microbial biomass and production, respectively. Dynamics of extracellular microbial enzymes involved in the acquisition of detrital C, N, and P were similar to observed patterns in fungal biomass and production, suggesting that litter-inhabiting fungal decomposers are responsible for a large portion of extracellular enzymes produced (see also Romani et al. 2006). Kuehn et al. (2000) also found that rates of fungal production associated with *J. effusus* litter greatly exceeded corresponding rates of bacterial production throughout early stages of submerged litter decomposition. Estimates of fungal and bacterial contributions indicated that fungal assimilation could explain a substantial portion (68%) of the observed *J. effusus* litter mass loss, providing evidence that fungi are also important players in the transformation and decay of plant litter under submerged or surface sediment conditions (Kuehn et al. 2000).

5.3.4 Factors Affecting Fungal Activities on Emergent Plant Litter

Plant litter decomposition in freshwater marshes involves a complex array of biotic and abiotic processes that result in the production of decomposer biomass (microbial and invertebrate consumers), release of CO_2 and nutrients (N and P) through mineralization, and also release of DOM and FPOM (Gessner et al. 1999; Kuehn 2008). From a microbial perspective, the rates of these processes are strongly influenced by the response of microbial communities to the environmental conditions and the intrinsic quality of the detrital resources they metabolize (Gessner et al. 2007). Being osmotrophic, both bacteria and fungi acquire detrital resources via the extracellular digestion of complex organic matter by degradative enzymes. The production and release of extracellular enzymes represent a substantial energy cost for microorganisms. As a result, microbes will regulate the production and release of extracellular enzymes in accordance with detrital resource availabilities (C, N, P), which serves to optimize their assimilatory return on investment (Sinsabaugh and Follstad Shah 2012; Sinsabaugh et al. 2014). Collectively, this regulation can affect outcomes related to fungal community performance (biomass, growth, reproduction) and hence the decomposition and mineralization of plant detritus.

Similar to deciduous leaf litter in streams, emergent plant litter in freshwater marshes typically has low nutrient concentrations and, hence, high C:nutrient ratios (e.g., $C:N > 80$, $C:P > 4000$). The nutrient stoichiometry (C:N:P) of fungal biomass is typically lower than that of the plant litter they colonize. As a consequence, the biomass and activities of fungi are often limited by N and P availability. These nutrients can be obtained either from the substrate they inhabit or from an exogenous source, such as dissolved nutrients in the water after the collapse of standing litter. For example, Kuehn et al. (2011) and Su et al. (2015) observed rapid increases in fungal biomass (ergosterol) during senescence and early standing leaf litter decay in *T. angustifolia* and *T. domingensis*. Despite similar patterns of increase, total fungal biomass accumulation in standing *T. angustifolia* leaf litter was notably higher than

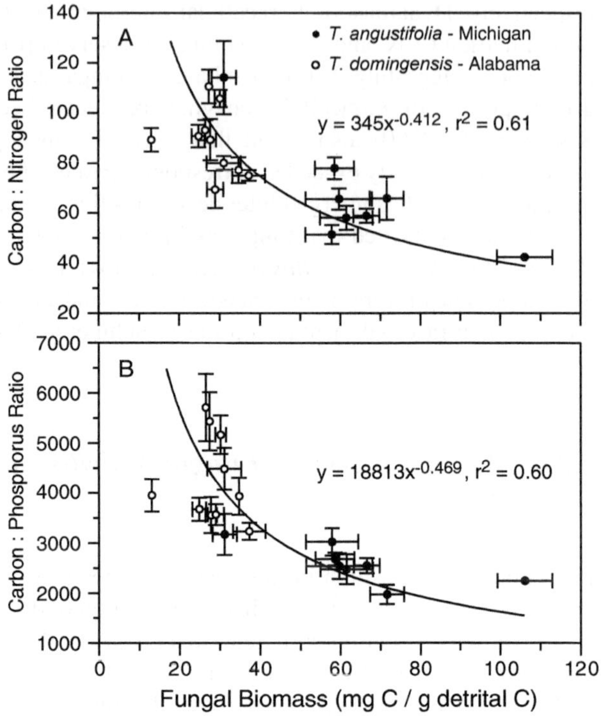

Fig. 5.11 Relationships between litter-associated fungal biomass and (**a**) detrital C:N ratios and (**b**) detrital C:P ratios of *T. angustifolia* (Michigan) and *T. domingensis* (Alabama) leaves throughout standing litter decomposition phase. Data were fitted using a power function

in *T. domingensis* (Fig. 5.8). Furthermore, daily and cumulative production of fungi in *T. angustifolia* were also markedly higher than in *T. domingensis*, which was consistent with the higher rates of mass loss observed in *T. angustifolia* (55%) versus *T. domingensis* (37%) leaf litter (Table 5.2). The contrasting performance and overall contribution of fungal decomposers to leaf decay in these studies may be explained by differences in litter nutrient content between these *Typha* species. Su et al. (2015) found that C:N and C:P ratios of *T. domingensis* leaf litter averaged 88 and 4352 throughout the post-senescent stages of standing litter decomposition, respectively. In contrast, Kuehn et al. (2011) observed much lower C:N and C:P ratios in standing *T. angustifolia* leaves, which averaged 67 and 2583, respectively. Fungal biomass in both *T. angustifolia* and *T. domingensis* leaf litter were negatively related to litter C:N and C:P ratios (Fig. 5.11), implying that fungal communities may have experienced considerable stoichiometric constraints on growth (Sinsabaugh et al. 2015) and, hence, litter decomposition rates.

A number of studies conducted in both subtropical and temperate freshwater marshes have also established that water availability is a key factor controlling the activity of microbial decomposers in the standing litter decay phase (Kuehn et al. 1998, 1999, 2004; Kuehn and Suberkropp 1998a; Welsch and Yavitt 2003). Kuehn et al. (2004) examined the effects of environmental conditions on rates of microbial respiration (CO_2 flux) from standing *P. australis* litter in two temperate freshwater marshes in Switzerland. Under field conditions, rates of microbial respiration from

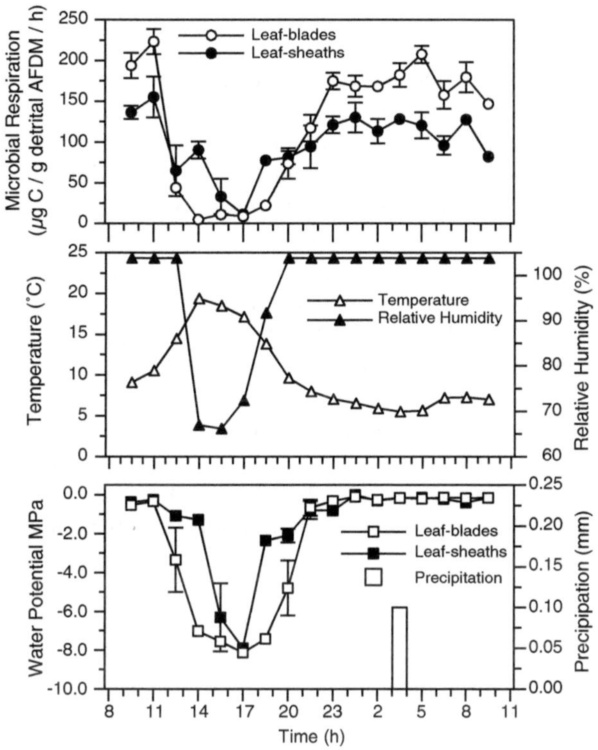

Fig. 5.12 Diel changes in rates of microbial respiration from standing-dead leaves and leaf sheaths of *Phragmites australis* within a temperate freshwater lake littoral marsh (top panel). Corresponding diel changes in temperature and relative humidity patterns and plant litter water potentials and precipitation are also illustrated (middle and bottom panels). Symbols indicate means ±1 SE ($n = 3$). Data from Kuehn et al. (2004)

standing litter exhibited a diel periodicity, with the highest rates occurring at night when water becomes available to litter-inhabiting microbial assemblages via dew formation (Fig. 5.12). In contrast, microbial respiration virtually ceased during the day as a result of increased desiccation stress. These results were remarkably similar to the diel microbial respiratory patterns reported earlier from standing litter in a subtropical freshwater marsh (Kuehn and Suberkropp 1998a), where temperature-driven increases in relative humidity and dew formation were identified as important mechanisms controlling nighttime increases in litter water potential (water availability) and associated microbial respiration. Additional research found that fungi inhabiting standing litter likely survive these diel periods of desiccation by accumulating intracellular compatible solutes (Kuehn et al. 1998).

Similar to observed patterns in fungal diversity, biomass, and production, distinct differences in rates of microbial respiration have also been observed among different plant litter substrates. Kuehn et al. (2004) reported significant differences in microbial respiration rates among standing litter fractions of *P. australis* (leaf blades vs. leaf sheaths vs. culms), which were consistent with differences in water absorption patterns, structural characteristics (lignocellulose), and the degree of fungal colonization. Maximum rates of respiration among litter fractions were positively correlated with litter-associated fungal biomass (ergosterol, $r = 0.72$). Similar findings by other researchers (Kuehn and Suberkropp 1998a; Su 2014; Su et al.

2015) lend support to the idea that respiratory activity in standing litter can be mostly attributed to fungal decomposers.

Under submerged or surface sediment conditions, rates of microbial respiration associated with emergent macrophyte plant litter are comparable to the maximum rates reported for standing-dead litter (Kominkova et al. 2000; Buesing and Gessner 2006; Su et al. 2007). Buesing and Gessner (2006) examined microbial respiration rates associated with submerged *P. australis* litter at the same littoral marsh site used by Kuehn et al. (2004). Respiration rates ranged between 31 and 319 μg CO_2–C g^{-1} litter organic mass h^{-1} throughout the year, with fluctuations closely following changes in lake water temperature. Since litter-associated microbial biomass is predominantly fungal (see above), it is plausible that a large portion of the respiratory C losses from collapsed plant litter may also be due to the metabolic activities of fungal decomposers.

In addition to chemical and physical factors described above, microbial performance and litter decomposition in freshwater marshes (and streams) may also be influenced by the myriad of biotic interactions that occur within the detrital landscape (Gessner et al. 1999, 2007). For example, under submerged conditions decaying plant litter often develops complex microbial biofilms (Battin et al. 2007), which can contain a diverse community of both autotrophic (algae) and heterotrophic (bacteria, fungi, protists) microorganisms. The close spatial proximity of these microbial groups on and within submerged plant litter suggests the potential for biotic interactions (e.g., Mille-Lindblom and Tranvik 2003; Francoeur et al. 2006; Mille-Lindblom et al. 2006; Danger et al. 2013; Kuehn et al. 2014). For example, Kuehn et al. (2014) provided intriguing evidence that photosynthetically active algae may elicit a "priming effect" on heterotrophic bacterial and fungal decomposers colonizing submerged emergent plant detritus (see also Danger et al. 2013). Well-established in terrestrial ecosystems, the priming effect describes a natural phenomenon whereby the microbial decomposition and mineralization of recalcitrant soil organic matter are enhanced by pulsed or continuous inputs of labile carbon, such as plant root exudates (Blagodatsky et al. 2010; Kuzyakov 2010). These carbon inputs produce hotspots and hot moments of soil-rhizosphere microbial activity, where microbial heterotrophs are provided energy-rich compounds that increase their metabolic capabilities (e.g., enzyme production) to degrade and mineralize refractory soil organic matter. In the presence of periphytic algae, Kuehn et al. (2014) and Francoeur et al. (2006) documented that short-term light exposure (400 μmol m^{-2} s^{-1} photosynthetically active radiation, PAR) rapidly stimulated rates of bacterial and fungal production (Fig. 5.13) and extracellular hydrolytic and oxidative enzyme activities in decaying plant litter, respectively. Furthermore, experimental incubations of decaying litter with ^{14}C- and ^{13}C-bicarbonate established that inorganic C fixed by algal photosynthesis was rapidly transferred to and assimilated by heterotrophic microbial decomposers (Kuehn et al. 2014), which was likely due to labile DOC exudation from co-occurring periphytic algae. These findings underscore that microbial interactions are important for our understanding of key ecological processes, such as plant litter decomposition. Furthermore, these findings also strengthen the contention that the priming effect phenomenon may be relevant within aquatic ecosystems (Guenet et al. 2010),

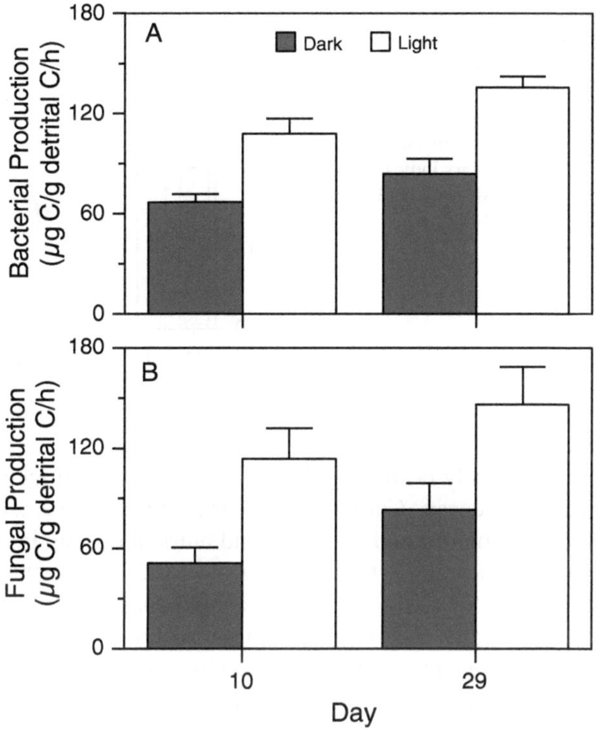

Fig. 5.13 Production rates of (**a**) bacteria (^{3}H-leucine incorporation into protein) and (**b**) fungi (^{14}C-acetate incorporation into ergosterol) in natural *T. angustifolia* detrital-periphyton communities when exposed to short-term light (400 μmol m^{-2} s^{-1} PAR, UV-free) or dark conditions in the laboratory. *T. angustifolia* detrital-periphyton was collected 10 and 29 days after litter submergence. Bars indicate means ±1 SE ($n = 5$). Data from Kuehn et al. (2014)

where algal presence and activity within the detrital microbial landscape may facilitate microbial decomposition and mineralization of detrital organic matter (see also Rier et al. 2007, 2014; Danger et al. 2013; Hotchkiss et al. 2014).

5.3.5 Fungal Importance at the Ecosystem Scale

To date, most studies examining fungal activities in freshwater marshes have focused on quantifying fungal biomass and production during specific litter decomposition periods. In this regard, plant litter of known age and type (e.g., leaves) was enclosed in litter bags or tagged in a standing position and sampled through time (Newell et al. 1995; Bärlocher and Biddiscombe 1996; Kuehn and Suberkropp 1998a, b; Kuehn et al. 1999, 2000, 2011; Komínková et al. 2000; Gessner 2001; Su et al. 2015). This approach has been useful in examining fungal dynamics during the litter decomposition process (e.g., Fig. 5.8) and for constructing partial litter decay budgets (e.g., Table 5.2) that focused on quantifying the contribution of fungal decomposers to litter mass loss. However, to assess the impact of fungi or other microbial assemblages on ecosystem-scale carbon and nutrient cycling, it is necessary to have areal (m^{-2}) estimates of microbial biomass and production associated

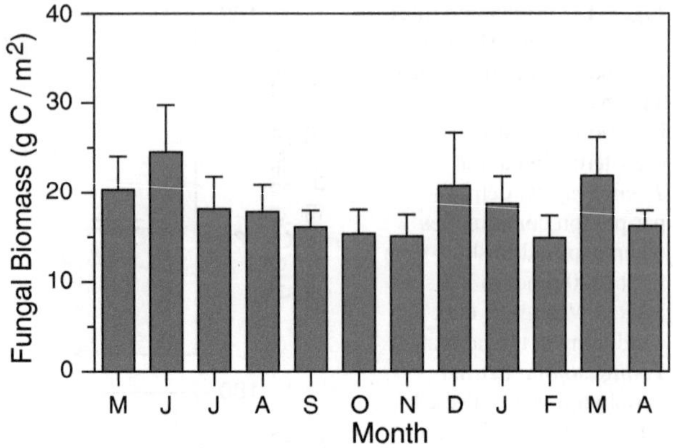

Fig. 5.14 Fungal biomass associated with standing-dead plant litter per m^2 of marsh surface area in a mixed *C. jamaicense* and *J. roemarianus* marsh stand (Alabama). Bars indicate means ±1 SE ($n = 6$). Data from Su (2014)

with naturally occurring plant litter, which would include plant detritus in various stages of decay. Currently, very few studies have estimated the ecosystem-scale contribution of fungi to carbon and nutrient cycling in freshwater marshes (Buesing and Gessner 2006; Ohsowski 2008; Su 2014; Kuehn and Gessner unpublished). Recently, Su (2014) estimated annual fungal biomass and production associated with naturally occurring standing-dead litter in a subtropical marsh dominated by *Cladium jamaicense*. Living fungal biomass per gram of plant litter carbon remained fairly constant over the study period, averaging 30 mg C g^{-1} detrital C. Because of appreciable standing litter accumulation, corresponding standing stock estimates of fungal biomass per square meter of marsh were considerable, averaging 18 g C m^{-2} over the annual study period (Fig. 5.14). Substantial rates of fungal production on areal basis were also observed. When integrated over the study period, estimated annual fungal production in standing litter totaled 90 g C m^{-2} $year^{-1}$ (Table 5.3), which was equivalent to ~14% of mean annual standing litter mass (643 g C m^{-2}). If annual estimates of microbial respiration are taken into account (CO_2 flux, 124 g C m^{-2} $year^{-1}$), which is most likely fungal, then total fungal assimilation could account for ~33% of mean annual standing litter mass. Comparable estimates of fungal production in standing-dead litter in other freshwater marsh systems (Table 5.3) suggest that fungal-mediated processes may be a significant pathway of carbon flow at the ecosystem scale.

High areal fungal production has also been reported for benthic plant detritus (Table 5.3; Buesing and Gessner 2006; Ohsowski 2008). Buesing and Gessner (2006) estimated annual production of both bacterial and fungal communities associated with naturally occurring benthic litter of *P. australis* in temperate lake's littoral marsh in Switzerland. Annual fungal production on submerged *P. australis* detritus totaled 93 g C m^{-2} $year^{-1}$, indicating that ~15% of the annual aboveground *P. australis* production (603 g C m^{-2} $year^{-1}$) was transformed into fungal biomass. However, in this study, despite certain reliance on dissolved organic carbon, annual bacterial production was considerably higher (661 g C m^{-2} $year^{-1}$), suggesting that litter-associated bacteria may play an important role in carbon transformation once standing litter collapses to the overlying marsh surface sediments.

Table 5.3 Estimates of annual production, production to biomass (P:B) ratios, and turnover times of fungi associated with standing-dead and benthic plant litter in freshwater marshes

Marsh	Annual fungal production (g C m^{-2})	Annual P:B	Turnover time (day)	Reference
Weeks Bay (Alabama)				
Cladium jamaicense/Juncus roemarianus marsh				
Standing-dead leaf and stem litter	90.3 ± 4.4	4.9	74	Su (2014)
Independence Lake (Michigan)				
Typha angustifolia marsh				
Standing-dead leaves	78.0 ± 5.5	4.0	90	Kuehn (unpublished)
Standing-dead stems	56.1 ± 3.8	8.2	44	
Paint Creek (Michigan)				
Typha angustifolia marsh				
Standing-dead leaves	7.9 ± 0.6	1.0	342	Ohsowski (2008)
Standing-dead stems	34.5 ± 2.2	2.7	137	
Benthic litter	112.3 ± 1.1	7.3	50	
Lake Hallwil (Switzerland)				
Phragmites australis marsh				
Standing-dead leaves	7.5 ± 4.3	3.8	103	Kuehn (unpublished)
Standing-dead sheaths	11.9 ± 6.2	7.2	53	
Standing-dead culms	2.5 ± 4	6.2	62	
Benthic litter	93.2 ± 36.0	10.0	37	Buesing and Gessner (2006)

Values for annual fungal production are means ±1 SD

Observations of considerable fungal production and biomass in emergent plant litter imply that fungi are also efficient in acquiring and immobilizing detrital N and P to meet their stoichiometric demands for growth and reproduction. This may be particularly true in standing litter decay systems where access to additional nutrients from exogenous sources is likely to be limited. Using a fungal stoichiometric ratio of $C_{100}N_9P_1$ (Gulis et al. 2017) and data from Su (2014), we estimated that fungal decomposers could immobilize ~25 and up to 100% of the detrital N and P, respectively, in naturally occurring standing litter in a *C. jamaicense* marsh. Similar observations in other freshwater marsh systems (Findlay et al. 2002b; Van Ryckegem et al. 2006; Kuehn et al. 2011; Su et al. 2015) suggest that fungal decomposers play an important role in the cycling of N and P in these systems.

5.4 Conclusions

Many freshwater environments, such as streams, rivers, and wetlands, depend on inputs of dead plant material as a source of carbon and other nutrients to fuel ecosystem processes. Fungi in these environments are key players in decomposition of plant litter, where they facilitate the transfer of energy and nutrients to higher trophic levels, such as aquatic invertebrates and ultimately vertebrate animals. It is not uncommon for fungal biomass to contribute around 10% of detrital mass, annual fungal production to reach tens to hundreds of g m^{-2}, and direct fungal assimilation of plant carbon to account for about one third of annual litter input. However, only relatively recent advances in methodology have allowed us to uncover the critical importance of fungi in decomposition of plant litter in these systems. Nevertheless, new approaches are needed to quantify the effects of global change (temperature, pollution, precipitation) on fungi and microbially driven processes in freshwaters. Since changes in fungal activity and community structure may affect ecosystem functioning, new molecular approaches, including the next-generation sequencing, hold promise to further our understanding of fungi in freshwaters.

Compliance with Ethical Standards

Funding Financial support from the National Science Foundation (DEB 0919054, DEB 1655797 to VG and DBI 0420965, DBI 0923063, DEB 0315686, DEB 1457217 to KAK) is gratefully acknowledged.

Conflict of Interest Vladislav Gulis declares that he/she has no conflict of interest. Rong Su declares that he/she has no conflict of interest. Kevin A. Kuehn declares that he/she has no conflict of interest.

Ethical Approval This article does not contain any studies with human participants or animals performed by any of the authors.

References

Abdullah SK, Taj-Aldeen SJ (1989) Extracellular enzymatic activity of aquatic and aero-aquatic conidial fungi. Hydrobiologia 174:217–223

Apinis AE, Taligoola HK (1974) Biodegradation of *Phragmites communis* Trin. by fungi. In: Kilbertus G, Reisinger O, Concela Da Fonseca JA (eds) Biodegradation et humification. Sarreguemines, Pierron, pp 24–32

Apinis AE, Chesters CGC, Taligoola HK (1975) Microfungi colonizing nodes and internodes of aerial standing dead culms of *Phragmites communis* Trin. Nova Hedwigia 26:495–507

Asaeda T, Nam L, Hietz P et al (2002) Seasonal fluctuations in live and dead biomass of *Phragmites australis* as described by a growth and decomposition model: implications of duration of aerobic conditions for litter mineralization and sedimentation. Aquat Bot 73:223–239

Baldy V, Gessner MO (1997) Towards a budget of leaf litter decomposition in a first-order woodland stream. C R Acad Sci Paris, Ser III 320:747–758

Baldy V, Gessner MO, Chauvet E (1995) Bacteria, fungi and the breakdown of leaf litter in a large river. Oikos 74:93–102

Baldy V, Chauvet E, Charcosset JY et al (2002) Microbial dynamics associated with leaves decomposing in the mainstem and floodplain pond of a large river. Aquat Microb Ecol 28:25–36

Baldy V, Gobert V, Guerold F et al (2007) Leaf litter breakdown budgets in streams of various trophic status: effects of dissolved inorganic nutrients on microorganisms and invertebrates. Freshw Biol 52:1322–1335

Bärlocher F (2005) Freshwater fungal communities. In: Dighton J, White JF, Oudemans P (eds) The fungal community. Its organization and role in the ecosystem, 3rd edn. CRC Press, Boca Raton, pp 39–59

Bärlocher F, Biddiscombe N (1996) Geratology and decomposition of *Typha latifolia* and *Lythrum salicaria* in a freshwater marsh. Arch Hydrobiol 136:309–325

Baschien C, Tsui C, Gulis V et al (2013) The molecular phylogeny of aquatic hyphomycetes with affinity to the Leotiomycetes. Fungal Biol 117:660–672

Battin T, Sloan W, Kjelleberg S et al (2007) Microbial landscapes: new paths to biofilm research. Nat Rev Microbiol 5:76–81

Batzer DP, Sharitz RR (2006) Ecology of freshwater and estuarine wetlands. University of California Press, Berkeley

Benstead JP, Rosemond AD, Cross WF et al (2009) Nutrient enrichment alters storage and fluxes of detritus in a headwater stream ecosystem. Ecology 90:2556–2566

Blagodatsky S, Blagodatskaya E, Yuyukina T et al (2010) Model of apparent and real priming effects: linking microbial activity with soil organic matter decomposition. Soil Biol Biochem 42:1275–1283

Bucher VVC, Pointing SB, Hyde KD et al (2004) Production of wood decay enzymes, loss of mass, and lignin solubilization in wood by diverse tropical freshwater fungi. Microb Ecol 48:331–337

Buesing N, Gessner MO (2006) Benthic bacterial and fungal productivity and carbon turnover in a freshwater marsh. Appl Environ Microbiol 72:596–605

Buesing N, Filippini M, Burgmann H et al (2009) Microbial communities in contrasting freshwater marsh microhabitats. FEMS Microbiol Ecol 69:84–97

Carter MD, Suberkropp K (2004) Respiration and annual fungal production associated with decomposing leaf litter in two streams. Freshw Biol 49:1112–1122

Cheever BM, Kratzer EB, Webster JR (2012) Immobilization and mineralization of N and P by heterotrophic microbes during leaf decomposition. Freshw Sci 31:133–147

Cheever BM, Webster JR, Bilger EE et al (2013) The relative importance of exogenous and substrate-derived nitrogen for microbial growth during leaf decomposition. Ecology 94:1614–1625

Christensen J, Crumpton W, van der Valk A (2009) Estimating the breakdown and accumulation of emergent macrophyte litter: a mass-balance approach. Wetlands 29:204–214

Chung N, Suberkropp K (2009) Contribution of fungal biomass to the growth of the shredder, *Pycnopsyche gentilis* (Trichoptera: Limnephilidae). Freshw Biol 54:2212–2224

Clemmensen K, Bahr A, Ovaskainen O et al (2013) Roots and associated fungi drive long-term carbon sequestration in boreal forest. Science 339:1615–1618

Cross WF, Wallace JB, Rosemond AD et al (2006) Whole-system nutrient enrichment increases secondary production in a detritus-based ecosystem. Ecology 87:1556–1565

Cross WF, Wallace JB, Rosemond AD (2007) Nutrient enrichment reduces constraints on material flows in a detritus-based food web. Ecology 88:2563–2575

Danger M, Chauvet E (2013) Elemental composition and degree of homeostasis of fungi: are aquatic hyphomycetes more like metazoans, bacteria or plants? Fungal Ecol 6:453–457

Danger M, Cornut J, Chauvet E et al (2013) Benthic algae stimulate leaf litter decomposition in detritus-based headwater streams: a case of aquatic priming effect? Ecology 94:1604–1613

Descals E (2005) Diagnostic characters of propagules of Ingoldian fungi. Mycol Res 109:545–555

Diez J, Elosegi A, Chauvet E et al (2002) Breakdown of wood in the Aguera stream. Freshw Biol 47:2205–2215

Duarte S, Pascoal C, Alves A et al (2010) Assessing the dynamic of microbial communities during leaf decomposition in a low-order stream by microscopic and molecular techniques. Microbiol Res 165:351–362

Duarte S, Seena S, Barlocher F et al (2013) A decade's perspective on the impact of DNA sequencing on aquatic hyphomycete research. Fungal Biol Rev 27:19–24

Fennessy M, Rokosch A, Mack J (2008) Patterns of plant decomposition and nutrient cycling in natural and created wetlands. Wetlands 28:300–310

Ferreira V, Chauvet E (2011) Synergistic effects of water temperature and dissolved nutrients on litter decomposition and associated fungi. Glob Chang Biol 17:551–564

Ferreira V, Elosegi A, Gulis V et al (2006a) Eucalyptus plantations affect fungal communities associated with leaf-litter decomposition in Iberian streams. Arch Hydrobiol 166:467–490

Ferreira V, Gulis V, Graça MAS (2006b) Whole-stream nitrate addition affects litter decomposition and associated fungi but not invertebrates. Oecologia 149:718–729

Ferreira V, Gulis V, Pascoal C et al (2014) Stream pollution and fungi. In: Jones EBG, Hyde KD, Pang KL (eds) Freshwater fungi and fungal-like organisms. De Gruyter, Berlin, pp 389–412

Ferreira V, Castagneyrol B, Koricheva J et al (2015) A meta-analysis of the effects of nutrient enrichment on litter decomposition in streams. Biol Rev 90:669–688

Findlay S (2010) Stream microbial ecology. J N Am Benthol Soc 29:170–181

Findlay S, Howe K, Austin H (1990) Comparison of detritus dynamics in two tidal freshwater wetlands. Ecology 71:288–295

Findlay S, Tank J, Dye S et al (2002a) A cross-system comparison of bacterial and fungal biomass in detritus pools of headwater streams. Microb Ecol 43:55–66

Findlay SEG, Dye S, Kuehn KA (2002b) Microbial growth and nitrogen retention in litter of *Phragmites australis* compared to *Typha angustifolia*. Wetlands 22:616–625

Fisher SG, Likens GE (1973) Energy flow in Bear Brook, New Hampshire: an integrative approach to stream ecosystem metabolism. Ecol Monogr 43:421–439

Francoeur SN, Schaecher M, Neely RK et al (2006) Periphytic photosynthetic stimulation of extracellular enzyme activity in aquatic microbial communities associated with decaying *Typha* litter. Microb Ecol 52:662–669

Gessner M (2001) Mass loss, fungal colonisation and nutrient dynamics of *Phragmites australis* leaves during senescence and early aerial decay. Aquat Bot 69:325–339

Gessner MO (2005) Ergosterol as a measure of fungal biomass. In: Graça MAS, Bärlocher F, Gessner MO (eds) Methods to study litter decomposition: a practical guide. Springer, Dordrecht, pp 189–196

Gessner MO, Chauvet E (1993) Ergosterol-to-biomass conversion factors for aquatic hyphomycetes. Appl Environ Microbiol 59:502–507

Gessner MO, Chauvet E (1994) Importance of stream microfungi in controlling breakdown rates of leaf litter. Ecology 75:1807–1817

Gessner MO, Chauvet E (1997) Growth and production of aquatic hyphomycetes in decomposing leaf litter. Limnol Oceanogr 42:496–595

Gessner MO, Van Ryckegem G (2003) Water fungi as decomposers in freshwater ecosystems. In: Bitton G (ed) Encyclopedia of environmental microbiology. Wiley, New York. https://doi.org/10.1002/0471263397.env314

Gessner MO, Chauvet E, Dobson M (1999) A perspective on leaf litter breakdown in streams. Oikos 85:377–384

Gessner MO, Gulis V, Kuehn KA et al (2007) Fungal decomposers of plant litter in aquatic ecosystems. In: Kubicek CP, Druzhinina IS (eds) The Mycota, vol IV. Environmental and microbial relationship. Springer, Berlin, pp 301–324

Gingerich R, Anderson J (2011) Litter decomposition in created and reference wetlands in West Virginia, USA. Wetl Ecol Manag 19:449–458

Grimmett I, Shipp K, Macneil A et al (2013) Does the growth rate hypothesis apply to aquatic hyphomycetes? Fungal Ecol 6:493–500

Guenet B, Danger M, Abbadie L et al (2010) Priming effect: bridging the gap between terrestrial and aquatic ecology. Ecology 91:2850–2861

Gulis V (2001) Are there any substrate preferences in aquatic hyphomycetes? Mycol Res 105:1088–1093

Gulis V, Bärlocher F (2017) Fungi: biomass, production, and community structure. In: Hauer FR, Lamberti GA (eds) Methods in stream ecology, vol 1. Academic Press, San Diego, pp 177–192

Gulis V, Suberkropp K (2003a) Effect of inorganic nutrients on relative contributions of fungi and bacteria to carbon flow from submerged decomposing leaf litter. Microb Ecol 45:11–19

Gulis V, Suberkropp K (2003b) Interactions between stream fungi and bacteria associated with decomposing leaf litter at different levels of nutrient availability. Aquat Microb Ecol 30:149–157

Gulis V, Suberkropp K (2003c) Leaf litter decomposition and microbial activity in nutrient-enriched and unaltered reaches of a headwater stream. Freshw Biol 48:123–134

Gulis V, Suberkropp K (2004) Effects of whole-stream nutrient enrichment on the concentration and abundance of aquatic hyphomycete conidia in transport. Mycologia 96:57–65

Gulis V, Rosemond AD, Suberkropp K et al (2004) Effects of nutrient enrichment on the decomposition of wood and associated microbial activity in streams. Freshw Biol 49:1437–1447

Gulis V, Marvanová L, Descals E (2005) An illustrated key to the common temperate species of aquatic hyphomycetes. In: Graça MAS, Bärlocher F, Gessner MO (eds) Methods to study litter decomposition: a practical guide. Springer, Dordrecht, pp 153–167

Gulis V, Ferreira V, Graça MAS (2006) Stimulation of leaf litter decomposition and associated fungi and invertebrates by moderate eutrophication: implications for stream assessment. Freshw Biol 51:1655–1669

Gulis V, Suberkropp K, Rosemond AD (2008) Comparison of fungal activities on wood and leaf litter in unaltered and nutrient-enriched headwater streams. Appl Environ Microbiol 74:1094–1101

Gulis V, Kuehn KA, Suberkropp K (2009) Fungi. In: Likens GE (ed) Encyclopedia of inland waters, vol 3. Elsevier, Oxford, pp 233–243

Gulis V, Kuehn KA, Shoettle LN et al (2017) Changes in nutrient stoichiometry, elemental homeostasis and growth rate of aquatic litter-associated fungi in response to inorganic nutrient supply. ISME J 11:2729–2739

Gutknecht J, Goodman R, Balser T (2006) Linking soil process and microbial ecology in freshwater wetland ecosystems. Plant Soil 289:17–34

Hagen E, McCluney K, Wyant K et al (2012) A meta-analysis of the effects of detritus on primary producers and consumers in marine, freshwater, and terrestrial ecosystems. Oikos 121:1507–1515

Hieber M, Gessner MO (2002) Contribution of stream detritivores, fungi, and bacteria to leaf breakdown based on biomass estimates. Ecology 83:1026–1038

Hotchkiss E, Hall R, Baker M et al (2014) Modeling priming effects on microbial consumption of dissolved organic carbon in rivers. J Geophys Res-Biogeosci 119:982–995

Jenkins CC, Suberkropp K (1995) The influence of water chemistry on the enzymatic degradation of leaves in streams. Freshw Biol 33:245–253

Kayranli B, Scholz M, Mustafa A et al (2010) Carbon storage and fluxes within freshwater wetlands: a critical review. Wetlands 30:111–124

Kominkova D, Kuehn KA, Busing N et al (2000) Microbial biomass, growth, and respiration associated with submerged litter of *Phragmites australis* decomposing in a littoral reed stand of a large lake. Aquat Microb Ecol 22:271–282

Kuehn KA (2008) The role of fungi in the decomposition of emergent wetland plants. In: Sridhar S, Bärlocher F, Hyde KD (eds) Novel techniques and ideas in mycology. Fungal Diversity Press, Hong Kong, pp 19–41

Kuehn KA (2016) Lentic and lotic habitats as templets for fungal communities: traits, adaptations, and their significance to litter decomposition within freshwater ecosystems. Fungal Ecol 19:135–154

Kuehn KA, Suberkropp K (1998a) Decomposition of standing litter of the freshwater emergent macrophyte *Juncus effusus*. Freshw Biol 40:717–727

Kuehn KA, Suberkropp K (1998b) Diel fluctuations in rates of CO_2 evolution from standing dead leaf litter of the emergent macrophyte *Juncus effusus*. Aquat Microb Ecol 14:171–182

Kuehn KA, Churchill PF, Suberkropp K (1998) Osmoregulatory responses of fungi inhabiting standing litter of the freshwater emergent macrophyte *Juncus effusus*. Appl Environ Microbiol 64:607–612

Kuehn KA, Gessner MO, Wetzel RG et al (1999) Decomposition and CO_2 evolution from standing litter of the emergent macrophyte *Erianthus giganteus*. Microb Ecol 38:50–57

Kuehn KA, Lemke MJ, Suberkropp K et al (2000) Microbial biomass and production associated with decaying leaf litter of the emergent macrophyte *Juncus effusus*. Limnol Oceanogr 45:862–870

Kuehn KA, Steiner D, Gessner MO (2004) Diel mineralization patterns of standing-dead plant litter: implications for CO_2 flux from wetlands. Ecology 85:2504–2518

Kuehn KA, Ohsowski B, Francoeur S et al (2011) Contributions of fungi to carbon flow and nutrient cycling from standing dead *Typha angustifolia* leaf litter in a temperate freshwater marsh. Limnol Oceanogr 56:529–539

Kuehn KA, Francoeur SN, Findlay RH et al (2014) Priming in the microbial landscape: periphytic algal stimulation of litter-associated microbial decomposers. Ecology 95:749–762

Kuzyakov Y (2010) Priming effects: interactions between living and dead organic matter. Soil Biol Biochem 42:1363–1371

Manning DWP, Rosemond AD, Kominoski JS et al (2015) Detrital stoichiometry as a critical nexus for the effects of streamwater nutrients on leaf litter breakdown rates. Ecology 96:2214–2224

Manning DWP, Rosemond AD, Gulis V et al (2016) Convergence of detrital stoichiometry predicts thresholds of nutrient-stimulated breakdown in streams. Ecol Appl 26:1745–1757

Methvin BR, Suberkropp K (2003) Annual production of leaf-decaying fungi in two streams. J N Am Benthol Soc 22:554–564

Mille-Lindblom C, Tranvik LJ (2003) Antagonism between bacteria and fungi on decomposing aquatic plant litter. Microb Ecol 45:173–182

Mille-Lindblom C, Fischer H, Tranvik L (2006) Antagonism between bacteria and fungi: substrate competition and a possible tradeoff between fungal growth and tolerance towards bacteria. Oikos 113:233–242

Mitsch WJ, Gosselink JG (2007) Wetlands. Wiley, New York

Moore J, Berlow E, Coleman D et al (2004) Detritus, trophic dynamics and biodiversity. Ecol Lett 7:584–600

Morrissey E, Berrier D, Neubauer S et al (2014) Using microbial communities and extracellular enzymes to link soil organic matter characteristics to greenhouse gas production in a tidal freshwater wetland. Biogeochemistry 117:473–490

Newell SY (2003) Fungal content and activities in standing-decaying leaf blades of plants of the Georgia Coastal Ecosystems research area. Aquat Microb Ecol 32:95–103

Newell SY, Fallon RD (1991) Toward a method for measuring instantaneous fungal growth rates in field samples. Ecology 72:1547–1559

Newell SY, Arsuffi TL, Fallon RD (1988) Fundamental procedures for determining ergosterol content of decaying plant material by liquid chromatography. Appl Environ Microbiol 54:1876–1879

Newell SY, Moran MA, Wicks R et al (1995) Productivities of microbial decomposers during early stages of decomposition of leaves of a freshwater sedge. Freshw Biol 34:135–148

Nikolcheva LG, Bärlocher F (2004) Taxon-specific fungal primers reveal unexpectedly high diversity during leaf decomposition in a stream. Mycol Progr 3:41–49

Nikolcheva LG, Cockshutt AM, Bärlocher F (2003) Determining diversity of freshwater fungi on decaying leaves: comparison of traditional and molecular approaches. Appl Environ Microbiol 69:2548–2554

Ohsowski BM (2008) Annual secondary production of fungal and bacterial decomposers associated with standing and benthic litter of the freshwater emergent macrophyte, *Typha angustigolia*. MS thesis, Eastern Michigan University

Pascoal C, Cassio F (2004) Contribution of fungi and bacteria to leaf litter decomposition in a polluted river. Appl Environ Microbiol 70:5266–5273

Pascoal C, Cassio F, Marcotegui A et al (2005) Role of fungi, bacteria, and invertebrates in leaf litter breakdown in a polluted river. J N Am Benthol Soc 24:784–797

Polunin NVC (1984) The decomposition of emergent macrophytes in fresh-water. Adv Ecol Res 14:115–166

Poon M, Hyde K (1998) Evidence for the vertical distribution of saprophytic fungi on senescent *Phragmites australis* culms at Mai Po marshes, Hong Kong. Bot Mar 41:285–292

Pugh GJF, Mulder JL (1971) Mycoflora associated with *Typha latifolia*. Trans Br Mycol Soc 57:273–282

Reddy KR, Delaune RD (2008) Biogeochemistry of wetlands: science and applications. CRC Press, Boca Raton

Rier ST, Kuehn KA, Francoeur SN (2007) Algal regulation of extracellular enzyme activity in stream microbial communities associated with inert substrata and detritus. J N Am Benthol Soc 26:439–449

Rier S, Shirvinski J, Kinek K (2014) In situ light and phosphorus manipulations reveal potential role of biofilm algae in enhancing enzyme-mediated decomposition of organic matter in streams. Freshw Biol 59:1039–1051

Romani AM, Fischer H, Mille-Lindblom C et al (2006) Interactions of bacteria and fungi on decomposing litter: differential extracellular enzyme activities. Ecology 87:2559–2569

Rosemond AD, Pringle CM, Ramirez A et al (2002) Landscape variation in phosphorus concentration and effects on detritus-based tropical streams. Limnol Oceanogr 47:278–289

Rothman E, Bouchard V (2007) Regulation of carbon processes by macrophyte species in a Great Lakes coastal wetland. Wetlands 27:1134–1143

Rousk J, Baath E (2007) Fungal and bacterial growth in soil with plant materials of different C/N ratios. FEMS Microbiol Ecol 62:258–267

Saccardo PA (1898) Sylloge fungorum omnium hucusque cognitorum 13, Index universalis. Fratres Borntrager, Lipsiae

Seena S, Wynberg N, Barlocher F (2008) Fungal diversity during leaf decomposition in a stream assessed through clone libraries. Fungal Divers 30:1–14

Shearer CA (1992) The role of woody debris. In: Bärlocher F (ed) The ecology of aquatic hyphomycetes. Springer, Berlin, pp 77–98

Shearer CA, Descals E, Kohlmeyer B et al (2007) Fungal biodiversity in aquatic habitats. Biodivers Conserv 16:49–67

Simon KS, Benfield EF (2001) Leaf and wood breakdown in cave streams. J N Am Benthol Soc 20:550–563

Sinsabaugh RL, Findlay S (1995) Microbial production, enzyme activity, and carbon turnover in surface sediments of the Hudson River estuary. Microb Ecol 30:127–141

Sinsabaugh RL, Follstad Shah JJ (2012) Ecoenzymatic stoichiometry and ecological theory. Ann Rev Ecol Evol Syst 43:313–343

Sinsabaugh RL, Belnap J, Follstad Shah JJ et al (2014) Extracellular enzyme kinetics scale with resource availability. Biogeochemistry 121:287–304

Sinsabaugh RL, Follstad Shah JJ, Findlay SG et al (2015) Scaling microbial biomass, metabolism and resource supply. Biogeochemistry 122:175–190

Spanhoff B, Gessner MO (2004) Slow initial decomposition and fungal colonization of pine branches in a nutrient-rich lowland stream. Can J Fish Aquat Sci 61:2007–2013

Stelzer RS, Heffernan J, Likens GE (2003) The influence of dissolved nutrients and particulate organic matter quality on microbial respiration and biomass in a forest stream. Freshw Biol 48:1925–1937

Su R (2014) Fungal contribution to carbon and nutrient cycling in a subtropical freshwater marsh. Ph.D. dissertation, University of Southern Mississippi

Su R, Lohner RN, Kuehn KA, Sinsabaugh R et al (2007) Microbial dynamics associated with decomposing *Typha angustifolia* litter in two contrasting Lake Erie coastal wetlands. Aquat Microb Ecol 46:295–307

Su R, Kuehn KA, Phipps SW (2015) Carbon and nutrient flow into decomposer fungi during standing-dead *Typha domingensis* decomposition in a subtropical freshwater marsh. Freshw Biol 60:2100–2112

Suberkropp K (1991) Relationships between growth and sporulation of aquatic hyphomycetes on decomposing leaf litter. Mycol Res 95:843–850

Suberkropp K (1992) Interactions with invertebrates. In: Bärlocher F (ed) The ecology of aquatic hyphomycetes. Springer, Berlin, pp 118–134

Suberkropp K (1995) The influence of nutrients on fungal growth, productivity, and sporulation during leaf breakdown in streams. Can J Bot 73:S1361–S1369

Suberkropp K (1997) Annual production of leaf-decaying fungi in a woodland stream. Freshw Biol 38:169–178

Suberkropp K (1998) Effect of dissolved nutrients on two aquatic hyphomycetes growing on leaf litter. Mycol Res 102:998–1002

Suberkropp K, Gessner MO (2005) Acetate incorporation into ergosterol to determine fungal growth rates and production. In: Graça MAS, Bärlocher F, Gessner MO (eds) Methods to study litter decomposition: a practical guide. Springer, Berlin, pp 197–202

Suberkropp K, Weyers H (1996) Application of fungal and bacterial production methodologies to decomposing leaves in streams. Appl Environ Microbiol 62:1610–1615

Suberkropp K, Gessner MO, Chauvet E (1993) Comparison of ATP and ergosterol as indicators of fungal biomass associated with decomposing leaves in streams. Appl Environ Microbiol 59:3367–3372

Suberkropp K, Gulis V, Rosemond AD et al (2010) Ecosystem and physiological scales of microbial responses to nutrients in a detritus-based stream: results of a 5-year continuous enrichment. Limnol Oceanogr 55:149–160

Tanaka Y (1991) Microbial decomposition of reed (*Phragmites communis*) leaves in a saline lake. Hydrobiologia 220:119–129

Tank JL, Webster JR, Benfield EF et al (1998) Effect of leaf litter exclusion on microbial enzyme activity associated with wood biofilms in streams. J N Am Benthol Soc 17:95–103

Tant CJ, Rosemond AD, First MR (2013) Stream nutrient enrichment has a greater effect on coarse than on fine benthic organic matter. Freshw Sci 32:1111–1121

Tsui CKM, Hyde KD (2003) Freshwater mycology. Fungal Diversity Press, Hong Kong

Van Ryckegem G, Verbeken A (2005a) Fungal diversity and community structure on *Phragmites australis* (Poaceae) along a salinity gradient in the Scheldt estuary (Belgium). Nova Hedwigia 80:173–197

Van Ryckegem G, Verbeken A (2005b) Fungal ecology and succession on *Phragmites australis* in a brackish tidal marsh. I. Leaf sheaths. Fungal Divers 19:157–187

Van Ryckegem G, Verbeken A (2005c) Fungal ecology and succession on *Phragmites australis* in a brackish tidal marsh. II. Stems. Fungal Divers 20:209–233

Van Ryckegem G, Van Driessche G, Van Beeumen J et al (2006) The estimated impact of fungi on nutrient dynamics during decomposition of *Phragmites australis* leaf sheaths and stems. Microb Ecol 52:564–574

Van Ryckegem G, Gessner MO, Verbeken A (2007) Fungi on leaf blades of *Phragmites australis* in a brackish tidal marsh: diversity, succession, and leaf decomposition. Microb Ecol 53:600–611

Wallander H, Ekblad A, Godbold D et al (2013) Evaluation of methods to estimate production, biomass and turnover of ectomycorrhizal mycelium in forests soils—a review. Soil Biol Biochem 57:1034–1047

Webster J (1992) Anamorph-teleomorph relationships. In: Bärlocher F (ed) The ecology of aquatic hyphomycetes. Springer, Berlin, pp 99–117

Webster JR, Benfield EF (1986) Vascular plant breakdown in freshwater ecosystems. Ann Rev Ecol Syst 17:567–594

Webster J, Descals E (1981) Morphology, distribution, and ecology of conidial fungi in freshwater habitats. In: Cole GT, Kendrick B (eds) Biology of conidial fungi, vol 1. Academic Press, New York, pp 295–355

Webster JR, Meyer JL (1997) Stream organic matter budgets. J N Am Benthol Soc 16:3–4

Weete JD, Abril M, Blackwell M (2010) Phylogentic distribution of fungal sterols. PLoS One 5: e10899

Welsch M, Yavitt J (2003) Early stages of decay of *Lythrum salicaria* L. and *Typha latifolia* L. in a standing-dead position. Aquat Bot 75:45–57

Weyers HS, Suberkropp K (1996) Fungal and bacterial production during the breakdown of yellow poplar leaves in two streams. J N Am Benthol Soc 15:408–420

Woodward G, Gessner MO, Giller P et al (2012) Continental-scale effects of nutrient pollution on stream ecosystem functioning. Science 336:1438–1440

Wurzbacher C, Bärlocher F, Grossart H (2010) Fungi in lake ecosystems. Aquat Microb Ecol 59:125–149

Wurzbacher C, Rösel S, Rychla A et al (2014) Importance of saprotrophic freshwater fungi for pollen degradation. PLoS One 9:e94643

Zare-Maivan H, Shearer CA (1988) Extracellular enzyme production and cell wall degradation by freshwater lignicolous fungi. Mycologia 80:365–375

Chapter 6
The Ecology of Methanogenic Archaea in a Nutrient-Impacted Wetland

Andrew Ogram, Hee-Sung Bae, and Ashvini Chauhan

Abstract Wetlands are the largest natural sources of methane, and many wetlands are subject to nutrient enrichment due to runoff from adjacent agricultural and urban lands. Methanogenic archaea are responsible for much of the methane produced in terrestrial wetlands and participate in a range of additional activities including nitrogen fixation and mercury methylation. Nutrient enrichment may impact the dominant metabolic groups of methanogens, such that the fundamental activities of methanogens may be associated with the nutrient status of the wetland. Regions of the Everglades, a large marsh in the southern part of Florida, in the USA, are subject to nutrient enrichment and are characterized by a gradient in available phosphorus, organic carbon, and sulfate concentrations. This marsh provides an outstanding system in which to study the impacts of nutrient enrichment on the distribution and activities of methanogens. Competition for acetate with sulfate-reducing prokaryotes plays an important role in structuring methanogenic consortia in nutrient-impacted regions, and the potentials for methanogenic nitrogen fixation and mercury methylation differ along the nutrient gradient.

6.1 Introduction

6.1.1 Background

Terrestrial wetlands are the largest natural source of the greenhouse gas methane and are responsible for the production of over 150 Tg of methane per year or approximately 20% of the total global methane budget (van Amstel 2012). Many of these wetlands are subject to nutrient enrichment resulting from runoff from agricultural or

A. Ogram (✉) · H.-S. Bae
Soil and Water Sciences Department, University of Florida, Gainesville, FL, USA
e-mail: aogram@ufl.edu; hsbae@ufl.edu

A. Chauhan
School of the Environment, Florida A&M University, Tallahassee, FL, USA
e-mail: ashivini.chauhan@famu.edu

C. J. Hurst (ed.), *The Structure and Function of Aquatic Microbial Communities*, Advances in Environmental Microbiology 7,
https://doi.org/10.1007/978-3-030-16775-2_6

urban systems. This nutrient enrichment may impact a range of ecosystem processes, including methane production. This review will use regions of the Florida Everglades, a large freshwater wetland in the Southeastern USA, as a model system to explore the impacts of nutrient enrichment on the microbial ecology of methane production and processes related to methane production, such as fermentation, sulfate reduction, and mercury cycling. Only the microbial ecology of methanogenesis, and not methane consumption or total methane flux, will be considered in this review. We also focus on the ecology of methanogens in peat within the Everglades, and not in other environmental compartments where methanogens may be active, such as the flocculent material and floating periphyton.

Most biological methane is the product of a strictly anaerobic metabolism conducted exclusively by archaea that are only capable of metabolizing relatively simple electron donors. However, the relationships that methanogenic archaea have with their environment and with other microorganisms can be quite complex. Methanogenic archaea respond to a variety of environmental conditions, including nutrient and electron donor availabilities and redox potentials. They are largely dependent on other microorganisms to provide electron donors and are typically outcompeted for those electron donors in many other freshwater systems by more efficient metabolic processes such as nitrate, iron, and sulfate reduction. Because the archaea are dependent on other microbial groups for electron donors, the fundamental microbial ecology of methane production is not only dependent on numerous environmental factors that affect the rates of methane production but is also dependent on a variety of fundamental interactions, including those between different functional groups at the cellular level, the nutrient limitations that control those interactions, and the pathways through which methane is produced. The Everglades provides a particularly good system for studying the relationships between nutrient availability, alternative terminal electron acceptors such as sulfate, and the ecology of methanogenic archaea.

6.1.2 The Everglades

The modern Everglades is comprised of 8000 km^2 (not including the upper Everglades that includes the Kissimmee River) of marshes and swamps that cover much of the southern tip of the state of Florida. The Everglades provides an array of ecological services, including habitat for numerous endangered species and migratory birds, and recreation and water for the growing human populations of South Florida. Approximately 3000 km^2 south of Lake Okeechobee were drained in 1948 to form the muck farms of the Everglades Agricultural Area, which are responsible for much of the domestic sugarcane crop. The Everglades was historically very low in available phosphorus due to the limestone substrata, which also maintained a relatively high pH (between 7 and 8). Fertilizers containing phosphate were applied to the land, as was elemental sulfur (S) to modulate the pH. Elemental S is converted to sulfate through the actions of sulfur oxidizing bacteria, which lowers the

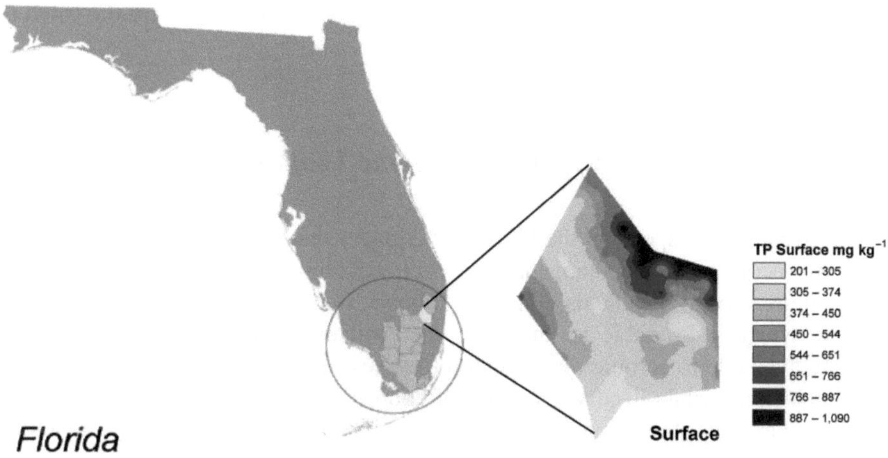

Fig. 6.1 Map showing location of the Everglades in the state of Florida, with Water Conservation Area 2-A identified and total soil P concentrations

pH. Canals drain the land to maintain crop production, which in turn carried P and sulfate from runoff into the Everglades through the Water Conservation Areas (WCAs, see Fig. 6.1). Gradients in the concentrations of P in water and soil were thereby established, with soil P concentrations ranging from 1.28 g per kg in the northern part of Water Conservation Area 2A (WCA-2A) to approximately 0.25 g per kg further south in WCA-2A (Bae et al. 2015).

The release of elevated concentrations of P into the northern Everglades effectively fertilized the originally very low P soils, leading to a shift in the dominant vegetation from sparse stands of sawgrass (*Cladium* sp.) to dense cattails (*Typha* sp.). Greater primary productivity accelerated peat development adjacent to the canals. Microbial activities are elevated in the P-impacted soils and include significantly higher production of methane and CO_2 (Bae et al. 2015), higher rates of nitrogen fixation (Inglett et al. 2011; Bae et al. 2018), and increased activities of many extracellular enzymes that are associated with microbial carbon and nitrogen acquisition than typically would be observed in soils that are not impacted by P (Penton and Newman 2007). The microbial communities of the P-impacted peat predictably exhibit characteristics of P-excess and mildly N-limited systems, whereas microbial communities in soils further south in the less P-impacted exhibit characteristics of highly P-limited systems (Morrison et al. 2016). A shallow gradient in surface water sulfate concentrations is also present in the WCAs, with concentrations ranging from 56 mM observed in the P-impacted regions of WCA-2A, down to less than 4 mM in regions of WCA-3A (Bae et al. 2015).

The amounts of P and sulfate entering the Everglades have been significantly reduced in recent years; however, there is sufficient P and sulfate stored in the peat to maintain significant internal cycling of these nutrients, such that these regions remain impacted by the elevated nutrients. Both P and sulfate are being gradually redistributed, however, with P and sulfate moving down stream toward the south of the system.

6.1.3 Decomposition in Methanogenic Peatlands

There are notable aspects of the Florida Everglades that impact methane production and the ecology of methanogens; however, decomposition of plant material in methane-producing peatlands generally follows a common trajectory. The energy that is typically available to anaerobic metabolism is significantly less than the energy available to aerobic metabolism processes, such that more complex organic molecules are metabolized via a series of coupled metabolisms. Decomposition of organic carbon is slow in anoxic peat systems due to a number of factors, including lack of terminal electron acceptors and the "peatland enzymic latch" (Freeman et al. 2001). The enzymic latch refers to the apparent importance of phenol oxidase, an enzyme that requires O_2, to decompose phenolics that would be inhibitory to extracellular hydrolases required for increasing the availability of N and P, which are necessary for organic matter decomposition. Peat therefore accumulates, with most of the available carbon coming from decomposition products from the aerobic zone above and the slow decomposition of polymers at greater depths. The general flow of carbon and electrons in a methanogenic peat accumulating wetland such as the Everglades is presented in Fig. 6.2, in which moribund and dead plant material as standing biomass is colonized by saprophytic fungi. Decomposition then proceeds through utilization of easily metabolized compounds and depolymerization of larger molecules such as starches, cellulose, and lignin. The primary genus responsible for depolymerization of cellulose to glucose and fermentation of those products in anoxic regions of the Everglades appear to belong to the *Firmicutes* and are represented by diverse members of the genus *Clostridium* (Uz et al. 2006). The dominant clades within *Clostridium* shift along the gradient, likely due to changes in carbon quality (Uz and Ogram 2006) as the dominant emergent vegetation shifts from cattail in the nutrient-impacted area to sawgrass in the nonimpacted area.

Fermentation of carbohydrates typically results in production of a range of products, including H_2 and small organic acids such as propionate, butyrate, acetate, and lactate. In the Everglades, as in many other wetlands, propionate is a dominant fermentation product (Bae et al. 2015). Propionate and butyrate contain significant energy, such that their fermentation is sufficient to provide energy for another group of microorganisms that function as secondary fermenters. These secondary fermenters are frequently referred to as syntrophs due to the requirement that their metabolisms be closely coupled with another group, such as methanogens, to successfully scavenge either H_2 or formate via interspecies electron transfer. The energy required to regenerate oxidized dinucleotides such as FADH requires that H_2 be maintained in very low concentrations (Schink 1997), such that H_2 must be removed quickly from the immediate vicinity of the syntroph. The great majority of syntrophs described to date are also capable of gaining energy via sulfate reduction, such that metabolism of

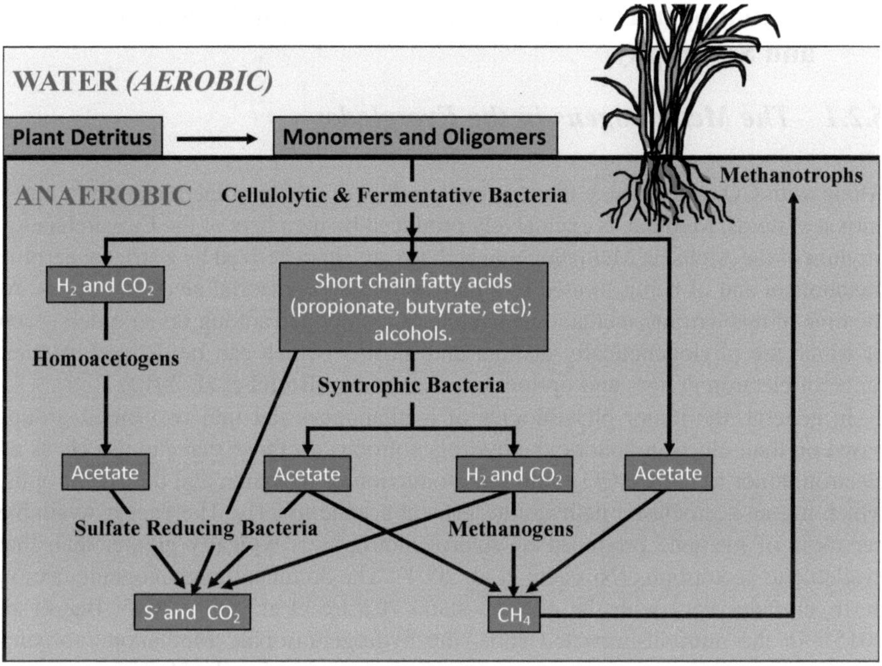

Fig. 6.2 The general pathway for decomposition of organic matter to methane in terrestrial wetlands

fermentation products by syntrophic interactions may be bypassed when sufficient sulfate is available. In addition to H_2, acetate and formate are common secondary fermentation products. In a typical methanogenic system, methanogenic archaea use H_2 either to reduce available CO_2 to methane or ferment acetate to methane in order to gain energy (Conrad 1999). While it is not known at this time how much of the methane produced is dependent on syntrophy, it is clear that both the physiologies of syntrophs and their methanogenic partners are intimately linked and that the relationships may be complicated by the presence of sulfate.

The physiologies and ecologies of methanogenic archaea in the context of the Everglades will be discussed in the sections below, with potential implications of the interactions between syntrophs and methanogens with respect to mercury cycling included.

6.2 Methanogenic Archaea, Sulfate-Reducing Prokaryotes, and Syntrophy

6.2.1 The Methanogens in the Everglades

Along with CO_2, methane is the terminal product of carbon mineralization in many anoxic systems. Methane is exclusively produced by members of the Euryarchaeota, phylum of the Archaea. Methanogenic archaea are characterized by a strict anaerobic metabolism and of being limited to a narrow range of potential electron donors. At the time of this writing, methanogenic archaea are divided among seven orders, each of which are phylogenetically distinct and most of which can be defined by their range of electron donors and optimum temperatures (Borrel et al. 2013).

In general, the major physiologies of methanogens fall into two broad groups based on their electron donors: the hydrogenotrophs, or those that can use H_2 as an electron donor to reduce CO_2 with the production of methane, and the acetotrophs, which use an acetoclastic pathway to ferment acetate to CH_4. The energy available per mole of methane produced by hydrogenotrophs is typically greater than that available to acetotrophs (Nüsslein et al. 2001). The dominant methanogenic taxa in the Everglades vary with the nutrient status (Castro et al. 2004, 2005; Bae et al. 2015). In the nutrient-impacted soils, the hydrogenotrophic Methanomicrobiales were consistently more numerous than the acetotrophic family Methanosaetaceae during 3 years of monitoring (Bae et al. 2015). Conversely, members of the Methanosaetaceae outnumbered the Methanomicrobiales during the same period. The redundancy analysis (RDA) presented in Fig. 6.2 indicates that numbers of Methanomicrobiales are inversely correlated with Methanosaetaceae and less strongly correlated with sulfate and total phosphorus concentrations. The differences in dominant methanogenic order and metabolism (hydrogenotrophic vs acetoclastic) suggest that the dominant pathway for methane production may also shift along the nutrient gradient, with a greater proportion of methane produced through the hydrogenotrophic pathway in the nutrient-impacted soils and less in the nonimpacted soils (Bae et al. 2015). This is supported by $\delta^{13}CH_4$ analysis of methane emitted from soils along the gradient, which indicated a shift from approximately 50% hydrogenotrophic in the nutrient-impacted soils to approximately 23% hydrogenotrophic in the nonimpacted soils (Holmes et al. 2014).

On a global basis, most terrestrial methane is thought to arise from acetate via acetoclastic methanogenesis. Stoichiometric conversion of carbohydrates to acetate and H_2 via primary fermentation (Fig. 6.2; fermentation of sugars to fermentation of fatty acids and alcohols) and secondary fermentation (a fermentation of primary fermentation products, such as fatty acids and alcohols) can only supply sufficient H_2 for reducing CO_2 to CH_4 to account for 1/3 of the total methane produced (Conrad 1999). Thus, it has generally been observed that acetate fermentation accounts for approximately 70% of the methane produced in terrestrial environments (Conrad 1999). A few notable exceptions to this paradigm have been described (Conrad 1999; Nüsslein et al. 2003), many of which favor the hydrogenotrophic pathway. As discussed above, the dominant pathway in the Everglades differs between the nutrient-impacted and nutrient-nonimpacted soils. As will be discussed below, the observed shift in methanogenic pathway is due to competition between methanogens and sulfate-reducing prokaryotes (SRPs) in the nutrient-impacted area.

6.2.2 *Interactions Between Methanogens and SRP Along the Nutrient Gradient*

The SRPs are distributed across a broad phylogenetic range of prokaryotes that spans domains and they exhibit both diverse metabolisms and diverse ecological roles (Castro et al. 2000). Sulfate concentrations at our study sites vary along the nutrient gradient, with sulfate concentrations consistently greater than 39 μM being observed in the nutrient-impacted soils and less than 4 μM in the nonimpacted soils (Bae et al. 2015). In a preliminary study utilizing terminal restriction fragment length polymorphism (TRFLP) analysis of the gene encoding dissimilatory bisulfite reductase (*dsrB*), a gene that is frequently used a phylogenetic marker of SRP, Castro et al. (2005) estimated that the dominant SRPs in WCA-2A soils are associated with the genus *Desulfotomaculum*. Members of this genus are capable of forming endospores, a trait that would be beneficial for strict anaerobes inhabiting soils with fluctuating redox conditions (Castro et al. 2000, 2005). The water levels in WCA-2A drop below soil surface each spring, and the surface soils shift from anoxic to oxic.

Usage of TRFLP analysis has methodological limitations (Prakash et al. 2014) in its ability to resolve closely related phylogenetic groups. However, Castro et al. (2005) noted that dominant *dsrB* TRFs consistently differed between the nutrient-impacted and nutrient-nonimpacted soils in monthly samplings conducted over the course of 1 year. The nature of SRP metabolism can be roughly divided into two types based on the ability to metabolize acetate as an electron donor and source of carbon, complete versus incomplete lactate oxidizers (Castro et al. 2000). Complete oxidizers are capable of metabolizing lactate to CO_2, while incomplete oxidizers metabolize lactate to acetate. Incomplete oxidizers are not capable of utilizing acetate as an electron donor. The genus *Desulfotomaculum* includes representatives of both complete and incomplete oxidizers. Significantly, Castro et al. (2005) reported that the majority TRFs associated with *Desulfotomaculum* sp. in nutrient-impacted soils are characterized by a predicted complete lactate oxidation metabolism, while predicted incomplete lactate oxidation dominated the *Desulfotomaculum* TRFs in the lower sulfate, nonimpacted soils. This suggested that SRP may compete with methanogens for acetate in the impacted soils and to a lesser degree in the nonimpacted soils. Potential competition between methanogens and SRP for acetate and the implications of this competition are inferred from the redundancy analysis (RDA) plot presented in Fig. 6.3, where the proportions of hydrogenotrophic Methanomicrobiales (MM) in WCA-2A soils are inversely

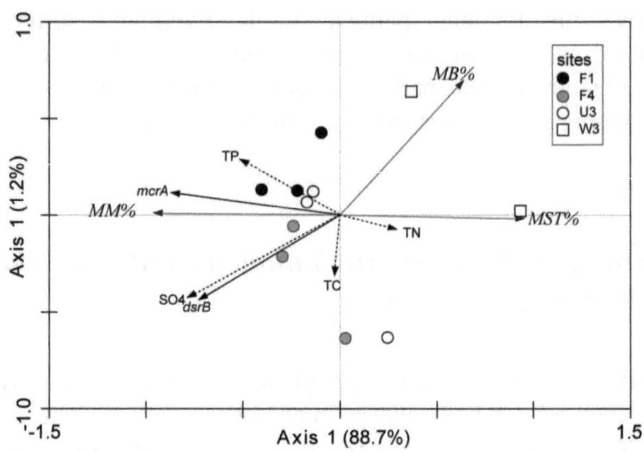

Fig. 6.3 Redundancy analysis (RDA) representing the correlation among *mcrA* and *dsrB* copies with geochemical parameters obtained from along the nutrient gradient. Arrows pointing in the same direction indicate positive correlations and arrows pointing in opposite directions indicate negative correlations. Arrow length corresponds to variance explained by the environmental variable. The first two axes explain 88.3% of the total canonical eigenvalues, with a significant Monte-Carlo test value ($P < 0.05$). From Bae et al. (2015)

correlated with the acetotrophic Methanosaetacae (MST) and weakly correlated with the proportions of SRP (represented by *dsrB*) and sulfate concentrations.

To assess whether competition for acetate between SRP and methanogens may control the dominant methanogenic pathways in the nutrient-impacted and nutrient-nonimpacted soils, laboratory incubations of soil from the impacted area with the sulfate reduction inhibitor MoO_4^{2-} indicated a shift in the $\delta^{13}CH_4$ toward the acetoclastic pathway and an increase in the mRNA transcribed from the acetotrophic family *Methanosaetaceae* (Bae et al. 2015). These results indicate that competition with SRP for acetate shifted the major pathway away from the acetoclastic pathway and toward the hydrogenotrophic pathway in the higher sulfate, nutrient-impacted soils.

In addition to competition, many SRP may cooperate with methanogens as syntrophs. Some SRP may shift from anaerobic respiration using sulfate as a terminal electron acceptor to a fermentative metabolism. Bae et al. (2014) found that the majority of *dsrB* transcripts clustered within the *Deltaproteobacteria* and were associated with SRP capable of fermentative metabolism and of syntrophy with methanogens. Depending on environmental factors such as carbon to sulfate ratios, SRP may either compete with or cooperate with methanogens.

6.2.3 Syntrophs in the Everglades

The term "syntrophy" has been used since the mid-twentieth century to describe a mutually beneficial relationship in anoxic systems, frequently (although not exclusively) between secondary fermentative bacteria (syntrophs) and their methanogenic archaeal partners (methanogens). The syntrophy allows the metabolic conversion of volatile fatty acids (VFAs) such as butyrate, propionate, and acetate into methane (Schink 1997). Thermodynamically, such conversions are not feasible because ΔG for the reaction is endergonic, but microorganisms facilitate such endergonic reactions by engaging in interspecies electron transfer (IET) (Schink 1997) to regenerate oxidized dinucleotides. The syntrophic members typically cluster with *Deltaproteobacteria* (e.g., *Syntrophus*, *Desulfovibrio*) or *Firmicutes* (e.g., *Syntrophomonas*, *Desulfotomaculum*) (McInerney et al. 2008). Notably, the deltaproteobacterial syntrophs are typically obligate anaerobes that ferment VFAs when grown in co-culture with a partner; however, if sulfate is bioavailable, many deltaproteobacterial syntrophs (e.g., *Desulfovibrionales* and *Desulfuromonadales*) may utilize VFAs. Representatives of syntrophic *Clostridia* and *Bacilli* may be either strict or facultative anaerobes; many are spore-formers and include typical syntrophic genera such as *Syntrophomonas* and *Pelotomaculum* (Morris et al. 2013). The prokaryotic groups that act as sinks to consume the by-products of syntrophic metabolism include the hydrogenotrophic methanogenic Euryarchaeota.

The compositions and architectures of the syntroph-methanogen partnerships are characteristic of the nutrient status of the site. Significant differences (*P*-values < 0.001) were observed by our research group in the phylogenies of syntrophs and archaea between microcosms with soil from nutrient-impacted, transitional, and nutrient-nonimpacted sites. Microcosms with soils from the different respective nutrient regions were spiked with different VFAs and then observed with regard to methane production, microscopic visualizations of the resultant syntrophic-methanogenic consortia, most probable number estimations of microbial population density, and ^{13}C-DNA-based stable isotope probing (SIP) to investigate the pathways of VFA metabolism and the dominant syntrophs. Most propionate-utilizing syntrophs in microcosms with soil from the nutrient-impacted and transition sites clustered with *Pelotomaculum* spp. and *Syntrophobacter* spp. and for butyrate, either *Deltaproteobacteria* or *Firmicutes* dominated, with *Syntrophospora* spp. and *Syntrophomonas* spp. (Chauhan and Ogram 2006). In the nonimpacted soils, *Pelospora* spp. and SRP dominated. Also noted in the nutrient-impacted and transition soils were 16S rRNA gene sequences clustering with the genus *Smithella*, members of which may convert propionate to butyrate. The mechanism used by *Smithella propionica* is to dismutate propionate to acetate and butyrate, followed by β-oxidation (de Bok et al. 2001; Liu et al. 1999). In addition to the typical syntrophic guilds, SRPs also constitute a significant proportion of the total relative phylotype abundance within the impacted soils. This situation was quite different in the nonimpacted microcosms, where SIP did not identify many known syntrophic

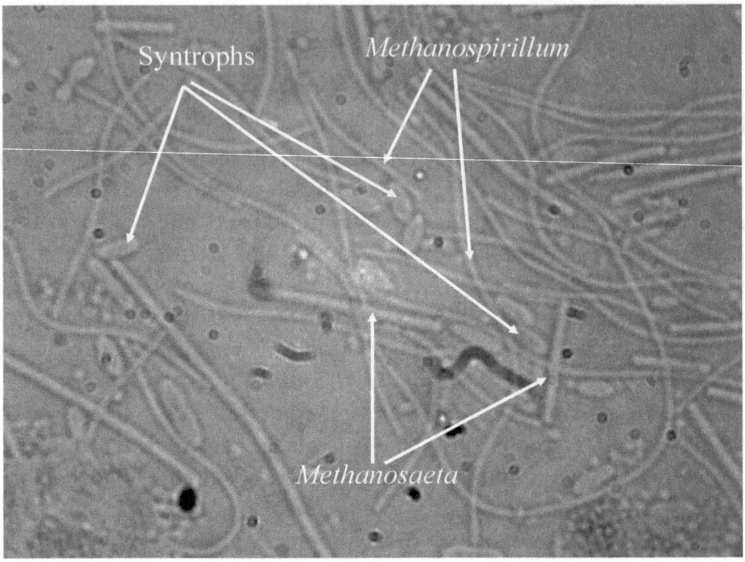

Fig. 6.4 Photomicrograph of syntroph-methanogen consortium isolated from enrichments of soil taken from the nutrient-impacted regions of WCA-2A. From Chauhan et al. (2004)

bacteria; however, 16S rRNA gene sequences clustered with various SRP and *Pelobacter* spp.

In lab-controlled microcosms spiked with VFAs, we observed an interesting feature: the presence of "tripartite" rather than a "bipartite" microbial syntrophic association which included a syntroph, hydrogenotrophic methanogen, and an acetotrophic methanogen (Fig. 6.4). The nest-like structures characteristic of these tripartite associations were only observed in enrichments of soil from the nutrient-impacted and transition soils and not in the nonimpacted soils (Chauhan et al. 2004). Similar associations have been reported in the literature and were primarily associated with enrichments in which VFAs were more rapidly consumed by tricultures of *Methanospirillum hungatei* (hydrogenotroph), *Methanothrix soehngenii* (acetate utilizer), and MPOB (mesophilic propionate-oxidizing bacteria) relative to bicultures consisting of *Methanospirillum* with MPOB (Voolapalli and Stuckey 1999), suggesting that the low acetate and H_2 concentrations were favorable for syntrophy in many systems. It is very likely that the dramatic increase in the methane production pathways and rates observed in the impacted Florida Everglades soils (Castro et al. 2004, 2005; Chauhan et al. 2004; Ogram et al., unpublished) occur because the microbiota in these systems have adapted and evolved into ecologically beneficial "tripartite" associations in nutrient-impacted soils. As noted above, estimations for methanogen numbers via quantitative PCR (qPCR) enumerations revealed that hydrogenotrophic methanogen populations numbers were significantly greater than those of acetotrophs, especially in the P-impacted soils (Chauhan et al.

2004; Castro et al. 2004, 2005; Bae et al. 2015), adding support to the conclusion that syntrophy may be a major route for methane production in the Everglades.

Sulfate concentrations fluctuate significantly in the Everglades (Bae et al. 2015), with porewater concentrations ranging from 39 to 56 μM determined in three samplings taken over 3 years. It is well-documented that SRP may outcompete methanogenic archaea for H_2 and acetate (Lovley et al. 1982; Schönheit et al. 1982; Robinson and Tiedje 1984), provided that sufficient sulfate is available. Little is currently known of the potential impact of fluctuating sulfate concentrations on syntrophy. As the level of bioavailable sulfate increases within a given habitat, we must ask "Does the syntrophic-methanogenic relationship breaks apart such that the syntrophs switch to sulfate reduction?" Most known syntrophs are capable of gaining energy via sulfate reduction metabolism, and our study has found that *Deltaproteobacteria* associated with the genus *Syntrophobacter* dominated in microcosms from the nutrient-impacted and transition soils and *Deltaproteobacteria* have been shown to metabolically perform dissimilatory sulfate reduction to produce energy. Evidence that members of this genus prefer to release the resultant reducing equivalents toward their syntrophic methanogenic partners or other SRPs has been reported (Wallrabenstein et al. 1995; Harmsen et al. 1998). Currently, it is unknown in what way fluctuating sulfate concentrations may impact the structures and functions of syntrophic consortia in the Everglades.

6.3 Associated Activities

6.3.1 Methanogenic Nitrogen Fixation

Elemental cycles are typically coupled, and the intersection points of cycles of environmental interest may interrelate within the metabolisms of individual populations. In addition to their importance in the production of methane, methanogenic archaea are also important contributors to the cycling of nitrogen in Everglades peat (Bae et al. 2018). The ability of some methanogens to fix N_2 has been recognized for over 30 years (Murray and Zinder 1984; Belay et al. 1984); however, the importance and potential for methanogenic nitrogen fixation to occur in wetlands has been understudied.

During peatland development, relatively young systems are typically limited in N, while N is put into the system by N-fixation, atmospheric deposition, and other sources (Anderson 1964). With time, a shift in nutrient limitation is frequently observed from N to P, such that older peatlands are typically limited by available P. Nitrogen limitation can also be imposed on peatlands through anthropogenic input of P, as has been observed in WCA-2A. In WCA-2A, nutrient runoff shifted the primary nutrient limiting primary productivity away from P and back to N (Corstanje et al. 2007). As expected, N-fixation rates are highest in the high P regions of WCA-2A peat and lower down the N-P limitation gradient (Inglett et al. 2011).

The highest rates of N-fixation in the Everglades are likely due to the activities of aerobic bacteria such as cyanobacteria in the water column and periphyton. The nitrogen fixed by these organisms is probably cycled within the periphyton and flocculent layer above the soil and within the top few centimeters of the peat, with little available nitrogen moving deeper into the peat where light is low, redox potentials drop dramatically, and anaerobic metabolisms such as methanogenesis dominate (White and Reddy 2003). Much of the available nitrogen at these depths is provided from the slow decomposition of organic matter and N-fixation by anaerobes (Inglett et al. 2011), with methanogens being among the dominant N-fixers at the Everglades peat (Bae et al. 2018). Sequence analysis of *nifH*, the gene encoding dinitrogenase reductase (critical to the nitrogen fixation process), showed that between 6 and 44% of the *nifH* copies at the 0–4 cm depth belonged to methanogens. It should be noted that these samples included surficial peat, such that many of the *nifH* copies originated from cyanobacteria or other aerobes that would not be present below the area where light could penetrate. Estimates of total copy numbers of *nifH* per gram of peat ranged from 5.5×10^8 to 1.9×10^{10} copies per gram of dry soil. The relative activities of methanogenic N-fixers in WCA-2A are inferred by the proportion of mRNA transcribed from methanogen *nifH*; up to 49% of the total *nifH* mRNA from these samples originated from methanogens, indicating that a significant amount of the N fixed in WCA-2A peat is due to the action of methanogenic Archaea. It is not known at this time how the nutrients that limit primary production in WCA-2A influence the rates of methanogenic nitrogen fixation in peat, and more work is required to fully understand the impacts of methanogenic N-fixation on methane production and peat decomposition along the nutrient gradient.

6.3.2 Mercury Methylation by Methanogens

Many wetlands, including the Everglades, are subject to atmospheric deposition of inorganic mercury, primarily in the form of Hg^{2+}. Once in anoxic environments, Hg^{2+} may be methylated to more toxic CH_3Hg^+, or methylmercury, by diverse anaerobic prokaryotes. Methylmercury is of particular concern because of its tendency to bioaccumulate through the food chain, potentially resulting in neurological damage and developmental problems. Fish in some areas of the Everglades are banned from human consumption due to methylmercury concentrations, and methylmercury has also been implicated in changes in the mating habits of wading birds that have led to declines in their numbers (Frederick and Jayasena 2011).

The SRPs were initially identified as being responsible for the majority of mercury methylation in the Everglades, with significant differences in methylation rates observed along the sulfate and P gradients in the WCAs (Gilmour et al. 1998). The known diversity of prokaryotic groups capable of mercury methylation was greatly expanded with the identification of the genes responsible for methylation, *hgcAB* (Parks et al. 2013; Gilmour et al. 2013). These genes encode a corrinoid protein that is involved in transferring methyl carbanions from one substrate to another. It is likely that mercury methylation is a gratuitous reaction that occurs during cellular metabolism and is not specifically directed toward mercury. All organisms known to carry *hgcAB* have been shown to methylate mercury (Gilmour et al. 2013), such that *hgcAB* may be used as a genetic marker for screening samples for the presence and phylogeny of mercury methylators. This work has led to identification of a great diversity of microbial groups representing different metabolic strategies as mercury methylators and include fermenters, syntrophs, and methanogens, as well as SRB.

Bae et al. (2014) reported that the dominant groups harboring *hgcAB* in the soils of the WCAs were affiliated with the deltaproteobacterial syntrophs (which may also function as sulfate reducers when in the presence of sufficient sulfate), *Firmicutes*, and methanogens. The relative proportions of methanogens ranged from 8% in the nutrient and sulfate-enriched soils of the northern WCA-2A to 27% in the low-sulfate soils of WCA-3A and include both acetotrophs (*Methanosarcinaceae*) and hydrogenotrophs (*Methanomicrobiales*). Subsequent investigations have found that the relative proportions of methanogenic *hgcAB* to other groups increase down the nutrient and sulfate gradients and may form a very significant proportion of mercury methylators in periphyton and floc in low-sulfate environments (Bae et al., manuscript in preparation). The rates of methanogenic methylation compared to methylation by sulfate reduction and other processes in the Everglades are not known at this time. However, methanogens were shown to be the dominant methylators in periphyton at a site in Canada (Hamelin et al. 2011), and mercury methylation by *hgcAB*-carrying syntrophic-methanogenic consortia may be significant. It is likely that the quality and availability of carbon, in addition to the range of available electron acceptors available to competitors of methanogens, will control the dominant processes involved in mercury methylation in this complex Everglades ecosystem.

6.4 Summary

Much of the microbial ecology, and hence biogeochemical cycling, in the Water Conservation Areas (WCAs) of the Everglades is impacted by the input of nutrients from the adjacent Everglades Agricultural Area (EAA). The Everglades was historically a very low-nutrient ecosystem, and input of nutrients (notably P) resulting from runoff from the adjacent EAA for many years significantly impacted a range of ecosystem processes in the Everglades. Notably, primary productivity increased dramatically, resulting in higher amounts of organic carbon put into the soils. In addition, sulfate runoff from the EAA resulted in a shallow gradient in sulfate concentrations and stimulated carbon mineralization through sulfate reduction in those areas. This in turn impacted the fundamental ecology of methanogens, which compete with sulfate-reducing prokaryotes (SRPs) for electron donors. Notably, SRP and methanogens compete for acetate in the nutrient and sulfate impacted soils of the northern WCA-2A, which increases the proportion of hydrotrophic methanogenesis relative to the typically more dominant acetoclastic pathway (Bae et al. 2015; Holmes et al. 2014).

Much of the carbon in the Everglades, as in most peatlands, is likely to be processed through methanogenic consortia involving secondary fermenters, or syntrophs, and hydrogenotrophic methanogens. The architecture of the syntroph-methanogen consortia involves three members: the syntroph, a hydrogenotrophic methanogen, and an acetotrophic methanogen. These consortia differ from those found in the nonimpacted soils, which include only the syntroph and a hydrogenotrophic methanogen.

In addition to methane production, methanogens are responsible for other significant processes, including nitrogen fixation and mercury methylation. The phylogeny of the dominant nitrogen-fixing methanogens in the WCAs falls primarily within the hydrogenotrophs, suggesting that the distribution of hydrogenotrophs and competition for electron donors along the nutrient gradient may impact methanogenic nitrogen fixation rates. The potential relationship between N-limitation of general microbial activities (Corstanje et al. 2007) and the rates at which methanogens fix N is unknown at this time, as is the potential importance of methanogenic N-fixation to the general microbial community. Methanogens may also play an important role in the methylation of mercury in the Everglades, and the relative proportions of methanogenic mercury methylators increases with decreasing sulfate inputs, which would promote mercury methylation by SRP.

In summary, the distributions and activities of methanogens are impacted by nutrient additions in the Everglades, much of which can be explained by the impacts that the availabilities of electron donors have on dominant metabolic groups of methanogens.

Compliance with Ethical Standards

Conflict of Interest Andrew Ogram declares that he has no conflicts of interest. Hee-Sung Bae declares that he/she has no conflicts of interest. Ashvini Chauhan declares that he/she has no conflicts of interest.

Ethical Approval This article does not contain any studies with human participants or animals performed by any of the authors.

References

Anderson JAR (1964) The structure and development of peat swamps of Sarawak and Brunei. J Trop Geogr 18:7–16

Bae HS, Holmes ME, Chanton JP et al (2015) Distribution, activities, and interactions of methanogens and sulfate-reducing prokaryotes in the Florida Everglades. Appl Environ Microbiol 81:7431–7442

Bae HS, Morrison E, Chanton JP, Ogram A (2018) Methanogens are major contributors to nitrogen fixation in soils of the Florida Everglades. Appl Environ Microbiol:e02222-17

Bae HS, Dierberg FE, Ogram A (2014) Syntrophs dominate sequences associated with the mercury methylation-related gene *hgcA* in the water conservation areas of the Florida Everglades. Appl Env Microbiol 20:6517–6526

Belay N, Sparling R, Daniels L (1984) Dinitrogen fixation by a thermophilic methanogenic bacterium. Nature 312:286–288

Borrel G, O'Toole PW, Harris HM et al (2013) Phylogenomic data support a seventh order of methylotrophic methanogens and provide insights into the evolution of methanogenesis. Genome Biol Evol 5:1769–1780

Castro HF, Newman S, Reddy KR, Ogram A (2005) Distribution and stability of sulfate reducing prokaryotic and hydrogenotrophic methanogenic assemblages in nutrient-impacted regions of the Florida Everglades. Appl Environ Microbiol 71:2695–2704

Castro HF, Ogram A, Reddy KR (2004) Phylogenetic characterization of methanogenic assemblages in eutrophic and oligotrophic areas of the Florida Everglades. Appl Environ Microbiol 70:6559–6568

Castro HF, Williams N, Ogram A (2000) Phylogeny of sulfate reducing bacteria. FEMS Microbiol Ecol 31:1–9

Chauhan A, Ogram A (2006) Stable isotope probing of fatty acid oxidizing guilds in the Florida Everglades. Appl Environ Microbiol 72:2400–2406

Chauhan A, Ogram A, Reddy KR (2004) Syntrophic-methanogenic associations along a nutrient gradient in the Florida Everglades. Appl Environ Microbiol 70:3475–3484

Conrad R (1999) Contribution of hydrogen to methane production and control of hydrogen concentrations in methanogenic soils and sediments. FEMS Microbiol Ecol 28:193–202

Corstanje R, Reddy KR, Prenger J et al (2007) Soil microbial eco-physiological response to nutrient enrichment in a sub-tropical wetland. Ecol Indic 7:277–289

de Bok FA, Stams A, Dijkema C, Boone DR (2001) Pathway of propionate oxidation by a syntrophic culture of *Smithella propionica* and *Methanospirillum hungatei*. Appl Environ Microbiol 67:1800–1804

Frederick P, Jayasena N (2011) Altered pairing behavior and reproductive success in white ibises exposed to environmentally relevant concentrations of methylmercury. Proc R Soc B 278: 1851–1857

Freeman C, Ostle N, Kang H (2001) An enzymic latch on a global carbon store. Nature 409:149

Gilmour CC, Podar M, Bullock AL et al (2013) Mercury methylation by novel microorganisms from new environments. Environ Sci Technol 47:11810–11820

Gilmour CC, Riedel GS, Ederington MC et al (1998) Methylmercury concentrations and production rates across a trophic gradient in the northern Everglades. Biogeochemistry 40:327–345

Hamelin S, Amyot M, Barkay T et al (2011) Methanogens: principal methylators of mercury in lake periphyton. Environ Sci Technol 45:7693–7700

Harmsen HJM, van Kuijk BLM, Plugge CM et al (1998) *Syntrophobacter fumaroxidans* sp. nov., a syntrophic propionate-degrading sulfate reducing bacterium. Int J Syst Bacteriol 48:1383–1387

Holmes BE, Chanton J, Bae HS, Ogram A (2014) Effect of nutrient enrichment on $\delta^{13}C$ and the methane production pathway in the Florida Everglades. J Geophys Res 118:1–11

Inglett P, Rivera-Monroy V, Wozniak J (2011) Biogeochemistry of nitrogen across the Everglades landscape. Crit Rev Env Sci Technol 41:187–216

Liu Y, Balkwill D, Aldrich H et al (1999) Characterization of the anaerobic propionate-degrading syntrophs *Smithella propionica* gen. nov., sp. nov. and *Syntrophobacter wolinii*. Int J Syst Bacteriol 49:545–556

Lovley DR, Dwyer DF, Klug MJ (1982) Kinetic analysis of competition between sulfate reducers and methanogens for hydrogen in sediments. Appl Environ Microbiol 43:1373–1379

McInerney MJ, Struchtemeyer CG, Sieber J et al (2008) Physiology, ecology, phylogeny, and genomics of microorganisms capable of syntrophic metabolism. Ann N Y Acad Sci 1125:58–72

Morris BE, Henneberger R, Huber H, Moissl-Eichinger C (2013) Microbial syntrophy: interaction for the common good. FEMS Microbiol Rev 37:384–406

Morrison E, Newman S, Bae HS et al (2016) Microbial genetic and enzymatic responses to an anthropogenic phosphorus gradient within a subtropical peatland. Geoderma 268:119–127

Murray PA, Zinder SH (1984) Nitrogen fixation by a methanogenic archaebacterium. Nature 312: 284–286

Nüsslein B, Eckert W, Conrad R (2003) Stable isotope biogeochemistry of methane formation in profundal sediments of Lake Kinneret (Israel). Limnol Oceanogr 48:1439–1446

Nüsslein B, Chin KJ, Eckert W, Conrad R (2001) Evidence for anaerobic syntrophic acetate oxidation during methane production in the profundal sediment of Lake Kinneret (Israel). Environ Microbiol 3:460–470

Parks JM, Johs A, Podar M et al (2013) The genetic basis for bacterial mercury methylation. Science 339:1332–1335

Penton CR, Newman S (2007) Enzyme activity responses to nutrient loading in subtropical wetlands. Biogeochemistry 84:83–98

Prakash O, Pandey PK, Kulkami GJ et al (2014) Technicalities and glitches of terminal restriction fragment length polymorphism (T-RFLP). Indian J Microbiol 54:255–261

Robinson J, Tiedje J (1984) Competition between sulfate-reducing and methanogenic bacteria for H_2 under resting and growing conditions. Arch Microbiol 137:26–32

Schink B (1997) Energetics of syntrophic cooperation in methanogenic degradation. Microbiol Mol Biol Rev 61:262–280

Schönheit P, Kristjansson JK, Thauer RK (1982) Kinetic mechanism for the ability of sulfate reducers to out-compete methanogens for acetate. Arch Microbiol 132:285–288

Uz I, Ogram A (2006) Cellulolytic and fermentative guilds in the Florida Everglades. FEMS Microbiol Ecol 57:396–408

van Amstel A (2012) Methane. A review. Integr Environ Sci 9:5–30

Voolapalli RK, Stuckey DC (1999) Relative importance of trophic group concentrations during anaerobic degradation of volatile fatty acids. Appl Environ Microbiol 65:5009–5016

Wallrabenstein C, Hauschild E, Schink B (1995) *Syntrophobacter pfennigii* sp. nov., a new syntrophically propionate-oxidizing anaerobe growing in pure culture with propionate and sulfate. Arch Microbiol 164:346–352

White JR, Reddy KR (2003) Nitrification and denitrification rates of Everglades wetland soils along a phosphorus-impacted gradient. J Environ Qual 32:2436–2443

Chapter 7
Briefly Summarizing Our Understanding of *Vibrio cholerae* and the Disease Cholera

Christon J. Hurst

Abstract *Vibrio cholerae* is a naturally existing aquatic bacteria that lives in association with the chitinous exoskeletons of crustaceans including copepods. Cholera is an infectious disease of humans which is caused by ingesting those strains of the bacteria *Vibrio cholerae* that carry both of two disease related factors, a toxin gene coded by the bacteriophage CTXΦ which produces the cholera toxin, and the toxin-coregulated pilus which both facilitates attachment of the bacteria to host cells and also serves as the CTXΦ receptor. Cholera is considered a waterborne infection, with the primary route of infection being ingestion of fecally contaminated water and secondary transmission being caused by ingesting fecally contaminated food. Development of mathematical modeling frameworks may help to provide an essential lead time for strengthening intervention efforts to either prevent or ameliorate outbreaks of cholera in regions where the disease is endemic.

7.1 Introduction

The Vibrionaceae are a family of heterotrophic bacteria found in oceanic environments. A few members of the species *Vibrio* have extended their range to occur in brackish and freshwater environments (Takemura et al. 2014). The growth and concentrations of *Vibrio* naturally found in coastal waters increases with warmer water temperatures, often leading to a seasonal distribution of *Vibrio* infections in temperate regions with most of those infections occurring from summer through early autumn (Sinatra and Colby 2018).

Vibrio cholerae naturally exists in association with the chitinous exoskeletons of crustaceans including copepods, and not surprisingly *Vibrio cholerae* produces chitinases (Nalin 1976). Indeed, it is presumed that all *Vibrio* species produce

C. J. Hurst (✉)
1814 Woodpine Lane, Cincinnati, OH 45255, USA

Universidad del Valle, Cali, Colombia

C. J. Hurst (ed.), *The Structure and Function of Aquatic Microbial Communities*,
Advances in Environmental Microbiology 7,
https://doi.org/10.1007/978-3-030-16775-2_7

chitinases (de Magny et al. 2011) which degrade the crustaceans insoluble exoskeleton chitin polymer into soluble chitin oligosaccharides (Hayes et al. 2017). The chitinases of *Vibrio cholerae* allow that bacterial species to use chitin as a sole carbon and nitrogen source (Mondal et al. 2014) and, correspondingly, *Vibrio cholerae* bacteria that are colonized onto the chitinous surface of copepods are able to utilize the copepods chitin as a sole carbon and nitrogen source (Mondal et al. 2014). One of the two known *Vibrio cholerae* chitinase enzymes, ChiA2, also is secreted by *Vibrio cholerae* within the mammalian intestine and that enzyme hydrolyzes intestinal mucin to release N-Acetylglucosamine. The released sugar then is utilized by *Vibrio cholerae* for growth within the host intestine (Mondal et al. 2014).

Cholera is an infectious disease of humans which is caused by ingesting only those strains of the bacteria *Vibrio cholerae* that carry both of two disease related factors. The pathogenicity factors for *Vibrio cholerae* are the cholera toxin which is encoded by a lysogenic bacteriophage CTXΦ, and a toxin coregulated pilus which is encoded by a pathogenicity island (Faruque and Mekalanos 2012) and also serves as the CTXΦ receptor (Krebs and Taylor 2011).

Cholera primarily is considered to be a waterborne disease with secondary transmission routes that include fecally contaminated food. Vaccination is one of our newest tools in fighting cholera. However, preventing deaths due to cholera will require that we both reduce the risk of exposure to pathogenic strains of the bacteria and that we successfully treat those individuals who contract the illness. Community understanding, along with awareness and intervention, will allow us to resolve this health problem. The goal is that John Snows wish (Snow 1849, 1854) of eliminating the health risk asssociated with cholera finally will be realized.

7.2 Ecology of the Vibrios that Cause Infections in Humans

Understanding a pathogenic microbes ecology is a key factor in comprehending its associated disease.

7.2.1 The Four Horsemen of Vibrio *Disease*

Human health concerns regarding the genus *Vibrio* encompass four environmental species and those are *Vibrio anguillarum*, *Vibrio cholerae*, *Vibrio parahaemolyticus*, and *Vibrio vulnificus*.

Vibrio anguillarum usually is considered to be a pathogen of fish, but it has a wide host range that also naturally encompasses bivalve molluscs plus crustaceans and extends to include larvae of the wax moth *Galleria mellonella* (McMillan et al. 2015). *Vibrio anguillarum* causes economic losses in the fishing and aquaculture industries, and has produced bacteremia in humans as a consequence of people consuming either fish or crustacean seafood which contained the bacteria (Sinatra and Colby 2018).

Both *Vibrio parahaemolyticus* (Letchumanan et al. 2014) and *Vibrio vulnificus* are found in estuaries and coastal waters, where these bacterial species naturally are concentrated into the tissues of filter feeding bivalve molluscs and cause an initially gastrointestinal disease when the contaminated shellfish are consumed either raw or undercooked (Raszl et al. 2016). The resulting human gastrointestinal infections caused by *Vibrio parahaemolyticus* and *Vibrio vulnificus* (Park and Lee 2018) can spread to the bloodstream producing bacteremia. *Vibrio vulnificus* additionally causes disease via wound infections (Raszl et al. 2016).

An examination of archived formalin-preserved plankton samples that included noting the presence of *Vibrio cholerae*, *Vibrio parahaemolyticus*, and *Vibrio vulnificus*, supplemented by using generalized additive models, revealed that long-term increase in *Vibrio* abundance from 1958 to 2011 in the North Atlantic was promoted by increasing sea surface temperatures and positively correlated with the Northern Hemisphere Temperature (NHT) and Atlantic Multidecadal Oscillation (AMO) climatic indices (Vezzulli et al. 2016).

7.2.2 Focusing on the Horseman Named "Pestilence" by Understanding the Ecology of **Vibrio cholerae**

The natural habitat for *Vibrio cholerae* is the chitinous shells of crustaceans (Grimes 1991) including copepods. Attachment of *Vibrio cholerae* to the surface of live copepods has been suggested as being important for ecological persistence of *Vibrio cholerae* in natural water (Huq et al. 1983) and that attachment beneficially affects multiplication of *Vibrio cholerae* (Huq et al. 1984). The beneficial effect of *Vibrio cholerae*'s association with copepods has been determined to be greatest at an alkaline pH of 8.5 with lesser pH levels of 6.5 and 7.5 also having been examined, at 30 °C with lower water temperatures of 5 through 25 °C having been examined, and with maximum attachment to copepods noted at 15 g/kg salinity as compared against salinities of 5 and 10 g/kg (Huq et al. 1984). It should be remembered that full salinity sea water is considered to be around 35 g/kg. *Vibrio cholerae* is present throughout the year both in the gut and on the surface of its zooplankton copepod hosts and in addition this bacteria has an association with some species of phytoplankton at least in Bangladeshi waters (de Magny et al. 2011). There additionally have been findings of *Vibrio cholerae* associated with *Acanthamoeba castellanii*, and of *Vibrio cholerae* attached to the mucilaginous sheath of the cyanobacteria *Anabaena* (Almagro-Moreno and Taylor 2013). *Vibrio cholerae*, along with copepods which are its presumed primary hosts, has been shown autochthonous to riverine, estuarine, and coastal waters. Copepods have been found as part of the microflora in drinking water distribution systems of even economically well developed countries (van Lieverloo et al. 2012) and this knowledge suggests that microcrustaceans may represent a potentially supportive environment for *Vibrio* contaminants that gain entrance to water distribution systems. Association of *Vibrio* organisms with the chitinous shells of micro crustaceans such as copepods, and macrocrustaceans such as crabs, may facilitate survival of the microbes during passage through the stomach, and that potential increase in gastric survival could be an important factor in waterborne disease if the *Vibrio cholerae* in source waters were attached to microcrustacean copepods when these microcrustaceans inadvertently get ingested along with the water (Nalin 1976; Nalin et al. 1978).

7.3 Characteristics of the Disease Cholera

Cholera is an acute diarrheal disease caused by toxin-producing strains of the bacterial species *Vibrio cholerae* (Lippi and Gotuzzo 2014). More than 200 serotypes of *Vibrio cholerae* have been identified, with several of those serotypes known to be capable of causing mild to serious gastroenteritis including local outbreak situations of diarrheal illnesses with "cholera-like" symptoms. However, it is the toxigenic strains of *Vibrio cholerae* serogroups O1 and O139 that have been identified with cholera epidemics (de Magny et al. 2011).

Cholera disease is noted for its severe watery diarrhea that can lead to dehydration and, if untreated, the disease can result in death. Azman et al. (2013) have provided a good general summary of the incubation period and symptomatology for cholera. It takes between 12 h and 5 days for a person to show symptoms after ingesting either contaminated water or food, and the disease then can kill within hours if left untreated. Susceptibility to infection by *Vibrio cholerae* seems to be determined by a combination of immunologic, nutritional, and genetic characteristics (Harris et al. 2008). Presumably most of those people who are infected with *Vibrio cholerae* will display no symptoms and yet those people still may have the bacteria present in their feces and thus potentially represent a source of infection for other people. Most of the people who do demonstrate symptoms have either mild or moderate illness that can be treated with oral rehydration solution (World Health Organization 2006). A minority of cholera cases will develop acute watery diarrhoea with severe dehydration that can lead to death if left untreated.

Cholera is considered to be a waterborne disease, principally acquired by ingesting fecally contaminated water, with there also being fecally associated secondary transmission (Snow 1849, 1854). It is fortunate that improved personal hygiene standards and improved drinking water quality largely have resulted in the disappearance of cholera from developed countries. Unfortunately, the menace of this disease does continue in many regions of the world due to poor sanitation accompanied by lack of important infrastructure and the complicating factor that flooding results in fecal contamination of water supples (Almagro-Moreno and Taylor 2013). The remaining endemnicity of cholera is in regions of the world where inadequate sanitary practices commonly are associated with consumption of contaminated water and food. It has been estimated that approximately 1.3 billion people are at risk for cholera in endemic countries, with perhaps 2.86 million cholera cases (uncertainty range: 1.3 m–4.0 m) and a possible 95,000 deaths occurring annually (Ali et al. 2015).

7.4 The Association of Cholera Incidence and Climatic Conditions

The incidence rate of cholera disease has been found to correlate both negatively and positively with rainfall patterns. In some studies is has seemed that dry weather periods may result in people using riskier sources of drinking water. Oppositely, surface runoff might overwhelm drinking water treatment systems and flood wells during wet weather seasons. In the study by Camacho et al. (2018) there was a positive correlation between the spring rainy season in Yemen and cholera incidence.

Rainfall, in addition to temperature and salinity, has proven to be an important factor in the ecology of *Vibrio cholerae* (de Magnya et al. 2008). Rainfall and daily hours of sunlight, along with water temperature, water depth, and conductivity, are characteristics that generally affect plankton populations. Phytoplankton blooms naturally are followed by blooms of zooplankton, including copepods, and that has accounted for the observation of an approximate 8 week lag time between the point when a phytoplankton bloom peaks and when cholera cases subsequently appear (Huq et al. 2005).

The results of a study by Fu et al. (2012) have suggested that a predictive model based upon growth curve functions which combined environmental temperature and organic nutrient levels might help to reliably predict *Vibrio cholerae* in the aquatic environment. Ecological analysis research by Jutla et al. (2013) has shown that the use of satellite based radiance observations of the difference between blue (412 nm) and green (555 nm) wavelengths helps to assess the timing and presence of organic matter in coastal waters which in turn can be related to likely incidence of seasonal cholera. Development of such modeling frameworks may help to provide an essential lead time for strengthening intervention efforts to either prevent or ameliorate outbreaks of cholera in regions where the disease is endemic.

7.5 The Role of Crustaceans in Some Other Diseases

Crustaceans are not associated with the transmission of only *Vibrio cholerae*. Several species of crustaceans including freshwater copepods have long been recognized as intermediate hosts of helminths (Leiper 1936). Particularly notable among those is the guinea worm *Dracunculus medinensis* whose larvae develop within a copepod's digestive tract before being transmitted to humans, and that transmission occurs when humans ingest freshwater copepods infested with

Dracunculus medinensis resulting in the disease dracunculiasis also called Guinea worm disease (Centers for Disease Control and Prevention 2012). Marine copepods and other aquatic crustaceans similarly also serve as primary hosts for many helminth species, and the consumption of those infested crustaceans by fish then produces a resultant infestation of the fish (Zander et al. 1994). Fish lice seem paticularly notable as intermediate crustacean hosts for nematodes (Moravec et al. 1999). Additionally, parasitic crustaceans including copepods serve as both hosts as well as vectors for viruses, and those vector relationships may play a role in transmitting the pathogenic agents to other economically valuable crustaceans among which are the penaeid shrimp (Overstreet et al. 2009).

7.6 Additional Transmission Routes for Cholera

In addition to *Vibrio cholerae* causing massive outbreaks of disease associated with the consumption of contaminated water, food also has served as a vehicle for transmission of cholera (Rabbani and Greenough 1999). *Vibrio cholerae* is particularly known to cause illness associated with the consumption of contaminated macrocrustaceans such as crabs (Finelli et al. 1992), and indeed that could be expected as a possible natural contamination of food given that the native habitat of this bacterial organism is the chitinous shells of crustaceans (Grimes 1991). The possibility that association of *Vibrio cholerae* with chitin may provide the bacteria with a natural means of protection from stomach acids (Nalin et al. 1978) indirectly may facilitate the spread of this diease. The natural contamination that occurs with crustaceans is not avoidable but can be ameliorated by adequately cooking food to kill the bacteria before consuming the food.

As with the disease typhoid (Hurst 2018) which is caused by *Salmonella enterica subsp. enterica serovar Typhi*, previously named *Salmonella typhi*, it should be presumed that for cholera ingestion of water contaminated with feces and sewage serves as the primary, or initial, route of transmission (Snow 1849). Careless fecal contamination of food likely is a secondary transmission route (Hurst 2018; Snow 1849). The transmission of cholera by food, aside from the instances of natural contamination as mentioned above for macrocrustaceans such as crab, is an outcome that easily could be avoided by adequate sanitation. Unfortunately, people often just simply do not wash their hands even when adequate means are available for doing that washing. The fact that cholera transmission can be associated with a failure to wash hands and the consequent contamination of food, has been understood at last since the time when it was explained by John Snow nearly two centuries ago (Snow 1849). Social behaviours relating to the sharing of food clearly may increase the risk of gastrointestinal disease transmission. As an example of this, Camacho et al. (2018) found that an increased risk of cholera transmission in Yemen seemingly had been associated with events related to celebration of Ramadan, a month of the Islamic calendar when there are large gatherings for meals in which people share food. That connection of social activity with cholera transmission occured following spring rains which naturally would have increased the chance for *Vibrio cholerae* contamination of water as a primary transmission route and associatively enhanced the risks of food associated cholera as a secondary transmission route.

There also has been a suggestion that synanthropic flies could serve as mechanical transmission vectors for cholera equally as they likey also transmit all other infections that are spread by the fecal-oral route. Vectoring of disease by flies was suspected as contributing to the transmission of the viral disease polio (Cirillo 2016) and has been hypothesized for both bacteria (Graczyk et al. 2001; Junqueira et al. 2017) as well as protozoa (Graczyk et al. 2001).

Once an outbreak of cholera has occurred, *Vibrio cholerae* strains can be traced geographically to understand the development of cholera disease outbreaks (Kiiru et al. 2013). Efforts also have been made by Chao et al. (2014) and Nishiura et al. (2017) to utilize disease transmission models for understanding the dynamics of cholera outbreaks. A basic explanation of compartment modeling and risk estimation for primary waterborne disease transmission accompanied by secondary routes of disease transmission can be found with examples in Hurst (2018).

7.7 Concurrent Infections Can Worsen Gastrointestinal Bacterial Disease

There always is the concern that malnutrition and also concurrent infections caused by other pathogenic organisms, such as those which produce malaria and measles and are known to increase the risk of severe outcome from other gastrointestinal infections, similarly could worsen the severity of cholera. My suggestion of concern relating to malaria is due to the fact that while non-typhoidal *Salmonella* serotypes (NTS) often are associated with gastroenteritis in immunocompetent individuals, individuals with severe pediatric malaria can develop bacteremic infections with NTS during which symptoms of gastroenteritis are commonly absent (Mooney et al. 2014). My concerns about malnutrition and measles were summarized in 2018 (Hurst 2018). The probability of human death from gastrointestinal llness caused by *Cryptosporidium parvum* in the general population is 0.0002 (100 deaths per 403,000 cases of illness) (Hurst 2018). Numbers from a publication by Crawford and Vermund (1988) indicate that the probability of cryptosporidial illness leading to death can be 0.14 (14%, or 1 in 7, or 2/14) if there is underlying malnutrition. The risk of death from *Cryptosporidium* can be 0.20 in the case of underlying measles (20%, or 1 in 5) (Crawford and Vermund 1988) which represents a 2000-fold increase over the rate of death from *Cryptosporidium* infections in the general population. By itself, illness caused by the measles virus *Morbillivirus measles morbillivirus* is almost never fatal but it may be one of the most immunosuppressive viruses that infect humans.

7.8 Prevention of Cholera Including Vaccination and "WASH"

Vaccination and community awareness ultimately will help us to defeat the pestilence known as cholera.

7.8.1 Immunity to Cholera and Vaccination Against the Disease

Vibriocidal antibody is an immunologic marker associated with protection from *Vibrio cholerae*. It has been found that knowledge of the levels of serum IgA to three specific antigens: the B subunit of cholera toxin, lipopolysaccharide, and the major subunit of the toxin-coregulated pilus TcpA that induces mucosal and systemic immunoglobulin A immune responses in patients with cholera caused by *Vibrio cholerae* types O1 and O139, can be used to predict protection in household contacts of patients infected with *Vibrio cholerae* O1 (Harris et al. 2008). Two types of killed-cell vaccines currently are available for helping to prevent cholera (World Health Organization 2017).

7.8.2 International Efforts to Control Cholera

The Global Task Force on Cholera Control (2017) has a goal of reducing cholera deaths by 90% and expresses the hope that as many as 20 of the 47 countries currently affected by cholera could completely eliminate cholera disease transmission by 2030. The task force strategy consists of interventions that include oral cholera vaccines plus a group of water related activities to which they have assigned the acronym "WASH". That acronym is said to represent several points, although the exact number and wording of those points seems to vary.

Firstly, "Basic water supply" is a concept that means having access to safe drinking water, and could be represented by a community supplied potable water distribution network which pipes safe drinking water to each household. Among the alternative possibilities that are considered reasonable for providing an adequate water supply are having access within a 30-minute round-trip to either a public standpipe, borehole, protected dug well, protected spring, or rainwater collection system. If the available water is not potable, meaning safely drinkable, then there should be a provision for either household or community disinfection of the water. That concept of basic water supply helps with preventing both primary transmission and also secondary transmission of waterborne infections. I have addressed elsewhere in this book the subject of providing microbiologically safe drinking water for populations ranging from municipalities to individual households (Chap. 9, "Microbiome of Drinking Water Distribution Systems" pp. 261–311).

Secondly, "Basic sanitation", meaning access to improved sanitation facilities as represented by either having households connected to a public sewer, or a septic system, or a pour-flush latrine, or a pit latrine, will help to prevent secondary transmission.

Thirdly, "Basic hygiene", meaning that every household has access to a hand-washing station with soap and water. Appropriate care when preparing food, including safely either peeling or washing ingredients, cooking things thoroughly, and then protecting the prepared food against inadvertent fecal contamination, will help to avoid secondary transmission. It also is helpful to establish effective community engagement programs to promote safe hygiene.

One of the important aspects of public health programs is to carefully manage the necessary monetary and physical resources that are required for achieving success in disease elimination. There also is a necessity for creating adequate governmental health agencies that will be assigned and comply with the task of monitoring the quality of community water supplies including piped water distribution networks. Disease surveillance and reporting, including the important step of confirming *Vibrio cholerae* infections at the peripheral level for suspected cholera cases, and conducting a monitoring program for identifying outbreaks, represent expensive but helpful efforts that require access to laboratory culture capacity and rapid diagnostic tests. Having an ability to test for antibiotic susceptibility of the causative bacteria during oubreaks, and the ability to geographically track the bacterial strains associated with outbreaks, also are helpful (Camacho et al. 2018; Kiiru et al. 2013) because accurate surveillance data can advance efforts towards achieving the goal of prioritizing preparedness.

The types of advance preparations which can be made for confronting cholera outbreaks include pre-positioning supplies of oral rehydration salt (ORS) solution for performing oral rehydration therapy, which is the administration of fluid by mouth to prevent or correct the dehydration caused by the diarrhoea (World Health Organization 2006), plus having available intravenous fluids for rehydration when instances of diarrhea are so severe that oral rehydration would not be sufficiently effective for saving life. Hypochlorite disinfectant solution should be available for sanitation to prevent iatrogenic infections inadvertently associated with medical examinations during the treatment of cholera patients. The health care system also can be benefited by advance training of health workers to improve patient care and benefited by education about reducing iatrogenic infections. Having dedicated health care facilities which might be either Cholera Treatment Centers or Cholera Treatment Units could further help to reduce the likelihood of nosocomial cholera infections (Global Task Force on Cholera Control 2017).

Compliance with Ethical Standards

Conflict of Interest Christon J. Hurst declares that he has no conflict of interest.

Ethical Approval This article does not contain any studies with human participants or animals.

References

Ali M, Nelson AR, Lopez AL et al (2015) Updated global burden of cholera in endemic countries. PLoS Negl Trop Dis 9(6):e0003832. https://doi.org/10.1371/journal.pntd.0003832

Almagro-Moreno S, Taylor RK (2013) Cholera: environmental reservoirs and impact on disease transmission. Microbiol Spectr 1(2). https://doi.org/10.1128/microbiolspec.OH-0003-2012

Azman AS, Rudolph KE, Cummings DA et al (2013) The incubation period of cholera: a systematic review. J Inf Secur 66(5):432–438. https://doi.org/10.1016/j.jinf.2012.11.013

Camacho A, Bouhenia M, Alyusfi R et al (2018) Cholera epidemic in Yemen, 2016–18: an analysis of surveillance data. Lancet Glob Health 6(6):e680–e690. https://doi.org/10.1016/S2214-109X(18)30230-4

Centers for Disease Control and Prevention (2012) Progress toward global eradication of Dracunculiasis - January 2011–June 2012. Morb Mortal Wkly Rep 61(42):854–857

Chao DL, Longini IM Jr, Morris JG Jr (2014) Modeling cholera outbreaks. Curr Top Microbiol Immunol 379:195–209. https://doi.org/10.1007/82_2013_307

Cirillo VJ (2016) "I am the baby killer!" house flies and the spread of polio. Am Entomol 62 (2):83–85. https://doi.org/10.1093/ae/tmw039

Crawford FG, Vermund SH (1988) Human cryptosporidiosis. Crit Rev Microbiol 16:113–159

de Magny GC, Mozumder PK, Grim CJ et al (2011) Role of zooplankton diversity in *Vibrio cholerae* population dynamics and in the incidence of cholera in the Bangladesh Sundarbans. Appl Environ Microbiol 77:6125–6132

de Magnya GC, Murtugudde R, Sapiano MRP et al (2008) Environmental signatures associated with cholera epidemics. Proc Natl Acad Sci USA 105:17676–17681

Faruque SM, Mekalanos JJ (2012) Phage-bacterial interactions in the evolution of toxigenic *Vibrio cholerae*. Virulence 3(7):556–565. https://doi.org/10.4161/viru.22351

Finelli L, Swerdlow D, Mertz K et al (1992) Outbreak of cholera associated with crab brought from an area with epidemic disease. J Infect Dis 166:1433–1435. https://doi.org/10.1093/infdis/166.6.1433

Fu S, Shen J, Liu Y et al (2012) A predictive model of *Vibrio cholerae* for combined temperature and organic nutrient in aquatic environments. J Appl Microbiol 114(2):574–585. https://doi.org/10.1111/jam.12058

Global Task Force on Cholera Control (2017) A global roadmap to 2030. World Health Organization, Geneva. https://www.who.int/cholera/publications/global-roadmap.pdf. Accessed 7 Dec 2018

Graczyk TK, Knight R, Gilman RH et al (2001) The role of non-biting flies in the epidemiology of human infectious diseases. Microbes Infect 3(3):231–235

Grimes DJ (1991) Ecology of estuarine bacteria capable of causing human disease: a review. Estuaries 14(4):345–360

Harris JB, LaRocque RC, Chowdhury F et al (2008) Susceptibility to vibrio cholerae infection in a cohort of household contacts of patients with cholera in Bangladesh. PLoS Negl Trop Dis 2(4): e221. https://doi.org/10.1371/journal.pntd.0000221]

Hayes CA, Dalia TN, Dalia AB (2017) Systematic genetic dissection of chitin degradation and uptake in *Vibrio cholerae*. Environ Microbiol 19(10):4154–4163. https://doi.org/10.1111/1462-2920.13866

Huq A, Small EB, West PA et al (1983) Ecological relationships between *Vibrio cholerae* and planktonic crustacean copepods. Appl Environ Microbiol 45:275–283

Huq A, West PA, Small EB et al (1984) Influence of water temperature, salinity, and ph on survival and growth of toxigenic *Vibrio cholerae* serovar 01 associated with live copepods in laboratory microcosms. Appl Environ Microbiol 48:420–424

Huq A, Sack RB, Nizam A et al (2005) Critical factors influencing the occurrence of *Vibrio cholerae* in the environment of Bangladesh. Appl Environ Microbiol 71:4645–4654

Hurst CJ (2018) Understanding and estimating the risk of waterborne infectious disease associated with drinking water. In: Hurst CJ (ed) The connections between ecology and infectious disease,

Advances in environmental microbiology, vol 5. Springer, Cham, pp 59–114. https://doi.org/10.1007/978-3-319-92373-4_3

Junqueira ACM, Ratan A, Acerbi E et al (2017) The microbiomes of blowflies and houseflies as bacterial transmission reservoirs. Sci Rep 7:16324. https://doi.org/10.1038/s41598-017-16353-x

Jutla A, Akanda AS, Huq A et al (2013) A water marker monitored by satellites to predict seasonal endemic cholera. Remote Sens Lett 4(8):822–831. https://doi.org/10.1080/2150704X.2013.802097

Kiiru J, Mutreja A, Mohamed AA et al (2013) A study on the geophylogeny of clinical and environmental *Vibrio cholerae* in Kenya. PLoS One 8(9):e74829. https://doi.org/10.1371/journal.pone.0074829

Krebs SJ, Taylor RK (2011) Protection and attachment of *Vibrio cholerae* mediated by the toxin-coregulated pilus in the infant mouse model. J Bacteriol 193:5260–5270. https://doi.org/10.1128/JB.00378-11

Leiper RT (1936) Crustacea as helminth intermediaries. Proc R Soc Med 29(9):1073–1074

Letchumanan V, Chan K-G, Lee L-H (2014) *Vibrio parahaemolyticus*: a review on the pathogenesis, prevalence, and advance molecular identification techniques. Front Microbiol 5:705. https://doi.org/10.3389/fmicb.2014.00705

Lippi D, Gotuzzo E (2014) The greatest steps towards the discovery of *Vibrio cholerae*. Clin Microbiol Infect 20(3):191–195. https://doi.org/10.1111/1469-0691.12390

McMillan S, Verner-Jeffreys D, Weeks J et al (2015) Larva of the greater wax moth, *Galleria mellonella*, is a suitable alternative host for studying virulence of fish pathogenic *Vibrio anguillarum*. BMC Microbiol 15:127. https://doi.org/10.1186/s12866-015-0466-9

Mondal M, Nag D, Koley H et al (2014) The *Vibrio cholerae* extracellular chitinase ChiA2 is important for survival and pathogenesis in the host intestine. PLoS One 9(9):e103119. https://doi.org/10.1371/journal.pone.0103119

Mooney JP, Butler BP, Lokken KL et al (2014) The mucosal inflammatory response to non-typhoidal Salmonella in the intestine is blunted by IL-10 during concurrent malaria parasite infection. Mucosal Immunol 7(6):1302–1311. https://doi.org/10.1038/mi.2014.18

Moravec F, Vidal-Martínez V, Aguirre-Macedo L (1999) Branchiurids (*Argulus*) as intermediate hosts of the daniconematid nematode *Mexiconema cichlasomae*. Folia Parasitol 46:79

Nalin DR (1976) Cholera, copepods, and chitinase. Lancet 2(7992):958

Nalin DR, Levine RJ, Levine MM et al (1978) Cholera, non-vibrio cholera, and stomach acid. Lancet 2(8095):856–859

Nishiura H, Tsuzuki S, Yuan B et al (2017) Transmission dynamics of cholera in Yemen, 2017: a real time forecasting. Theor Biol Med Model 14:14. https://doi.org/10.1186/s12976-017-0061-x

Overstreet RM, Jovonovich J, Ma H (2009) Parasitic crustaceans as vectors of viruses, with an emphasis on three penaeid viruses. Integr Comp Biol 49(2):127–141. https://doi.org/10.1093/icb/icp033

Park J, Lee CS (2018) *Vibrio vulnificus* infection. N Engl J Med 379(4):375. https://doi.org/10.1056/NEJMicm1716464

Rabbani GH, Greenough WB III (1999) Food as a vehicle of transmission of cholera. J Diarrhoeal Dis Res 17(1):1–9

Raszl SM, Froelich BA, Vieira CRW et al (2016) *Vibrio parahaemolyticus* and *Vibrio vulnificus* in South America: water, seafood and human infections. J Appl Microbiol 121:1201–1222. https://doi.org/10.1111/jam.13246

Sinatra JA, Colby K (2018) Fatal *Vibrio anguilarum* infection in an immunocompromised patient-maine, 2017. Morbid Mortal Wkly Rpt 67(34):962–963

Snow J (1849) On the mode of communication of cholera. John Churchill, London

Snow J (1854) On the mode of communication of cholera, 2nd edn. John Churchill, London

Takemura AF, Chien DM, Polz MF (2014) Associations and dynamics of Vibrionaceae in the environment, from the genus to the population level. Front Microbiol 5:38. https://doi.org/10.3389/fmicb.2014.00038

van Lieverloo JH, Hoogenboezem W, Veenendaal G et al (2012) Variability of invertebrate abundance in drinking water distribution systems in the Netherlands in relation to biostability and sediment volumes. Water Res 46(16):4918–4932

Vezzulli L, Grande C, Reid PC et al (2016) Climate influence on *Vibrio* and associated human diseases during the past half-century in the coastal North Atlantic. Proc Natl Acad Sci USA 113 (34):E5062–E5071. https://doi.org/10.1073/pnas.1609157113/-/DCSupplemental

World Health Organization (2006) Oral rehydration salts production of the new ORS. WHO/FCH/CAH/06.1. World Health Organization, Geneva

World Health Organization (2017) Cholera vaccines: WHO position paper – August 2017. Wkly Epidemiol Rec 92:477–500

Zander CD, Groenewold S, Strohbach U (1994) Parasite transfer from crustacean to fish hosts in the Lübeck Bight, SW Baltic Sea. Helgol Meeresunters 48:89–105. https://doi.org/10.1007/BF02366204

Chapter 8
Options for Providing Microbiologically Safe Drinking Water

Christon J. Hurst

Abstract This chapter describes how to provide microbiologically safe drinking water for different population groups, ranging from large municipalities to small communities and households. Information is presented about potential sources of treatable water including surface water, springs and seeps, groundwater, rainwater and fog. Storage of rainwater harvested by ground catchments and roof catchments also is explained. The appropriate processes for treating water are sedimentation, coagulation, flocculation, filtration and disinfection. Those processes often are achieved differently depending upon the population size that is being served. For example, disinfecting water by boiling works well to centrally treat water at the household level but trying to centrally boil enough water to supply a population of one million people would be unmanageable. I also have included information about safely distributing drinking water including the use of municipal plumbing networks, tanker trucks, prepackaged bottled water, water vending machines, and water dispensers. As a final topic, I explain how water is collected and recycled aboard the International Space Station along with images which show the developmental testing of that technology.

8.1 Introduction

Having safe drinking water available in our homes, accessible by simply opening a faucet, is something which many of us have the good fortune of being able to take for granted. Sadly, there are a great many people in the world who can only dream of having such luxury. This chapter examines the approaches which may allow us to reach the goal of successfully providing microbiologically safe drinking water for all households and communities.

C. J. Hurst (✉)
1814 Woodpine Lane, Cincinnati, OH 45255, USA

Universidad del Valle, Cali, Colombia

C. J. Hurst (ed.), *The Structure and Function of Aquatic Microbial Communities*,
Advances in Environmental Microbiology 7,
https://doi.org/10.1007/978-3-030-16775-2_8

Perhaps the best starting point for discussing any aspect of aquatic microbiology is a reminder of two basic truths about water and microbes. First of those would be acknowledging that water and microorganisms naturally co-occur. Second of those would be recognition that by its nature aquatic microbial life is persistent even when we might wish it not to be. When we speak of either finding or creating microbiologically safe drinking water, we almost cannot intend that no microbes would be present. Sometimes sustaining health does indeed require being able to medically administer microbe free solutions, and we are able to do that for relatively small quantities of liquid. Our practical goal with drinking water is to remove those microorganisms that are pathogenic. Unfortunately even that goal becomes difficult when we think about not only creating microbially safe drinking water but then also needing to distribute the water without degradation of its quality.

Sometimes, what we consider to be microbial contaminants in water are environmental microorganisms that will naturally be present in even the most pristine of water sources, and many of those microorganisms are of health concern for us. More often, however, the aquatic microbial burden with which we need to be concerned originates from humans and animals that naturally reside either in or near to the water. We acquire some waterborne infections by ingestion of contaminated water and most often those infections are gastrointestinal. Many other water associated infections are acquired from body surface contact with contaminated water, and that category includes partially as well as fully immersive activities such as bathing. Bathing activities typically result in infections of the skin, eyes, ears, and nose. Bathing activities can result in gastrointestinal infections because water often gets swallowed either accidently or intentionally when we bathe. Inhalation of aerosolized water tends to cause respiratory illnesses, with those being a potential result of shower bathing and also by encountering infectious aerosols arising from waterfalls, vaporizers, cooling towers, and air conditioning units.

The infections which humans acquire from water recently have been summarized elsewhere in this series (Hurst 2018). Table 8.1 lists a few of those diseases along with their respective causal microorganisms. Many of the waterborne pathogens found in source water, such as *Legionella*, *Naegleria*, and *Vibrio vulnificus*, presumably are naturally present environmental organisms and the diseases which they cause in humans are unnecessary to the ecology which sustains the natural existence of those microbes. Some of the other microbes on that list, such as *Cryptosporidium parvum* and *Giardia intestinalis*, seem to exist naturally as pathogens of other animals and while infection of humans by those microbes may be a part of the microbes natural existence those infections of humans would be coincidental and not a core part of the microbes existence. A few among the viruses, such as the human enteroviruses, have an ecology that centers upon their infectiousness for humans. All of these microbes successfully must be either removed from the source water or rendered harmless in order for consumption of the water to become free from risk.

The most recently available estimates from the World Health Organization (2018b) about availability of usable drinking water for human populations are that:

> In 2015, 71% of the global population (5.2 billion people) used a safely managed drinking-water service—that is, one located on premises, available when needed, and free from contamination.

Table 8.1 Examples of infectious diseases associated with waterborne pathogens

Exposure route	Pathogen group	Type of disease	Causative microorganism(s)[a]
Ingestion (includes contaminated drinking water and other beverages plus ice and water-associated contamination of foods)			
	Bacterial		
		Enteric fever	*Salmonella* (especially *Salmonella enterica* subsp. enterica serovar Typhi, which causes typhoid fever)
		Enteritis	*Campylobacter*, *Shigella* (causes bacterial dysentery), *Vibrio* (especially *V. cholerae*, which causes cholera)
		Febrile syndrome	*Francisella tularensis*
		Septicemia	*Vibrio vulnificus*
	Protozoan		
		Enteritis	*Cryptosporidium parvum*, *Entamoeba histolytica* (causes amebic dysentery), *Giardia intestinalis* (includes the formerly named Giardia lamblia)
	Viral		
		Encephalitits	*Enterovirus*
		Gastroenteritis	*Alphacoronavirus*, *Mamastrovirus*, *Norovirus*, *Rotavirus*, *Vesivirus*
		Hepatitis	*Vesivirus*, *Hepatovirus*
		Meningitis	*Enterovirus*
Body Surface Contact (usually associated with either personal bathing, recreational, aquatic activities or occupational activities)			
	Bacterial		
		Enteritis	*Vibrio cholerae*
		Nephritis	*Leptospira interrogans*
		Wound infections	*Vibrio parahemolyticus*, *Vibrio vulnificus*
	Metazoan		
		Worm infestation	*Schistosoma*
	Protozoan		
		Encephalitis	*Naegleria*

(continued)

Table 8.1 (continued)

Exposure route	Pathogen group	Type of disease	Causative microorganism(s)[a]
		Enteritis	*Entamoeba histolytica*
	Viral		
		Encephalitis	*Enterovirus*
		Gastroenteritis	*Alphacoronavirus*, *Mamastrovirus*, *Norovirus*, *Rotavirus*, *Vesivirus*
		Meningitis	*Enterovirus*
		Pharyngoconjunctival fever	*Mastadenovirus*
Inhalation			
	Bacterial		
		Pneumonic fever	*Legionella pneumophila*

[a]If an organism is indicated by both its genus and species names, then the disease association is with that particular species. If only a genus name is given, then the disease association is with more than one species belonging to that genus

89% of the global population (6.5 billion people) used at least a basic service. A basic service is an improved drinking-water source within a round trip of 30 min to collect water.

844 million people lack even a basic drinking-water service, including 159 million people who are dependent on surface water.

By 2025, half of the world's population will be living in water-stressed areas.

In low- and middle-income countries, 38% of health care facilities lack an improved water source, 19% do not have improved sanitation, and 35% lack water and soap for handwashing.

Those estimates about safely managed water services could allow us to feel overly optimistic about how we are progressing towards the goal of reducing waterborne human health risks, when in fact the product water which those services provide very often contains microorganisms that would cause disease if the water were consumed (Daud et al. 2017; Ghaderpoori et al. 2009; Hashmi et al. 2012; Hurst 2018; Payment et al. 1991, 1997).

Community distributed water usually comes from sources of surface freshwater where that is available. Otherwise, potentially drinkable water may be obtained from either groundwater sources via wells, collection of rainwater where it is available in sufficient quantity, or by desalinating brackish and marine water. Sometimes communities are able to supply drinking water but not deliver it to individual households. If it is not possible to install a pressurized water piping network for distributing the water to individual households, then a good choice for managing the situation can be a pressurized distribution network that makes the water available to community standpipes. If a piping network is entirely not available then developing community cisterns with multiple taps may be the answer for safely achieving storage and community distribution of the treated water. Either way, if the community is served by community standpipes or cisterns, then people in the community will need to bring their own water carrying containers to those standpipes and cisterns, and the people then must transport their drinking water home with them.

There are times and places where water successfully is delivered to households through community plumbing systems but the quality of that distributed water is inadequate, such that further household treatment of the delivered water is necessary for microbial safety. All too frequently, the task of finding potentially drinkable source water and then treating that water to make it safe for human consumption is left entirely to the individual consumers, an aspect that is covered below in Sect. 8.5 of this chapter. Both of those scenarios result in our needing to rely upon drinking water treatment techniques that have been developed for household usage. There also are times when the only available source for generating potable water is directly recycled wastewater, and that need will be particularly obvious in the section of this chapter which addresses water management on the International Space Station.

One of the best documents that I have found on the subject of developing available water sources is a publication on rural water supply which seems to have been designed as the text for an engineering course (United States Agency for International Development et al. 1982). It gives guidance for developing safe water supplies, along with notes for providing drinking water treatment to small communities and households. It also presents a very thorough compilation of technical notes helpfully containing basic planning, design and construction information. The surface water sources considered for utilization are ponds, lakes, reservoirs, streams, and rivers. It also mentions using spring boxes for collecting water from springs, and using seepage collection systems to obtain water from seeps, both of which represent groundwater sources that either occur very near to the surface or emerge onto the land surface. Information additionally is provided for constructing different types of vertical and horizontal wells to obtain ground water that is not emerging onto the nearby land surface. Rainfall harvesting by either roof catchment or ground catchment also is included. That publication then explains how to treat the obtained water for household and small community usage. Means for accomplishing that water treatment include small community sedimentation basins, slow sand filters, small community disinfection units represented by chlorinators, boilers, household sand filters and other household treatment techniques, in addition to household water storage. That document now is a rather old publication, but much of the newer information available on the internet draws from that 1982 publication and reproduces its illustrations.

The need for using treatment technology for producing microbiologically safe drinking water is generally understood and accepted by the people who live in communities that range in population size from large to moderately small. Part of the challenge in trying to provide microbiologically safe drinking water to very small communities and individual households in poorer and rural areas relates to such factors as the people sometimes having received limited previous education regarding the risks associated with unsafe water, overall community economic factors including their possibly limited ability to pay for drinking water treatment and perhaps a question of their willingness to pay, their level of water demand and usages, the availability of chemicals and spare parts for sustaining treatment technology, plus raw water characteristics such as availability, source quantity and source quality. All of those factors will together determine what level of treatment technology can be accepted and used (Parr and Shaw 2019).

8.2 The Three Technical Tasks Set Before Us

There are three tasks to be accomplished. The first is to find usable water. The second is treating the water as necessary to make certain that it will be safe to drink. The third is, if possible, to help by safely delivering the treated water to those are in need of it.

8.2.1 *Finding a Suitable Water Source*

Collection of source water often is done as a municipal function and delivering that water to its users may also be a municipal function. Private companies both large and small, often limited to being one-person businesses, do also play a part in obtaining and delivering water. There are several options for finding suitable source water, whether that be surface water, ground water, rain or fog. The most technologically challenging choice for a water source would be either brackish or saline water that needs to be desalinated prior to its consumption. The information that I am presenting in this chapter presumes desalination is not necessary, although that technology is indeed used to create safe drinking at the both the level of communities and households. My suggestions for reference information to understand desalination and the microbial hazards associated with inadequately maintained desalination process equipment are the publications by Nagaraj et al. (2017), World Health Organization (2011b), and Yari et al. (2018).

8.2.1.1 Surface Water

The first of our tasks is to find a source which can provide a sufficient quantity of water. Surface supplies of freshwater containing minimal sewage pollution are the best options but can be hard to find in areas of high population and dry climates. Figure 8.1 shows three of my students holding sampling bottles by the shoreline of the Ohio River, at the Public Landing for Cincinnati, Ohio. The Ohio provides plenty of raw, albeit turbid and sewage polluted, source water both for Cincinnati and for numerous other towns and cities between that rivers official beginning in Pittsburgh, Pennsylvania, and the point in Cairo, Illinois, where the Ohio River joins with the Mississippi River.

Fig. 8.1 This image shows the Ohio River near the public landing in Cincinnati, Ohio. Slightly east of that location is the water supply inlet for Cincinnati, Ohio. Use of this image is courtesy of the author

8.2.1.2 Groundwater

The second best option would be a hopefully uncontaminated and non-saline groundwater. Figure 8.2 shows two places near Cincinnati where groundwater emerges and flows onto the surface. The top of that figure shows icicles where water emerges between limestone layers at a roadcut. The bottom of that figure shows a creek where the water is clear but always served as our positive control for viruses, because during periods of low flow most of the water carried by that creek comes from household septic tanks. Figure 8.3 shows an artesian well in Colombia, South America, and also two bucket wells in southwestern Ohio. Hand operated pumps serve for providing well water to very small communities and individual households when the wells used are shallow and only low flow rates are required. Deeper wells and wells intended to provide high rates of water flow will need reliable power sources for motorized pumping. Figure 8.3 additionally shows three pump wells in Southwest Ohio that provide different capacities. All of the wells in Fig. 8.3 have their sides protected against contamination from surface flow and five of the wells also have their top openings protectively covered The well head shown in the bottom right image of Fig. 8.3 serves the city of Hamilton, Ohio and that well head further is enclosed by a steel shed with a locking door. Figure 8.4 shows a protected pump well which is used to fill an elevated storage tank, providing flowing water for a nursing home facility near Cali, Colombia, that pump is enclosed and protected by a locked gate. Figure 8.5 shows an unprotected shallow well in the

Fig. 8.2 This image shows groundwater returning to the land surface. The upper image is of icicles that resemble a frozen waterfall showing where groundwater has emerged between layers of limestone at a roadcut in Highland Heights, Northern Kentucky. The lower image shows Clough Creek in Hamilton County, Ohio slightly east of Cincinnati, Ohio. The water in the creek is aesthetically pleasing due to its very low turbidity, but the water has a high virus level because it receives leachate from buried septic tanks. Use of these images is courtesy of the author

Fig. 8.3 This figure shows wells that provide water for groups with different levels of need and the wells show different levels of sanitary protection. Upper left is an artesian flow well in the Cauca River valley region of Colombia. Center left is a covered bucket well under a roof shelter near Oxford, Ohio, USA. Lower left is a capped bucket well near Cincinnati, Ohio, USA. Upper right is a sealed well with a hand powered pump near Cincinnati, Ohio, USA. Center right is a sealed well operated by the drive shaft of a farming tractor in Hamilton county, Ohio, USA. Lower right is a sealed well within an enclosure in Butler County, Ohio, USA, and this view shows that water in the well head was being sampled for possible viral contaminants. Use of these images is courtesy of the author

seasonally dry bed of an African river. Figure 8.6 shows a spring box which collects groundwater that emerges on a hillside near Cincinnati and, as with the creek in Fig. 8.2, water being collected into and distributed from that spring box largely

Fig. 8.4 This image shows the water supply system for a nursing home near Cali, Colombia. The water source is a protected well. The well casement extends above the ground surface and the opening to that casing is covered with a concrete lid. A large stone provides additional weight on top of the concrete lid. A pipe leads horizontally above ground from the well casing to a pump that is located in the bottom of the tower. That pump is protectively housed behind a locked gate. The water is stored in a covered reservoir tank on top of the tower. Use of this image is courtesy of the author

originates from septic tanks. Figure 8.7 shows a spring that has been intercepted by means of a horizontal well in Ukraine. Figure 8.8 shows the collection of a water from a seep in the United States. The publication by Meuli and Wehrle (2001) discusses different types of springs, legal rights, water quality, water quantity, design and construction of gravity fed catchments, plus the design and construction of artesian catchments. The publication by United States Agency for International Development et al. (1982) describes different types of wells plus catchment systems for springs and seeps.

8.2.1.3 Rainwater

We often can capture rainwater when it is available and then store that water for use. Figures 8.9 and 8.10 show ground collection systems. Water collected by the drainage system of ITESM Campus, Ciudad de México, shown in Fig. 8.9 does

Fig. 8.5 This figure shows collection of turbid seepage which is being used as the source of drinking water for Mwamanongu Village in Tanzania. This water has seeped into an open pit that is called a seep well. It this instance, the well was dug in the bed of a dry river. The image is titled "Mwamongu water source" by Bob Metcalf and is in the public domain. Seep wells also can be dug a slight distance away from bodies of surface water so that water subsequently collected from the seep well will have been filtered to some extent by passage through the ground, and that process is similar to the engineered collection and processing of water by riverbank filtration

not seem to be purposefully used, and descriptive text published with that photograph suggests the circular structure into which the water drains may be the modified entryway of a sinkhole. Figure 8.10 shows a cistern located in Rajasthan, India, which stores rainwater collected by ground catchment. The ground water collection system for the cistern in Fig. 8.10 seems to show no protection against entrance of contaminants except for a concrete labyrinthine entry weir which allows some provision for settling. The water in that cistern serves as a source of drinking water and is accessed by using a bucket that would be lowered through the cisterns top opening. You will notice that the top opening is covered and has of been elevated with protective sides to help preclude contamination of the cistern opening by any water that may collect on the cisterns lid.

Figure 8.11 shows two rooftop rainwater collection systems and their storage containers. Environmental contaminants including animal feces can get washed into the rainwater storage containers and cisterns. The water storage tanks that are used with rooftop collection systems can be protected against entry of contaminants by any of several methods. One option is to wait until perhaps 5–10 min after a rainfall event has begun before manually diverting flow of rooftop collected water into the storage container. There is an option of using automatic 'first flush' diverters, which

Fig. 8.6 These images show a spring water cistern, sometimes called a "Spring box", which intercepts and stores for distribution the water that emerges from a hillside spring near Cincinnati, Ohio. Use of these images is courtesy of the author

Fig. 8.7 This image shows water from a spring which is being conducted by means of a horizontal well pipe into a collection pool. This spring is the water source for the village Dzyhivka in Ukraine. The image is titled "Siphot" by USchick and being used under a Creative Commons Attribution-share alike unported license. 3.0 Generic license

attach to the collection downspouts and assure that the initial water collected, which hopefully carries the greatest amount of contaminants such as animal feces, will not enter the storage tank. There also are filtration units that can be mounted at the top of the storage tanks and those filters will remove much of the contaminants that are washed down from the roof. The publication by United States Agency for International Development et al. (1982) describes catchment systems for rainwater. Some additional suggested references for rainwater catchment would be: Despins et al. (2009), de Kwaadsteniet et al. (2013), Karim et al. (2005), Manitoba Conservation and Water Stewardship (2014), PennState Extension (2016) which is particularly well detailed, Rahman et al. (2014), Skinner and Shaw (2019a).

8.2.1.4 Fog

There even are some places where we similarly can capture the water contained in dense fog (Dodson and Bargach 2015) as shown in Fig. 8.12.

Fig. 8.8 This image shows collection of water from a seep. The water was collected using a perforated polyvinyl chloride drainage pipe. Usage of the image is courtesy of Shawn and Beth Dougherty https://onecowrevolution.wordpress.com/2018/05/13/spring-water-catching-a-seep/. Accessed 10 Jan 2019

8.2.2 Understanding the Basic Processing Steps for Treating Water

The second task is to assure that the water we do find and will then provide is safe for human use, including its being drinkable. Very often the found water needs to be treated before it safely can be consumed. Our concerns about the safety of drinking water include not only the microbes which cause disease, but also turbidity, toxic metals, and toxic chemicals (World Health Organization 2017). This chapter deals with the first two of those, microbes and turbidity. The recent lead crisis in Flint, Michigan, of the United States has made it very obvious that hazardous contaminants other than microbes can even originate from the materials with which we construct water distribution systems (Ruckart et al. 2019; Zahran et al. 2017). Hurst (2001) offers a good general summary of turbidity removal and water disinfection. The publication by Parr and Shaw (2019) presents some discussion regarding the process used for plain sedimentation which is simple gravitational sedimentation done without adding either coagulants or flocculants, and also mentions roughing filters, slow sand filters, rapid sand filters, aeration, coagulation, disinfection.

Fig. 8.9 This figure shows the receiving basin for a rainwater ground catchment system. The lower image is a magnification of the upper image, and better shows the water inlet pipe. This image is titled "03242012Taller sostenibilidad lore037" by Talento Tec and used under the Creative Commons Attribution 3.0 Generic license. The text described this as being a cenote, which suggests that it may be an adapted sink hole

Fig. 8.10 This cistern is located in Rajasthan, India, and it is a traditional tank used for storing rainwater collected by ground catchment. The image is titled "Rainwater harvesting tank, India" by Spiritualfade and is in the public domain

8.2.2.1 The Processes of Sedimentation, Coagulation and Flocculation

It often is necessary to treat surface water, and sometimes to treat groundwater, in ways that will aesthetically improve the waters acceptability. We can improve both the visual appearance and taste of the water by reducing its turbidity. Some microbes will be associated with the particles that contribute to turbidity of the water, and thus removing those particles can by itself help to reduce the waters microbial burden. Many of the things that cause turbidity can complicate the task of disinfection, and so lowering turbidity for the sake of facilitating disinfection provides yet another reason for turbidity reduction.

The simplest approach for turbidity reduction is plain sedimentation and an easy procedure for performing that on a household basis is called the three pot technique. The three pot technique consists of allowing water to rest stationary in containers so that the suspended solids can settle to the bottom by gravity. Clearer water subsequently can be collected from the containers either by decanting, which is pouring off liquid in a way that will not carry over the sedimented solids, or by collecting the settled water using a siphon. This settling technique is performing by placing source water into a storage container and allowing the water to sit there for one day. The water then is carefully transferred from that first container into a second storage

Fig. 8.11 These images show receiving containers for rainwater roof catchment systems. The upper image is a rain barrel of which the lid contains a hole for the downspout pipe, courtesy of the author. The lower image titled "Rainwater harvesting tank (5981896147)" is by SuSanA Secretariat (The Sustainable Sanitation Alliance) and used under the Creative Commons Attribution 2.0 Generic license

container where the water will be allowed to settle for a second day. On the third day, the settled water is carefully collected from the second container into a third container and then can be disinfected for use.

We can speed up the process of sedimentation by assisted it with coagulation and flocculation. It also is important to note that while some natural turbidity components

Fig. 8.12 This image shows a fog harvesting net above a water collection trough. The water which condenses on fog nets drains into the collection troughs which are mounted beneath those nets. That collected water then drains into either a storage tank or a piping system by which the water is distributed to a community. The image is titled "Atrapanieblas en Alto Patache" by Pontificia Universidad Católica de Chile and used under the Creative Commons Attribution 2.0 Generic license

will sediment easily by gravity, other components of turbidity must be coagulated and even flocculated to facilitate their removal. Although these two terms, coagulation and flocculation, often are used synonymously they technically are distinct. A coagulating agent promotes particle collision by neutralizing the electrostatic charge of suspended particles. Flocculation is considered to be a physical process by which fine particulates are caused to clump together into a floc without involving the neutralization of charge. Typically, for the purpose of drinking water treatment, a polycationic compound is used as a coagulant to help manage the removal of negatively charged particles such as silt, clay, and bacteria. Flocculation often can be achieved by adding polymers that cause small destabilized particles to aggregate, in turn facilitating separation of those particles from the water. If the floc floats to the top of the liquid (creaming) then it can be collected by skimming. If the floc settles to the bottom of the liquid (sedimentation) then it can be collected by decanting the overlying liquid. The floc also can be filtered from the liquid.

Salts of aluminium and iron are the most commonly used coagulants for treatment of drinking water. Aluminium sulfate, which is a chemical compound with the formula $Al_2(SO_4)_3$, often is used for that purpose in the form of potassium aluminum sulfate dodecahydrate and is called alum. Ferric sulfate is a form of iron commonly and similarly used for drinking water treatment. Plant based coagulants used in drinking water treatment include that made from *Moringa oleifera*. *Moringa* works as a coagulant due to its positively charged, water-soluble proteins.

Powdered activated carbon is added to the water being treated for two purposes, to bind organic contaminants and because it also may help with the removal of fine particulate material during subsequent sedimentation and filtration. This type of activated carbon can be added directly in its powdered form, although often the powder is added in the form of a slurry. Powdered activated carbon is used at the municipal level and also has been tried at the household level. Drinking water treatment plants need to add powdered activated carbon to the water prior to adding coagulants because coagulants can coat the powdered activated carbon and that coating will reduce the effectiveness with which the carbon removes organic compounds.

8.2.2.2 The Process of Filtration

Traditionally, filtration was used to remove suspended particulates. We still do use it for that purpose, but we also apply the term filtration in reference to many other selective removal processes.

Paper filtration is based upon particle size retention and uses the same technique that is presented as part of the school experiments done by chemistry students. In chemistry classes, a flat round sheet of filter paper is folded into a pleated paper cone and that paper cone placed into a funnel shaped holder. Paper filtration, when used as a water treatment technique, relies upon a rectangular flat sheet of filter paper which has been pleated and packaged as an inline cartridge. Figure 8.13 shows a paper filter cartridge in its protective housing that has designed for home use. These filters must be discarded once they have clogged, as the filter paper cannot successfully be washed and reused.

Cloth filtration, which consists of pouring water through cloth, is based upon particle size retention and works for removing copepods and correspondingly reduces health risks associated with both the bacterial species *Vibrio cholera* that causes cholera (Chap. 7: "Briefly Summarizing Our Understanding of *Vibrio cholerae* and the Disease Cholera" by Christon J. Hurst, pp. 173–184), and the nematode species *Dracunculus medinensis* that causes *dracunculiasis*. Cloth filtration usually is done to produce small volumes of treated water for satisfying limited needs. Sari cloth has been recommended with success for reducing the risk of cholera (Colwell et al. 2003; Huq et al. 2010). Polyester 100 μm mesh has been recommended for reducing the risk of *dracunculiasis*. Figure 8.14 presents the idea of cloth filtration. Cloth filters can be washed and reused. Cloth filtration also can be used to remove precipitated floc and natural suspended solids as a part of other treatment processes noted below.

Sand filtration is based upon particle size retention called mechanical straining, plus physical adsorption, and it is a drinking water treatment option that we can scale in size to accommodate the needs of any human community. Figure 8.15 shows the basic concept of rapid sand filtration using a filter created from a plastic bucket, some

Fig. 8.13 This image shows a water filtration unit designed for whole house installation, meaning that it would be installed as part of the household water piping just inside of the point where the water line enters the house. This version utilizes a pleated paper filter for removing particulates. Other water filtration units can be more complex, incorporating an activated charcoal cartridge, perhaps a reverse osmosis cartridge, and additionally use ultraviolet light. Some under the counter water treatment units also incorporate water storage tanks. Use of this image is courtesy of the author

pipe fittings, plus fine gravel and fine sand. The filter is created as a layered filtration bed. Gravel is used as a bottom layer to facilitate draining water from the bottom of the filter, the sand is on top of the gravel. Either a diffuser plate or another layer of fine gravel needs to be on top of the sand so that addition of water onto the sand will not result in the upper layer of sand being agitated, because agitation would cause fine particulates trapped in the upper sand to be resuspended and the outflow water consequently become turbid until the filtration stabilizes to again produce clear water. Sand filters can be washed by agitation which may be achieved either from the top or by rapid upflow of filtered water through the filtration bed. Additional materials such as crushed garnet, anthracite, and zeolites may be added to the filtration medium either to supplement the sand or as partial substitutes for the sand. When performed as rapid sand filtration at a municipal level, the goal usually is for the filter to remove fine particles that did not settle during a preliminary coagulation and flocculation treatment of the water. Slow sand filtration treatment plants typically do not use preliminary coagulation and flocculation treatment, but may use gravel filtration of the water to achieve large particle removal prior to actually sand filtering the water. Those preliminary gravel filters are known as roughing filters. Slow sand filtration relies upon a hypogeal biological layer called a schmutzdecke which forms in the upper layer of the sand for achieving the removal of fine particulates including some capability for removing bacteria. Rapid sand

Fig. 8.14 This image helps present the concept of filtering water through clean cloth to remove large natural particulate material and coagulated fine particulate material. The cloth would be folded several times and then placed over the top of the bottle. Safely filling the bottle would be done by pouring water into it through the cloth. Use of this image is courtesy of the author

filtration usually is performed for larger communities, and slow sand filtration usually is performed for smaller communities.

The use of sand filtration is very effectively for treating drinking water at the municipal level. Table 8.2 uses data from a publication by Johnson (1916) to show how effectively the implementation of municipal sand filtration for treating drinking water reduced the typhoid fever death rate early in the twentieth century. The publication of Yang et al. (2018) describes limitations faced when trying to complete the task by developing effective vaccination against typhoid.

Riverbank filtration was perhaps the precursor of municipal sand filtration plants and will again be mentioned later in this chapter. Riverbank filtration still is being used municipally and consists of establishing wells with pumps to collect groundwater from a nearby surface water source. The basic idea is similar to what is shown in Fig. 8.5.

Ceramic filters and porous pot filters are a concept that goes back to at least colonial times in the Americas. The earliest filters seem to have been made in the

Fig. 8.15 This image shows a sand filter constructed from a 5 US gallon plastic bucket using some brass plumbing fittings, fine gravel and fine sand. The image on top shows the components used for creating the filter. The lower left image shows the filter with naturally turbid water on top of the sand, the glass jar on the left holds a sample of that naturally turbid water before filtration, and the glass jar on the right shows a sample of that same water after having passed through the sand filter. The lower right image shows a frontal view of the filter, plus the samples of naturally turbid water and filtered water. Use of these images is courtesy of the author

form of porous volcanic stone bowls and also in the form of ceramic bowls. This process is based upon particle size retention. Water is placed into the bowl or pot and slowly permeates through the porous material. The filtered water is collected in a bowl which has less permeability. Modern porous ceramic filtration pots are made using clay that has been augmented with fine organic material such as sawdust. The organic matter burns away and leaves voids or pores when the clay pot is fired to create ceramic. The upstream surface of the filter, which for a filter pot is the inside surface, will become coated with a layer of solids which slows the filtration process. That layer can be scrubbed away. Eventually, however, the internal pores of the filter material will clog and the ceramic filter or stone pot filter will need to be replaced.

Table 8.2 Relation between increase in percentage of United States city population supplied with filtered drinking water and decrease in overall city population typhoid death rates

Year	Percentage of city population served by water filtration utilities	Typhoid fever[a] death rates for same cities
1900	8.7	36
1901	10.8	34
1902	11.9	37
1903	13.3	38
1904	16.0	35
1905	17.4	30
1906	20.5	33
1907	23.2	32
1908	23.3	25
1909	30.1	21
1910	34.6	24
1911	37.2	20
1912	42.4	16
1913	48.0	16

[a]Death rates given per 100,000 population per year
Data is presented for those United States cities having accurate registration as to cause of human death
Source: Johnson (1916)

Ceramic filters also are made as hollow tubes called ceramic candles that are closed at one end with the other end serving as a drain. Filtration of water using ceramic candles is based upon water flowing from outside of the filter to the inside, and the outside of the filter is scrubbed to remove excluded particulates.

Reverse osmosis filters use thin membranes that often are manufactured as sealed cartridges. Reverse osmosis filters act by passing water through the filter membrane using hydrostatic pressure while the membrane excludes passage of charged ions. The filters are used both commercially and in homes for water desalination. It is necessary to remove turbidity from the water by some other type of preliminary filtration before processing the water with reverse osmosis filters. Otherwise, without preliminary turbidity removal, the reverse osmosis membranes quickly will clog. After having once being wetted, the reverse osmosis membrane filters are not allowed to dry because they may develop micro cracks which will allow contaminants to pass through the filter membrane. There also is a hazard of biofilm growth on the downstream side of the reverse osmosis membranes and microbes contained in that biofilm can contaminate the product water.

Activated carbon filters use either granular or powdered carbon which has been made by charring cellulosic materials. The carbon has been treated in a way called activation that produces micropores which increase its surface area. Activated carbon filtration is used for retaining organic material by adsorption. There is a problem in that microorganisms can create a biofilm growing on the activated carbon, and that biofilm may contaminate the filtered water. Granular activated carbon can periodically be reactivated, and large water treatment facilities successfully do use reactivation as a means of saving cost. Powdered activated carbon is discarded instead of being recovered and reactivated.

8.2.2.3 The Process of Disinfection

Chlorination is the most commonly used method for disinfecting large quantities of water. The options most relied upon include: chlorine gas, sodium hypochlorite liquid, calcium hypochlorite powder, and sodium dichloroisocyanurate powder. Community drinking water treatment facilities sometimes substitute chloramination in place of chlorination to reduce the formation of carcinogenic byproducts. Chlorine and chloramines temporarily provide at least some residual disinfectant capacity which reduces regrowth of those microbes that survived the physical water treatment processes, and that residual disinfectant capacity also protects the treated water against recontamination by microbes that enter the water through plumbing mishaps. Electrolytic generation of mixed oxidants from sodium chloride brine has been used for disinfecting water (Bajszár and Dekonenko 2010) with the disinfecting products created by that process largely consisting of chlorine, hypochlorite, and hypochlorous acid.

Ozone is popular in some regions as a drinking water disinfectant and the ozone is generated on-site. Ozonation can improve both the taste and odor of the water. Using ozonation in place of chlorination also avoids the problem of creating hazardous chlorinated organic byproducts.

Ultraviolet light disinfection by germicidal lamps is used at some large scale community water treatment facilities and also serves in many home treatment units. Successfully achieving ultraviolet light disinfection requires that the water not contain suspended particulates which could shield the microorganisms against germicidal exposure. A disadvantage is that water treated by ultraviolet light can be easily recontaminated because that water retains no residual disinfection capacity.

Thermal treatment is effective and commonly used at the household level but it is not practical for use at large scale drinking water treatment facilities. Boiling water is the safest thermal treatment from a disease prevention perspective and it is a very commonly used technique for home treatment of drinking water even in some areas of the world where community treated drinking water is delivered to individual households by a distribution network. Pasteurization is achieved by heating water to temperatures that are below the boiling point. Distillation also is effective as a means of disinfecting drinking water, although distillation has limited practicality even at small population scales. The design of a solar distillation unit will be described later in this chapter. The fact that thermal treatments provide no disinfection residual is critically important.

Skinner and Shaw (2019b) presents a summary of information about straining water with cloth, the three pot system, boiling, chlorine disinfection, and solar disinfection which is pasteurization that uses sunlight as a source of heat.

8.2.3 Generating and Distributing Drinkable Water at the Municipal Level

The methodology used for preparing safe drinking water at the municipal level will vary depending upon the source and quality of the raw water, and must be appropriately designed to meet the needs of the community's population size.

Water from surface sources may have high turbidity, and for that reason surface water often is sand filtering prior to distribution. Filtration can make the water visually more acceptable, make successful disinfection easier, and avoids sedimentation in the water distribution system. The options for sand filtration are either rapid sand filtration, which mostly removes turbidity, or slow sand filtration which has the additional benefit of offering better bacterial removal capability. Aeration and coagulation may be incorporated into the treatment process, and I generally think of those two processes as being used in conjunction with rapid filtration. Slow sand filtration may include the use of prefiltration through roughing filters which are composed of gravel. After either rapid sand filtration or slow sand filtration, the water should be disinfected prior to its distribution. Parr and Shaw (2019) have presented a discussion on plain sedimentation, roughing filters, rapid sand filters, aeration, coagulation, and disinfection. A third option is riverbank filtration which uses pumps located some distance away from a surface water source to pull water from that surface source through an aquifer and the pumps then discharge that water either into a public distribution system or into a water treatment facility. Riverbank filtration is used as an alternative to constructing a sand filtration drinking water treatment plant.

Groundwater sources and springs produce water that may have a very low turbidity and therefore water from those sources might not be sand filtered prior to distribution. Low turbidity surface water similarly may not be sand filtered prior to distribution. Hopefully, some disinfectant such as either chlorine or ozone always will have been added to the water before that water enters a distribution system.

Community distribution networks for drinking water may include pipelines that extend to connections for individual households, or more simply deliver water only to communal standpipes. The treated water in municipal drinking water distribution systems is not always microbially safe (Hashmi et al. 2012). Using Iran as an example, Ghaderpoori et al. (2009) reported on community distribution systems in rural areas that often provide water from springs and from deep and semi-deep wells. It was estimated that perhaps eighty eight percent of the water was safe. However, much of the rural population drank water that was not disinfected for reasons that included lack of chlorine availability. In one instance, the level of fecal coliforms was indicated to have been 1100 MPN/100 ml (Ghaderpoori et al. 2009) when there should be no fecal coliforms in the water.

Fig. 8.16 This image shows the entrance to the Barranquilla, Colombia rapid sand filtration drinking water treatment plant. Use of this image is courtesy of the author

8.2.3.1 Municipal Conventional Treatment

Conventional treatment of water means many different things. For purification of surface waters, conventional treatment usually consists of treatment with alum to achieve coagulation and flocculation, followed by sedimentation, rapid sand filtration, and disinfection. Conventional treatment including rapid sand filtration can be adapted to serve populations numbering in the hundreds of thousands to millions of people.

Figure 8.16 shows the entrance to the drinking water treatment plant "Estación de Tratamiento de Agua Potable" in Barranquilla, Colombia which uses Rio Magdalena as its source. It is a conventional treatment plant. Figure 8.17 shows the coagulation, flocculation and sedimentation section of the treatment plant, where the alum treatment causes coagulation of the suspended solids and the floc then is allow to settle prior to filtration. Powdered activated carbon sometimes is added as a slurry to aid with sedimentation of fine particulates and the powdered activated carbon also will help with removal of organic compounds. Figure 8.18 shows the rapid sand filters for that treatment plant. The normal flow of water through the filters is vertically downward during which the water on top of the filters has a very calm surface as shown in upper image of Fig. 8.18. When the rate of water flow through a filter has decreased too greatly due to clogging of the filtration media, the filter is cleaned by backwashing which means having filtered water flow rapidly upward through the filter material which fluidizes the filter material and washes out suspended particulates that were trapped in the filter material. During backwashing,

Fig. 8.17 This image shows the basins for flocculation and sedimentation, also termed clarification, at the Barranquilla rapid sand filtration drinking water treatment plant. Use of this image is courtesy of the author

the clean filtered water flowing vertically up through the filter causes the top of the filter to appear as if it were raining. The backwash water and suspended particulates either may be discharged or returned to the beginning of the plants water treatment process. The bottom image in Fig. 8.18 shows the appearance of a filter that is being backwashed. The sand in the rapid sand filters periodically is removed and replaced. Figure 8.19 shows a rapid sand filter tank at the Rio Cauca treatment plant near Cali, Colombia, which draws its source water from the Rio Cauca. Most of the filtration medium, plus the gravel support which underlays the filter medium, had been removed from that filter tank for maintenance which allows us to see the water collection system installed at the bottom of the filter. Chlorine gas is used to disinfect the product water of the Barranquilla treatment plant, just prior to that water entering the municipal distribution network. The Rio Cauca plant also uses chlorination of their product water. The distribution networks of both cities provide water directly to individual residences. A good brief explanation of rapid sand filtration accompanied by illustrations which demonstrate its engineering has been prepared by Bruni and Spuhler (2018a).

Fig. 8.18 These images show rapid sand filters at the Barranquilla drinking water treatment plant. The upper image shows filters in which the surface of the water is calm, and that indicates the water in those filters is flowing downwards as normal. The lower image of a single filter shows a rippling appearance on the surface of the water as if it either were raining or the water were boiling. That appearance of rippling shows water is flowing upwards through the filter during its backwashing cycle. Use of these images is courtesy of the author

8.2.3.2 Municipal Slow Sand Filtration

Slow sand filtration typically is used for small to medium size communities. This technology can serve large communities, although doing so may involve prohibitive land costs because the filter surface area required to process the same volume of water by slow sand filtration is greater than the surface area needed for rapid sand filtration.

The images that I am providing are for the "El Retiro" slow sand filtration plant which serves a community of that same name located near Cali, Colombia. That treatment plant uses the Rio Pance as its source. Prefiltration at the El Retiro plant is by means of roughing filters which utilize a combination of downward and horizontal flow of the water through filter beds that are composed of gravel. Figure 8.20 shows the first course of prefiltration at top, and the second course at prefiltration at bottom. After passing through the roughing filters, the water at this treatment plant is slow sand filtered. Figure 8.21 shows the slow sand filters and you can see a reflection of the sky in the bottom image of that figure. These slow sand filters are

Fig. 8.19 This image shows maintenance being performed on a rapid sand filter at the Rio Cauca water treatment plant in Cali, Colombia. Most of the filtration media has been removed from this filter tank which allows viewing of the system of perforated pipes through which water would flow when it collects and drains beneath the layer of sand, gravel and other filtration material that otherwise would be in this tank. Seeing this empty tank allows us to understand the relative surface area and depth of a rapid sand filter. Use of this image is courtesy of the author

covered with metal roofing to reduce potential growth of algae and cyanobacteria on top of the filter material.

Cleaning of slow sand filters when they clog is done by removing the uppermost centimeters of sand in which the biological layer has developed. Most of the filtration effectiveness occurs in that biological layer, but eventually the top layer clogs and must be replaced with fresh sand. It can become necessary to entirely replace the filter material. Figure 8.22 shows a closer view of a slow sand filter in the top image, and the bottom image in that figure shows replacement of the sand in one of the filters. In the bottom image of Fig. 8.22 the workman is standing on the floor of the filtration tank, and by comparing that with Fig. 8.19 you can notice the difference in relative depth of municipal rapid sand filters versus municipal slow sand filters.

The filtered water at this treatment plant is disinfection with sodium hypochlorite. Figure 8.23 shows the clearwell at the El Retiro sand filtration drinking water treatment plant. The treated water from the El Retiro plant enters a municipal distribution network which delivers water directly to individual residences. Figures 8.24 and 8.25 show

Fig. 8.20 This image shows roughing filters at the El Retiro slow sand filtration drinking water treatment plant near Cali, Colombia. Use of these images is courtesy of the author

examples of municipal distribution networks in other places that deliver water to community standpipes rather than extending the networks to individual homes.

The design and operation characteristics of the El Retiro treatment plant and also of two additional slow sand filtration drinking water treatment plants are described in

Fig. 8.21 This image shows sand filters at the El Retiro slow sand filtration drinking water treatment plant near Cali, Colombia. Use of these images is courtesy of the author

the reference authored by Federación Nacional de Cafeteros de Colombia et al. (1988) along with some history of using slow sand filtration for water treatment. The multi-stage slow sand filtration process used at the El Retiro treatment plant has been described by Galvis Castaño et al. (1999). Chan et al. (2018) has studied the microbial community in full scale slow sand filters. A good brief explanation of

Fig. 8.22 These images show sand filters at the El Retiro slow sand filtration drinking water treatment plant near Cali, Colombia. The lower image shows replacement of the sand from one of the filters. The workman seen in this image is standing on the floor of the filter tank. An important detail to notice is the difference in depth of this slow sand filter as compared to the depth of the rapid sand filter shown in Fig. 8.19. Use of these images is courtesy of the author

Fig. 8.23 This image shows the clear well at the El Retiro slow sand filtration drinking water treatment plant near Cali, Colombia. Use of this image is courtesy of the author

slow sand filtration including illustrations which demonstrate its engineering has been prepared by Bruni and Spuhler (2018b).

8.2.3.3 Municipal Riverbank Filtration

Riverbank filtration uses the suction provided by a pumping station to collect surface water by pulling that water through a natural aquifer. The objective of riverbank filtration is to use the subsurface matrix as a filter and this avoids collecting directly from a potentially risky body of surface water. Jeyakumar et al. (2017) reviewed studies of river bank filtration in India and discussed how this process of filtration is affected by geological conditions and well types, plus the effect of bank filtration on physical, chemical and biological qualities of the water. Gutiérrez et al. (2017) addresses the engineering issues associated with river bank filtration and also

Fig. 8.24 This image shows a public standpost that provides safe drinking water for a community. The image is titled "A drinking fountain in Saint-Paul-de-Vence" by Claritas, and used under the Creative Commons Attribution 3.0 Generic license

mentions its presumed historical role as the precursor technology for other types of drinking water sand filtration methods.

8.2.3.4 Health Risk Associated with Those Microbial Contaminants that Are Not Successfully Removed Before Drinking Water Is Municipally Distributed

This section refers to the study of conventionally treated drinking water. Conventional treatment is good at protecting human health although it is not perfect in terms of either removing or destroying all of the microorganisms that are present in source water. The residual level of microorganisms in treated water represents a quantifiable risk of infectious disease for people who consume that water. This section is not meant to imply that the other treatment options described in this chapter are more effective than is conventional treatment for reducing the risk associated with residual levels of microorganisms in the treated water.

One part of the health risk from water is due to microorganisms that are neither removed nor successfully rendered harmless during municipal treatment processes

Fig. 8.25 This image shows a public standpost that is used for providing drinking water to a community. The image is titled "Communal tap (standpost) for drinking water in Soweto, Johannesburg, South Africa (2941729790)" by author SuSanA Secretariat and is being used under the Creative Commons Attribution 2.0 Generic license

(Hurst 2018) as mentioned above and in Table 8.1. Two epidemiological studies have been performed in Montreal, Canada to assess that risk. Results of a 1991 epidemiological study (Payment et al. 1991) indicated that the annual risk to consumers of acquiring gastroenteritis from ingesting microorganisms contained in the community-distributed conventionally treated tap water of Montreal, Quebec, Canada, was 0.26, equivalent to 26%, or roughly one illness per person every 4 years. The results of that study have been evaluated by Hurst (2018) using a risk estimation technique with knowledge of the estimated level in the treated water of virus potentially pathogenic for humans, plus the levels of *Cryptosporidium* and *Giardia*. The corresponding bacteriological data available for the 1991 study by Payment et al. were not for pathogenic bacteria, and thus the risk of illness due to bacteria in the community distributed water was presumed to be all residual risk not accounted for by virus and protozoa. The summary of findings for that validation of epidemiologically determined annual risk of illness (designated validation exercise 1 in Hurst 2018) per individual consumer were: together all microbial contaminants in the municipally supplied tap water had been responsible for an annual risk of 0.26 (100.0%), the estimated amount of that risk caused by virus in the municipally supplied tap water was 0.20861 (80.2%), the estimated amount of that risk attributable to *Cryptosporidium* in the municipally supplied tap water 0.00267 (1.0%), and the estimated amount of that risk attributable to *Giardia* in the municipally supplied tap water 0.00674 (2.6)%. The difference, or residual level of risk, between the epidemiologically determined annual risk of illness and the level of risk predicted by this estimation technique using data for levels of virus and protozoa in the water was 0.04198 (16.1%) and presumed to have been the bacterial risk.

Several years later, after efforts to improve the efficiency of that community's drinking water treatment facilities, Payment and coworkers performed a second epidemiological study (Payment et al. 1997) during which they found that individuals ingesting the conventionally treated, community-distributed tap water had an overall 0.66 annual risk for incidence of gastrointestinal illness. That contrasted with an annual illness risk of 0.58 among a control group of consumers who either drank bottled water or else drank tap water that originated from the same community distribution system but which had gone through an in-home treatment system including filtration prior to the water being ingested. The difference in annual risk thereby attributed to microorganisms in the tap water during the second epidemiological study was 0.08. The validation exercise presented by Hurst (designated validation exercise 2, in Hurst 2018) for the second epidemiological study of Payment et al. (1997) suggested that of the 0.08 annual risk of illness per individual consumer, the amount which could have been due to virus in the drinking water was 0.068749 (85.9%), with protozoa of the genera *Cryptosporidium* and *Giardia* respectively accounting for approximately 0.00118 (1.5%) and 0.00337 (4.2%) of the observed illnesses. The data which were published by Payment et al. (1997) concerning bacterial concentrations in the tap water for that study represented indicator bacterial groups rather than known waterborne bacterial pathogens. As such, the only way to estimate the proportion of observed infectious disease which would have been attributable to bacterial pathogens contained in the water during the time of the second epidemiological study again was by subtraction. Thus, pathogenic bacteria presumably accounted for an annual risk of 0.00670 representing 8.4% of the observed cases of illness for the second study by Payment et al. (1997).

8.2.3.5 Health Risk Associated with Microbial Contaminants that Are Acquired by Municipally Treated Water During Distribution of the Water

Having made water safe for human consumption we then need to protect its quality. Unfortunately, the need for moving water from where it was treated and made safe to where it will be consumed does involve a possibility that quality of the water will degrade.

8.2.3.5.1 Contamination Occurring in the Water Distribution Network

Part of the risk associated with community distributed drinking water is caused by microbial contaminants that are introduced after the water begins its path through the water pipeline distribution network. Within that distribution network, microorganisms originating from the source water will be joined by biological contaminants that accidentally enter the distribution system in association with infiltrations related to piping engineering failures. Accidental cross contamination of drinking water by septage collection systems represents yet another source of microorganisms that will be entering the drinking water distribution system. The total accumulation of biomass and its ecosystem structure that exists inside of a drinking water distribution system must be understood and controlled, because the biomass in drinking water interferes with those chemical disinfectant processes upon which we rely for delivering safe product water to consumers. The complex microbiology of drinking water distribution networks is discussed in Chap. 9 of this volume "Microbiome of Drinking Water Distribution Systems" by authors "Laurence Mathieu, Tony Paris and Jean-Claude Block, pp. 261–311".

8.2.3.5.2 Microbial Safety Issues Associated with Delivery of Drinking Water by Tanker Truck

If safe water is available but must be delivered in containers, then options for that delivery include the use of either tanker trucks or bottling the water. The microbiology of bottled water will be discussed below in Sect. 8.3.1.

Tanker trucks do offer an economical means of delivering treated drinking water to small communities and groups of people. Water deliveries by tanker may be used to fill community cisterns that have multiple taps, from which people subsequently will be expected to fill their own water containers and then carry that water away for household usage. Tanker trucks also can deliver drinking water for community systems that include distribution pipelines. Such distribution systems may include water transmission lines that feed only community standposts with multiple taps or the distribution systems may have household connections that deliver water to individual dwellings.

Sule et al. (2014) found that the water coming out of delivery tanks in Nigeria generally had higher levels of bacterial contamination, including total and fecal coliforms, than did the water going into the tanks, and the authors suggested that this contamination may have represented the tanker trucks having being used for other hauling instead of being reserved for transporting only treated drinking water. Mendonça et al. (2017) performed bacteriological analysis of drinking water marketed by pipe trucks in Brazil and determined that the water delivered by this means did not meet portability requirements as assessed by total coliforms, thermotolerant coliforms, *Pseudomonas aeruginosa*, and heterotrophic bacteria. Mendonça et al. (2017) found that the microbial quality of the water decreased during the time that it was in the truck tank, and that plastic containers were the worst for water storage perhaps because they are more favorable to microbial attachment which will start the development of a biofilm. The State of Queensland (2015) has offered typical guidelines with their requirement that tanker trucks which are used for transport of potable water cannot be used for hauling other materials, and stating that the trucks as well as their pipes, hoses and fittings, must not only be cleaned but also disinfected with chlorine. The Texas Commission on Environmental Quality (2018) has classified water haulers as transient noncommunity public water systems because they have the potential to provide drinking water to individuals in various locations at a variety of times. Texas considers that such a transient noncommunity public water system may be treated as a consecutive system for purposes of compliance under Texas Commission on Environmental Quality rules. Their regulations require that the water must be from an approved source, the tank can only be used for transporting drinking water and can never have been used for transporting anything other than potable liquids. They also have regulations requiring the water carrying tank to have a manhole, vent and drain. They additionally require that the hoses and pump, as well as the tank itself, must be disinfected at least monthly, microbiological sampling of water from the truck must be done at least monthly, and the hauler must maintain a specified level of disinfectant in the water that is being transported.

8.2.3.5.3 Microbiology of Water Dispensers

We often reply upon public water fountains as well as beverage dispensers in restaurants and other institutions to provide individuals with drinking water. However, even in places that are considered quite modern, the quality of water from those dispensers may fail to meet microbiological standards and the dispensed water can containing opportunistic pathogens (Al Moosa et al. 2015).

8.3 Bottled Water and Water Vending Machines

The use of bottled water and water vending machines are common options but a sense of care must accompany our reliance upon them.

8.3.1 Bottled Water

There are two main source categories of commercially distributed bottled water. The first of those would be water collected from springs. Spring water may be captured where the spring naturally emerges at the land surface. At other times, a spring's water is obtained by using a well to intercept the flow of that water underground at a point before the water can reach the surface. The other source for bottled water is conventional tapwater that has been further purified.

The quality of bottled water varies widely. Fewtrell et al. (1996) sampled natural mineral water and other bottled water, finding that some of the water contained total coliforms although none of the water which they sampled contained either *Escherichia coli*, faecal streptococci or aeromonads. Leclerc and Moreau (2002) examined the microbial ecology of natural mineral water and did find the water to have contained some groups of bacteria which are pathogenic in mammals.

Saleh et al. (2008) examined commercial bottled water from 35 sources, including purified drinking water and natural spring water. Nine of the 35 sources examined by Saleh et al. (2008) contained bacteria and among those the authors found: *Acidovorax delafieldii*, *Acidovorax temperans*, (some species of *Acidovorax can cause sepsis in human*), *Agrobacterium rhizogenes*, *Burkholderia glumae* (is pathogenic in human), *Moraxella caviae* (pathogenic in guinea pig), *Parageobacillus thermoglucosidasius* (formerly *Bacillus thermoglucosidasius*) which produces spores that are used as a challenge for sterilization, *Ralstonia pickettii* (formerly *Klebsiella terrigena* which is pathogenic in human causing mainly respiratory infections), and *Raoultella terrigena* (causes sepsis in human). Totaro et al. (2018) examined spring water and bottled mineral water by polymerase chain reaction (PCR) and found that, even in the absence of cultivable bacteria, the bottled water contained evidence of naturally free living human pathogenic amoeba and bacteria. The organisms detected by Totaro et al. (2018) were *Acanthamoeba polyphaga*, *Vermamoeba vermiformis*, the amoeboid genus *Vahlkampfia* of which at least some species are pathogenic for humans, the bacterial genus *Legionella*, and non-tuberculous mycobacteria.

Amenu (2014) performed a review of literature on the subject of bottled water and found a lack of extensive quality standards for packaged water, as compared to the type of standards which exist for municipal water supplies. The United States Government Publishing Office (2018) did publish regulations which indicate that the water which is bottled must be from an approved source, and that publication included regulations regarding the containers used for the water, the water processing equipment, and the facilities where the bottling is performed.

8.3.2 Water Vending Machines

Water vending machines purify tap water on site and vend it. Such machines are a common approach for purchasing drinking water in a form which is less expensive than is purchasing commercially distributed prepackaged bottled water. The typical series of processes used in those machines would be activated carbon filtration, followed by particle filtration, reverse osmosis, a second carbon filter, then ultraviolet light. Figure 8.26 shows the appearance of a water vending machine. Some states in the United States have both performance and inspection requirements for these machines. The Florida Department of Agriculture and Consumer Services (2017) has inspection requirements for vending machines that sell either water or packaged ice. The Florida requirements are for disinfection of the water with either ultraviolet, or ozone, or some equivalent means, and there also is a requirement of testing the product water for total and fecal coliforms with specified maximum allowed bacterial limits. The California Department of Public Health (2014) similarly requires inspection of water vending machines and disinfection of the water with either ultraviolet, or ozone, or some equivalent, but microbiological testing is not required by California. Figure 8.27 shows a row of water vending machines for

Fig. 8.26 This image shows a water vending machine which processes tap water and then dispenses the product. Such machines typically are coin operated. Water processing done within that machine involves a sequence including initial activated carbon filtration to remove organic compounds from the water, small pore size filtration to remove microorganisms including those which grow on organic compounds that accumulate in the activated carbon filter, reverse osmosis, a second carbon filtration to improve taste following passage through the reverse osmosis filter, then ultraviolet light immediately prior to dispensing

sale in Thailand, where there is a concern that such coin operated vending machines frequently distribute water which fails to meet microbial safety regulations (Fredrickson 2017).

8.4 Community Education Helps with Providing Background Knowledge of Waterborne Disease and Acceptance of the Need for Drinking Water Treatment

Public education can be a critically important aspect of planning water treatment for very small communities of a few families, and for individual households. Reasons for removing turbidity are easily understood because doing so improves both the visual appeal and taste of the water. Understanding the existence and nature of pathogenic microorganisms requires that people learn to believe in dangers

Fig. 8.27 This image is titled "E8661-Pattaya-water-vending-machines" by Vmenkov and shows water vending machines for sale in Pattaya, Thailand. It is being used under the Creative Commons Attribution 1.0 Generic license

associated with things that cannot easily be seen since most of the microbes in water are invisible without some form of technical assistance such as magnified viewing.

The goal of public health microbiology requires that we overcome the invisibility of pathogenic microorganisms. Achieving that goal includes convincing people of there being some benefit in filtering and disinfecting water before they drink it, even though reducing the risk of microbially associated illness may not have been a part of their earlier education. Understanding and utilizing options for destroying infectivity associated with those microorganisms that remain in the water after we have applied turbidity removal processes does sometimes nearly require a leap of faith.

Another issue is that people cannot be helped in ways that they either do not want or cannot afford. The people that we would help need to feel comfortable with the technology. Their use of that technology also must be both affordable and reliably continue to be available.

A community needs to feel ownership of its drinking water treatment processes and understand responsibility for maintenance of shared equipment. If people sense that they have a personal involvement with the water treatment techniques then they are more likely to use those techniques. Instilling a sense of personal involvement can be even more critical when trying to sustain the use of water treatment technologies that are being applied to small communities as compared to household treatments, because if no individual person or group of people within a community feels a sense of personal ownership for the responsibility of participating in maintenance of such structures as sand filters, rainwater cisterns, and chlorinators, then the maintenance will never get done. Brikké and Bredero (2003) have written a good general background document regarding the technologies of drinking water treatment including technical requirements for design, along with discussing the importance of assigned maintenance and helping people to sense participatory ownership of a project so that those people will feel they should care for the equipment. If you just simply give the information and water processing equipment such as sand filters to people without those people sensing ownership and evolvement, then no one in the community will take care of the equipment (Brikké and Bredero 2003).

8.5 Adapting Water Treatment Technologies for Usage by Very Small Communities

8.5.1 Turbidity Removal by Sedimentation, Coagulation and Flocculation

Source water may contain sand and other material that contributes to high turbidity. Turbidity makes the water unpleasant for drinking and the material that contributes to turbidity often will bind chlorine which complicates the task of chemical disinfection. The process of plain sedimentation is used to make water less turbid by allowing sand, silt, and clay as well as other particulates to settle by gravity. Sedimentation can be achieved either by constructing a small community sedimentation basin (United States Agency for International Development et al. 1982) or by allowing the water to settle in buckets. If gravitational sedimentation is not sufficient to lower the turbidity then it may be necessary to add compounds that will assist in achieving coagulation and flocculation.

The techniques used for adapting coagulation and flocculation, as well as filtration and disinfection, downward in scale from the volumes of water required at municipal levels to the volumes that will provide an adequate amount of microbiologically safe drinking water to small communities and households will vary depending upon the number of people being served. And, importantly it should be noted that some water treatment techniques which serve well at the household level will not work at larger scale. For example, the use of boiling as a water disinfection technique works at the level of an individual household but cannot practicably be done in a central location to supply the needs for a million people.

8.5.2 Adaptation of Slow Sand Filtration for Very Small Communities

Rapid sand filtration typically is used at facilities that treat surface freshwater for large communities. Rapid sand filtration occurs as the water flows through a layer of natural sand. Alternatively, sand may be the major component of a mixed media filtration bed. The sand or mixed media is underlain with a layer of gravel which provides for easy drainage of the filtered water. When this technique is used at large scale water treatment facilities, rapid sand filtration is done in combination with preliminary steps of coagulation, flocculation and sedimentation. Rapid sand filters typically will achieve turbidity removal but can accomplish only minimal removal of bacteria, and so the water processed by rapid sand filtration usually is disinfected using either chlorine or ozone prior to the water being distributed. Rapid sand filtration units can be adapted for use even at the household level and that concept will be presented at a later point in this chapter.

Slow sand filters serve as the standard for successfully achieving bacterial removal by sand filtration. Slow sand filtration relies upon importance of a hypogeal biological layer which forms in the uppermost section of the sand because that layer effects most of the bacterial removal. While rapid sand filters can be allowed to run dry, slow sand filters cannot be allowed to run dry because that would cause the sand bed to develop cracks and also risk damaging integrity of the biological layer. If a slow sand filter does run dry then it must go through a restart process and the turbidity of the output water should be monitored so as to determine when turbidity removal has again reached a satisfactory level, indicating that the feedwater is being filtered effectively, at which point the filter is said to be mature.

Slow sand filtration is one of the treatment technologies which might have seemed difficult to adapt for very small scale usage, but indeed success has been found in a form that commonly is termed a biosand filter. Biosand filters are very small community scale slow sand filtration drinking water treatment systems which can be used to provide intermittent flow, and they can work effectively after maturation of the filter (Nair and Ahammed 2014). Figure 8.28 illustrates the design of these filtration units. There is mention of these filters being used for individual households, but such filters often would be relatively expensive for household use. The containers used for housing these filters are about 90 cm meters tall and 30 cm square (Centers for Disease Control and Prevention 2014b). The two most important aspects to be noted are firstly that the outlet for the filter is above the top level of the sand because the filter must be kept submerged to prevent need for a technical restarting of the filter. The height of the outlet pipe is set to maintain 5–6 cm of water above the sand (Centers for Disease Control and Prevention 2014b). Secondly, a diffuser plate is present so that addition of water will not disturb the top layers of the sand. Water to be filtered is just simply poured on top of the diffuser plate. Figure 8.29 shows four of these filters being used to supply drinking water for a small community consisting of a few families in Guatemala. The New England Water Treatment Technology Assistance Center (2018) has recommended that the volume of each charge of raw water applied to a biosand filter not exceed 80% of the

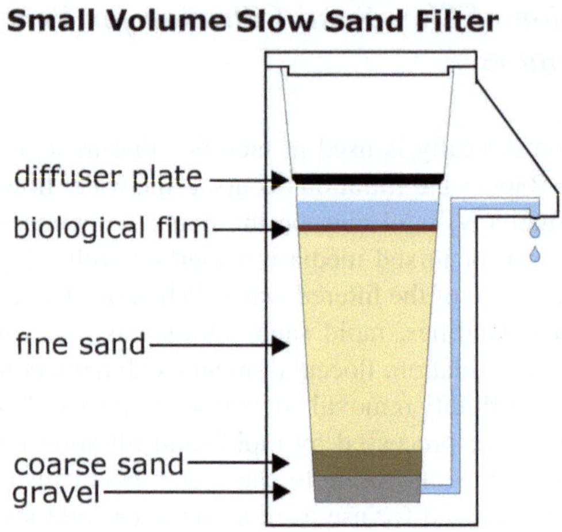

Fig. 8.28 This is a diagram of a small volume slow sand filter, also called a biosand filter. Use of this image is courtesy of the author

Fig. 8.29 This image shows biosand filters, they are small volume slow sand filters used to provide water for drinking and washing. The image is titled "Biosand Filters in Guatemala" by Nora. jeanine530 and used under the Creative Commons Attribution 3.0 Generic license

filter's void volume. Allowing turbid source water to settle as a pretreatment before that water is poured into the filter tank can help to reduce clogging of the filter and thus help to maintain the filters flow rate. When flow through the biosand filter begins to slow severely, the diffuser plate can be removed and the biological layer either agitated or scraped away. The diffuser plate then needs to be reinstalled. Eventually, as with all slow sand filters, it does become necessary to replace the sand.

Clark et al. (2012) has created effective slow sand filters using polyvinyl chloride buckets and summarized information on both the biofilm layer and fineness of the sand. Even though most of the filtration effectiveness for slow sand filters occurs in the biofilm layer, the depth of the filter still is an important consideration because a filter with greater initial sand depth can be scraped and cleaned more times before additional sand is needed. The size and uniformity of the sand particles importantly alters both the flow speed and effectiveness of a slow-sand water filter. Water can filter faster through coarser sand, and yet fine sand with uniform grain size will provide more effective although slower filtration. Having finer grain size unfortunately also means that a filters media will clog more easily. Uniformity in size of the sand particles is helpful for maximizing effectiveness of the filter because having great variability in grain sizes can result in the smaller sand particles filling in the spaces between the larger particles and thus contributing to clogging of the filter.

Two negative aspects of biosand filters are that flow limitations can require excessive waiting, and that inadequate water purification occurs when high flow rates are imposed. Tellen et al. (2010) examined both increasing the sand's effective size and adding zero-valent iron into the media as a disinfectant. Those researchers found that after 65 days of filter operation, the percent reductions in total coliform, fecal coliform, and fecal streptococci averaged 98.9% for traditional versus 99% for their improved biosand filter. Both of those modifications proved to be statistically significant filter improvements. Therefore, the suggestions of Tellen et al. (2010) were that increased sand size and addition of zero-valent iron may counter some drawbacks of traditional biosand filters.

Mwabi et al. (2012) examined biosand filters including those supplemented with zeolite, versus the use of a sand and gravel bucket filter, a ceramic candle filter, and silver-impregnated porous pot filters that had been manufactured using plastic buckets with removable lids. The silver-impregnated porous pot was superior in effectiveness and consistently produced bacterially acceptable drinking water regardless of the quality of the source water treated, given sufficient contact time after filtration so that the silver could take effect against the bacteria. Second best were the biosand filters, which were similar and sometimes slightly better than the ceramic candle filter. The bucket filter, which is a rapid filtration technique, demonstrated the worst efficiency at bacterial removal.

8.6 Water Treatment for Individual Households

There are many options for preparing microbiologically safe drinking water at the household level, and the choices vary based upon the location where the techniques might be used. Summaries of information on household drinking water treatment techniques have been published by Agrawal and Bhalwar (2009), Laurent (2005), Skinner and Shaw (2019c), and the World Health Organization (2011a).

The household level technologies for removing suspended solids include natural settling and filtering, plus the possibility of assisting sedimentation by using chemically induced coagulation and flocculation with subsequent precipitation (Oxfam 2012). Filtration can be accomplished by many techniques, with some of those options such as ion exchange and reverse osmosis requiring a pressurized source of water and thus having been designed primarily for households which receive community supplied tapwater. Other filtration technologies such as cloth filters, granular media filtration including sand, and ceramic filters, have been developed for emergency situations and for households that do not have community supplied tapwater, and correspondingly those techniques do not need a pressurized water supply. Disinfection of the water can be done by using any one of many techniques including chlorination, heating to pasteurize the water using either fuel or solar energy, boiling and distillation, as well as by the use of ultraviolet lamps. General information on performing filtration at the household level and on performing chlorine disinfection of typical household water volumes can be found in the reference by Centers for Disease Control and Prevention (2009). Disinfection of water with ozone is not practical at the household level.

Many of these methods for accomplishing successful drinking water treatment at the household level will be further explained later in this chapter. Using the methods in combinations performed either simultaneously or sequentially creates a multi-barrier approach, of which an example would be coagulation combined with disinfection. The reference pathogens cited by the World Health Organization (2011a) for drinking water treatment were: for bacteria *Campylobacter jejuni*, for viruses *Rotavirus*, and for protozoan *Cryptosporidium*. The targets which the World Health Organization has stated for microbial reduction to be considered "protective" were $2\log_{10}$ reductions in bacteria and protozoa, and a $3\log_{10}$ reduction in virus (World Health Organization 2011a). The targets which have been stated for microbial reduction to be considered "highly protective" were $4\log_{10}$ reductions in bacteria and protozoa, and a $5\log_{10}$ reduction in virus (World Health Organization 2011a). Having accomplished successful microbial removal it then becomes necessary to store the water safely to prevent recontamination. Safe storage requires keeping the water covered. And, because it is not safe to store water in a way that will allow people to wash their hands in the water that someone else subsequently will be drinking, drinking water storage containers must have narrow openings (International Network to Promote Household Water Treatment and Safe Storage 2007). People also must be taught not to drink water from those containers with their mouths directly on the containers opening!

8.6.1 *Coagulation and Flocculation*

Using the processes of coagulation and flocculation can be simply done by individual households as suggested in Fig. 8.30. Aluminium sulfate, commonly known as alum, and powder from *Moringa* seeds are two commonly used compounds for aiding the settling of suspended solids (Oxfam 2012) which then allows a clearer water to be decanted. Whenever possible, the decanted water should be filtered

Fig. 8.30 This image suggests the addition of coagulant to a bowl of water. Use of this image is courtesy of the author

through a piece of clean cloth to help remove the suspended solids because micro-organisms associated with those solids likely will remain infectious and represent a possible source for recontamination of the treated water.

Moringa treatment achieves water flocculation by using a powder prepared from the seeds of *Moringa oleifera* (Doerr and ECHO Staff 2005). Sánchez-Martín et al. (2012) evaluated optimizing the pH, stirring rate and stirring time for that technique. There have been comparative studies in which the effectiveness of *Moringa* was evaluated against aluminum sulfate (Alo et al. 2012; Tunggolou and Payus 2017) and it has been suggested that seeds of *Moringa stenopetala* may be more effective than is *Moringa oleifera* for purifying water (Abiyu et al. 2018). Habtemariam (2017) has described the process by which *Moringa* extracts succeed at water purification. It is important to note that *Moringa* treatment does not remove 100% of all microbial contaminants.

Alum also is used as part of a combined treatment that replies upon disinfection of the water by either heating or chlorination (Crump et al. 2004; Wrigley 2007). Ferric sulfate similarly is used as part of a combined treatment with chlorination, as will be described below.

8.6.2 Filtration

Household water filters are available in different capacities designed for either whole house use or only kitchen use. Commercially available wholehouse filters can be installed in the water supply line near to where the water line has entered the house. A whole house inline paper filter for removing sediments is shown in Fig. 8.13.

Some kitchen usage filtration units are designed to be faucet mounted, meaning that they attach at the end of a standard kitchen faucet and typically those filtration units include a bypass valve which allows for filtering only water that will be used for cooking and drinking. Faucet mounted units have small filters that because of their reduced size must be replaced frequently. Sink mounted kitchen water filters are connected into the kitchen plumbing. Some of the smaller sink mounted kitchen filtration units are designed to sit on the counter next to a sink. There are larger kitchen units designed for installation under the counter and many of those will include a reservoir tank that holds filtered water. The more common household filtration options include woven string, pleated paper, spun polypropylene, and activated carbon. Reverse osmosis filtration units which include several of these techniques are available for household kitchen use, they typically are installed under the counter and I briefly will discuss them later in this chapter.

For a general review of household filter units I would suggest the articles by Agrawal and Bhalwar (2009), and by Johnson and Scherer (2011). Most of the fabric, polypropylene and paper filters have effective pore sizes greater than the diameters of viruses and bacteria, and thus any effectiveness which such filters have against those groups of microorganisms would depend upon removing microorganisms that are associated with solids. Those filters may, however, serve to remove protozoa and metazoa. I will discuss later in this chapter the use of cloth filters, including folded sari cloth, to reduce the risk of cholera and *dracunculiasis* by removing copepods from water.

8.6.2.1 Household Rapid Sand Filtration

Rapid sand filters can easily be designed for household use. Figure 8.15 shows a rapid sand filter that was constructed using a 5 US gallon plastic bucket, and these often are described as being 'bucket filters'. Both the United States Agency for International Development et al. (1982) as well as Skinner and Shaw (2019c) have published designs for constructing rapid sand filters to serve the needs of households. Such filters are effective at removing turbidity but their capability for removing microorganisms is considered poor, and so achieving microbiological safety of the treated water would require that the filtered water be disinfected.

8.6.2.2 Porous Pot Filters

Porous pot filters function by allowing water to permeate through the sides and bottoms of either stone or ceramic pots into which the water has been placed. The filtered water then is retained in a less porous container located beneath the porous pot. If the water to be filtered has a high level of turbidity then you should first prefilter the water through a clean piece of fine cloth (Oxfam 2012) which will reduce clogging of the porous pot. That prefiltration easily can be done by placing a

Fig. 8.31 These images show a stone water filter at the Santa Catalina monastery in Arequipa, Peru. The left image is titled "Water filtration stone 2, Santa Catalina monastery, Arequipa, Peru", the right image is titled "Water filtration stone, Santa Catalina monastery, Arequipa, Peru", both are by Pethrus and being used under a Creative Commons Attribution-share alike 1.0 license. The ceramic pot beneath the filter serves as a receptacle for the filtered water

piece of clean cloth across the top of the porous pot filter and pouring the dirty water though that cloth into the pot.

8.6.2.2.1 Introduction to Porous Pot Filters

Porous pot filters also are called porous jar filters. They consist of a permeable pot which serves as the filter through which water slowly permeates, and there will be a less permeable container called a receptacle that sits below the filter and into which the filtered water collects. Historically these types of filters have been made of either carved stone or ceramic. There still is some usage of the stone porous pot filters Fig. 8.31 (Ramos 2018) although perhaps that mostly is done for a sense of nostalgic curiosity, and apparently new stone filters are fashionable as decorative items in Latin America. Figure 8.32 shows the historical development of ceramic porous pot filters, which today are manufactured in a more modern style that uses a plastic bucket as the container into which filtered water collects. A small spigot is fitted to the receptacle so that filtered water can be withdrawn without inserting either dirty hands or potentially contaminated implements into the receptacle.

The composition of the mixture used for creating modern ceramic filter pots consists of kaolin, sawdust and grog. In this sense, grog will be either ground ceramic or a ground mineral. The formulation and sintering temperature will together determine the removal efficiency that subsequently can be achieved by the filtration pot. Lower porosity ceramic will achieve a greater removal of bacteria

Fig. 8.32 These images show colonial, Victorian era and modern ceramic filters. The creation of household ceramic water filters with collection reservoirs for holding the processed water has been a developing concept for several centuries. The ceramic material used for the filter has a higher permeability than does the receptacle or reservoir into which the water collects. The upper left image shows a Colonial era Colombian ceramic filter courtesy of Benjamin Villegas, Villegas Editores, Bogota, Colombia. The upper right image is of a "germ proof" Victorian era ceramic water filter with a brass tap, titled "CeramicWaterFilter" by ClarkMills and used under the Creative

and turbidity, albeit the filtration rate of lower porosity ceramic correspondingly is slower. van der Laan et al. (2014) found that the burnt material content of the ceramic did not affect *Escherichia coli* removal efficiency. Zereffa and Bekalo (2017) have examined how the composition and sintering temperature influenced removal of Ca^{2+}, Mg^{2+}, iron, nitrite, and conductivity, as well as affecting pH of the filtered water.

Properly caring for a ceramic pot filtration unit should include pouring the raw water through clean cloth as a prefiltration to remove suspended particulates prior to placing that water into the filter pot. Prefiltration treatment will remove some suspended particulates from the water and thereby reduce clogging of the pores that are in the ceramic body. This prefiltration can easily be done by placing cloth over top of the ceramic filter pot, then pouring water through the cloth into the filter pot (Filtron 2018). Silver often is added to the pots and silver is effective as a disinfectant against some bacteria, fungi and virus, but there are toxicological limits for ingestion of silver (World Health Organization 2018a). Addition of the silver may be done by either applying colloidal silver to the ceramic pot or incorporating $AgNO_3$ into the clay, although incorporating $AgNO_3$ can result in the initial water which passes through the filter containing too high a level of silver (Mwabi et al. 2012). Silver is commonly used both in other types of domestic ceramic water filters and in powdered activated carbon filters for the stated purposes of reducing biofilm growth and potentially providing an additional level of water treatment. Copper and silver ionization often is used in swimming pool disinfection, also is used to prevent *Legionella* bacterial colonization of hot water plumbing systems, and may serve as a secondary disinfectant of drinking water supplies (World Health Organization 2018a).

8.6.2.2.2 Specific Details About Performance of Porous Pot Filters

Ceramic filters typically can remove most of the larger protozoan and bacterial organisms from water, but are ineffective at removing the much smaller viral organisms and the filtration effectiveness varies depending upon the production quality of the ceramic. Ceramic pot filters now often are coated with silver which leaches into the water that is being filtered, and if sufficient contact time is allowed between filtering of the water and the time when that water is consumed, then silver in the filtered water will show good effectiveness in achieving bacterial inactivation (Innovación para el Desarrollo y la Cooperación Sur-Sur 2018). Using silver as part of the pot construction unfortunately seems to decrease the effectiveness with which the filter removes viruses. Studies have also shown significant bacterial contamination of the filtered water either when poor-quality locally produced filters are used, or

Fig. 8.32 (continued) Commons Attribution-Share Alike 3.0 Unported license. The lower images are a photograph and cutaway drawing of a Filtrón ceramic water filter courtesy of Potters for Peace, Boulder, Colorado

when the receptacle is contaminated at the household level (Centers for Disease Control and Prevention 2012). Because there is no residual protection for the filtered water other than the potential presence of silver, it is important that users be trained to properly care for and maintain both the ceramic filter and receptacle. Given these precautions, an impressive 60–70% reduction in diarrheal disease incidence has been documented in users of ceramic pot filters (Centers for Disease Control and Prevention 2012).

van Halem et al. (2007) tested for 12 weeks colloidally silvered and unsilvered ceramic filter pots from three production locations, Cambodia, Ghana and Nicaragua, for removal of microorganisms from canal water and for presence of metals in the filtrate. They also studied the microstructure of the filter material using mercury intrusion porosimetry and bubble-point tests. The effective pore size of the ceramic was measured to have been a mean of 40 mm, which they noted is larger than many pathogenic microorganisms and that also indicated microbial removal by the filters to be dependent upon high tortuosities in the filter material facilitating a higher chance of water contaminants being removed by adsorption, diffusion and sedimentation.

van Halem et al. (2007) found that turbidity in the raw canal water was well removed by the filters, but turbidity removal corresponded with clogging of the filter material. Scrubbing the insides of the pots to remove accumulated layers of solids that had been retained on the surface of the ceramic was beneficial but did not stop the effect of long term clogging which occurs within the ceramic body. The filters without silver did not show different water flow reductions compared to filters with a silver coating. Notably, those authors do not seem to have prefiltered the water by any means such as perhaps passing it through a cloth to remove suspended particulates prior to placing that water into the ceramic filters.

van Halem et al. (2007) found that several metals leached from the ceramic pots, particularly in the first few weeks, and those were aluminium, antimony, arsenic, barium, copper, manganese, silicon and silver. For most of those compounds the effluent measurements eventually decreased to concentrations below World Health Organization guidelines. Arsenic was the only compound found to leach in concentrations above the acceptable level of 10 mgL. Although the arsenic levels did reduce over time, arsenic levels for the Cambodian filters during even week 12 still were high. The suggestion was that filter material should be tested for leaching of arsenic, especially if the ceramic was made in regions where high arsenic concentrations are known to be present in the soil.

van Halem et al. (2007) compared the filter pots for removal efficiency of total coliforms naturally present in the canal water, sulphite reducing *Clostridium* spores, *Escherichia coli* K12 and MS2 bacteriophage. Their summary microbiological results are that bacteria and bacterial spore removal was higher with the silvered pots, but contrastingly bacteriophage removal was lower with the silvered pots. The specific microbiological results are that silver impregnated ceramic pot filters successfully removed total coliforms, sulphite reducing *Clostridium* spores, and *Escherichia coli* K12. Filters manufactured in Nicaragua without application of silver showed the highest concentrations of *Escherichia coli* in the filtered water.

The Nicaraguan filters with silver usually did achieve the bacterial guideline level of no detectable coliforms in any 100 mL sample of filtrate. Water quality produced by the filters from Ghana and Cambodia was good but did not meet the bacterial guideline level. MS2 bacteriophage were only partially removed from the water, and viral removal was significantly better for the filters that lacked colloidal silver impregnation, a finding that was noted to be in contrast with previously reported studies. van Halem et al. (2007) concluded that the removal of *Escherichia coli* proved better by filters with a layer of silver but that impregnation of colloidal silver was not needed to remove significant concentrations of *Escherichia coli* as filters without silver also reached high $\log_{10}$ reduction values. They also concluded that colloidal silver was unnecessary for achieving efficient removal of *Clostridium* spores by the filter pots.

van der Laan et al. (2014) subsequently found that storage time of the filtered water in the receptacle, and not contact time with silver during the filtration phase, was the dominant parameter in achieving *Escherichia coli* inactivation by silver. The removal effectiveness for viruses is still of major concern for ceramic pot filtration. The initial hypothesis of van der Laan et al. (2014) that the absence of silver would enhance virus removal due to biofilm formation on the ceramic filter element could not be confirmed.

8.6.2.3 Ceramic Candle Filters

Ceramic candle filters typically are hollow cylinders with one end closed and rounded, and the other end having a sealed connection to an outflow drain. The ceramic usually is made from sintered diatomite (diatomaceous earth) which is the fossilized remains of diatoms. Water flows from outside the candle to the inside. These filters can use either gravity or another source of hydrostatic pressure to push

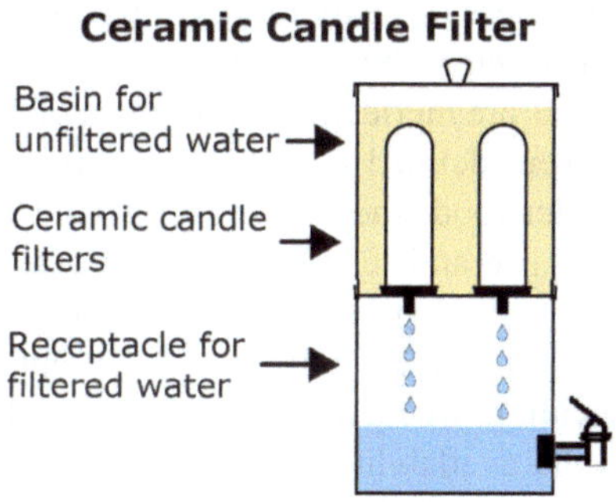

Fig. 8.33 Diagram of a ceramic candle filter that works by gravity. In this drawing there are two hollow ceramic candles which fasten to the bottom of the upper basin and they extend into the unfiltered water of the upper basin. The ceramic candles typically are made of sintered diatomaceous earth. Some filter candles now have an activated carbon core, contain exchange resins, and they may be silvered. The effective pore size of the ceramic candles is graded and seems to range from 0.5 to 0.9 microns. Use of this image is courtesy of the author

water through the ceramic material. Figure 8.33 shows how ceramic candle filters commonly are used to create a household water filtration device which uses gravity as the source of hydrostatic pressure. The ceramic candle filters are similar in concept to the ceramic porous pot filters, except that for the porous pot filters the upper basin is itself the filter (Oxfam 2012). The filter housing for ceramic candle filters can have a sealed basin and the system be designed to operate inline connected to the plumbing of a home water supply. There also are large volume commercial scale applications for ceramic candle filters. In terms of maintenance, it periodically is necessary to scrub away large solids that have collected in a layer on the outer side of the filter candles. Ceramic candle filters typically have effective pore sizes ranging from 0.5 to 0.9 microns and are able to remove most bacteria and protozoa. Importantly, not all bacteria will be removed by these ceramic filters as some bacteria have a diameter smaller than 0.5 microns. There are newer composition ceramic filters that can achieve partial removal of particulates down to 0.2 microns. The most modern designs for ceramic candle filters incorporate a first stage which consists of the ceramic body, a second stage anti-bacterial matrix integrated into the ceramic mix, a third stage which is an inner core of activated carbon block that can retain organic chemicals from the water that is being filtered, a fourth stage which is an ion exchange resin intended to reduce lead and other heavy metals but potentially also reducing any silver that was added for bacteriocidal quality, and a fifth stage which imparts silver to the water. Indeed, often now the ceramic filter candles are silvered as is done with ceramic porous pot filters to provide some antibacterial disinfectant capability for the filtered water.

The housing containers in which ceramic candle filters are installed generally now are made of either stainless steel or plastic. A common problem is that people often clean the outside surfaces of the ceramic candles and then clean the inside of the bottom container with the same cloth. The outside of the candle is likely to be contaminated and this approach to cleaning the unit will then cause contamination of the clean water in the bottom container (Oxfam 2012).

The quality and effectiveness of many locally produced types of ceramic candle filters may be doubtful, as often is the case for ceramic filter pots. It should not be surprising that ceramic filters are ineffective for removal of mycoplasma and viruses. Indeed, viruses initially were defined as “filterable viruses” because, before their biological nature was understood, they were recognized as being infectious and capable of passing through a ceramic candle filter. The pores in the ceramic material of filter candles eventually will clog just as will the pores in a porous pot filter, at which point the ceramic filter candles will need to be replaced. Oxfam (2012) has suggested that when deciding whether to distribute candle filters, consideration should also be given to the availability of replacement ceramic candle filters and taps. It is important to understand that significant training and follow up will be needed if people are not familiar with the filters. Another important understanding is that turbid water may need to be prefiltered through a clean cloth before using the ceramic candle filter. One means of using cloth to prefilter the water would be to place the cloth over top of the basin before adding water to the basin, then pour water into the basin through the cloth.

Ceramic filtered water lacks the residual protection that a disinfectant such as chlorine would provide, and so it is important that users of ceramic filters be trained to properly care for and maintain the ceramic filters and to not contaminate the filtered water receptacle.

8.6.2.4 Cloth Filtration

The natural ecology of *Vibrio cholerae* is its existence as a commensal on the chitinous shell of aquatic crustaceans including copepods. Ingesting some strains of that bacterial species will produce the disease cholera, and filtering copepods out of the source water before drinking that water can reduce the risk of cholera disease (see Hurst Chap. 7, "Briefly Summarizing Our Understanding of *Vibrio cholerae* and the Disease Cholera" pp. 173–184). Colwell et al. (2003) suggested the use of multiply folded Sari cloth as a filter method applicable at the household level for the purpose of removing copepods from intended drinking water so as to reduce the incidence rate of cholera. Household use of cloth filtration is present in Fig. 8.14. Colwell et al. (2003) also presented alternative suggestions for reducing the incidence of cholera, those being either the use of nylon netting for removing copepods or boiling of the water, although for some households boiling drinking water represents the use of expensive firewood. Regarding the Colwell et al. (2003) reference, there are notes on its first page about boiling plus instructions at the top of its page 1053 for using the cloth filtration technique. Huq et al. (2010) discovered by analysis of data that several years after initiation of a cloth filtering project 31% of the community's women still used filtration for household water, of which 60% used sari cloth filtration. Those results showed that sari cloth filtration not only was accepted and sustained by the villagers and benefited them in reducing the incidence of cholera, but that use of filtration also benefitted their neighbors who did not filter water. The benefit to neighbors was unexpected and presumably relates to filtration having reducing primary transmission via ingestion of contaminated water, and by having prevented some cases of primary transmission there would have been a subsequent benefit of reducing the chance that neighbors would acquire cholera by person-to-person secondary transmission. The reference by Hurst (2018) discusses primary and secondary transmission routes.

The two types of cloth suggested for this water filtration technique have a large effective pore size and cannot be expected to remove either bacteria or viruses, unless those categories of microorganisms are attached either to large suspended solids or to other particulates such as copepods. Cloth filtration is useful for straining out suspended natural solids to prevent clogging of ceramic filters. The technique of cloth filtration possibly can increase the effectiveness of solar disinfection and would do that mostly by eliminating solids that either can block or absorb ultraviolet radiation. Removal of solids additionally may reduce the chance of microbial regrowth following solar disinfection. Cloth filtration also can remove solids that are produced by the coagulation and flocculation processes which result from using

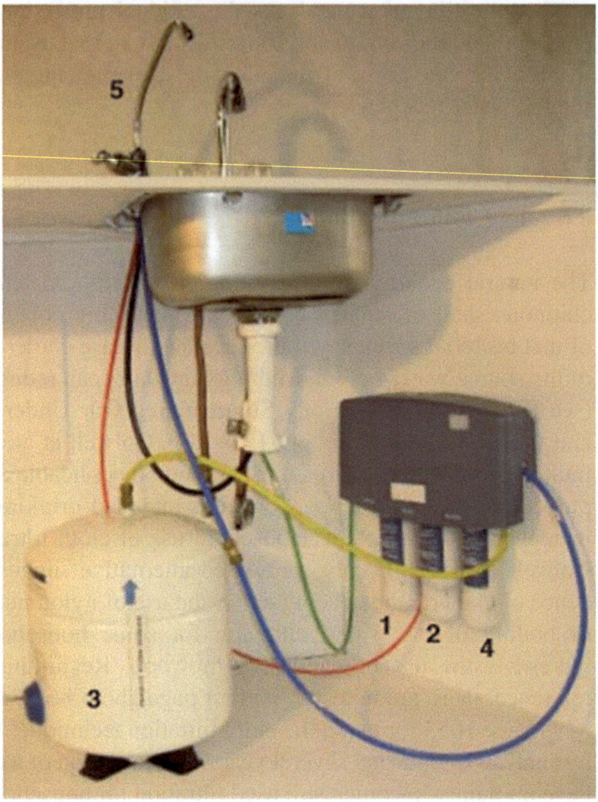

Fig. 8.34 This image is titled "A typical home RO system" including (1) particle filter, (2) reverse osmosis membrane unit, (3) pressurized treated-water storage container, (4) carbon adsorption post-filter and (5) separate treated-water tap. It is by North Dakota State University, authors Roxanne Johnson and Tom Scherer, and being used under a Creative Commons Attribution-share alike 3.0 license. https://www.ag.ndsu.edu/pubs/h2oqual/watsys/wq1047.pdf

such compounds as alum, ferric sulfate, and *Moringa* extracts to help clarify water prior to disinfection of the water.

8.6.2.5 Reverse Osmosis

Reverse osmosis filtration also is available as a household technology Fig. 8.34 (Johnson and Scherer 2013). Yari et al. (2018) studied the microbiological quality associated with household water desalination devices that included reverse osmosis membranes. Those devices were being purchased and used by community members because of the high levels of total dissolved solids in their community provided tap water. Yari et al. (2018) determined that the heterotrophic bacterial level in the output water of those household water desalination devices was greater than the bacterial level found in the input water, and presumably that increase was due at least in part to development of a biofilm in the filtration units.

8.6.3 *Disinfection*

There are several techniques that can be used to achieve disinfection of drinking water for individual households. Those most commonly suggested are chlorination, solar disinfection which is a thermal pasteurization, and boiling. Ion exchange disinfection also can be done (Agrawal and Bhalwar 2009) and that primarily uses iodine in the form of either tri-iodide or penta-iodide exchange resins. Portable and point-of-use devices that use iodine exchange resin have been developed and extensively evaluated for the inactivation of waterborne pathogens, primarily in developed countries. Most of these iodide exchange resin devices are designed as pour through cups, pitchers, and columns through which water is passed so that microbes come into contact with iodine on the resin. That technology is too expensive and complex for much of the world's household needs, although when cost is not an issue as with the water recycling that takes place on the International Space Station, iodine resins can become very important and that will be discussed later in this chapter.

8.6.3.1 Chlorine Disinfection

Chlorine disinfection is an important technique that, given sufficient contact time and disinfectant concentration, seems effective against all microorganisms although destruction of protozoan cysts and oocysts can require lengthy periods of chlorine exposure. It is important to leave a residual level of chlorine in treated water to protect against accidental recontamination of the water. The available options for household disinfection using chlorine are sodium hypochlorite which often is obtained as liquid chlorine laundry bleach (Centers for Disease Control and Prevention 2017; United States Environmental Protection Agency 2017), calcium hypochlorite which is a power and can be purchased for household usage as powdered bleach, and sodium dichloroisocyanurate which most commonly is available as a tablet (Clasen and Edmondson 2006).

Sodium dichloroisocyanurate is considered to be more stable that is the sodium hypochlorite in household bleach (Clasen and Edmondson 2006). Sodium dichloroisocyanurate is available as effervescent (self-dissolving) tablets which will disinfect two quarts of water per tablet. It is recommended that turbid water should first be allowed to settle and then filtered through either a clean cloth, paper towel, or a coffee filter, if such filters are available, prior to treating the water with sodium dichloroisocyanurate and then drinking the water (Federal Emergency Management Agency 2006). There is a general note of caution which must be considered when treating groundwater that contains reduced forms of arsenic, iron, manganese, and sulfur, which can react with chlorine-based disinfectants and effectively increase the waters chlorine demand. Naser et al. (2018) is a reference for this reaction, and in their study they noted that iron present in groundwater did affect the level of residual chlorine which remained after water was treated with sodium dichloroisocyanurate. Chlorine treatment (Laurent 2005; International Network to Promote Household Water Treatment and Safe Storage 2007) often is used as a disinfection technique in combination with coagulants and flocculants, and those combination techniques will be presented later in this chapter.

8.6.3.2 Pasteurization, Boiling, and Distillation

Boiling is a standard for effectiveness in destroying the infectivity of microorganisms in water. The World Health Organization's 2015 publication provides a summary table of microbial inactivation rates by temperature and time for bacteria, protozoa and virus and clearly shows that a good amount of safety can be accomplished by heating water to less than the boiling point. Heating liquid to a specified temperature that is below the boiling point is termed pasteurization, and the exact combinations of temperature and time that are used for achieving pasteurization vary quite a bit. It is better to store the thermally treated water within the same container in which it was either pasteurized or boiled because thermal treatment will not leave any residual disinfection capacity in the water as would an alternative treatment such as chlorination.

8.6.3.2.1 Solar Water Disinfection

Using sunlight to pasteurize water has been suggested and researched by many people. Pasteurization with sunlight, which is called solar water disinfection and known by the acronym SODIS, has been done by placing water into clear polyethylene terephthalate (PET) plastic bottles and then exposing those bottles of water to sunlight. A good general reference on solar water disinfection would be that by Luzi et al. (2016). Figure 8.35 shows examples of solar water disinfection. As seen in Fig. 8.35, the outside of PET bottles may be painted black to increase the rate at which the contained water can be heated by exposure to sunlight. Ciochetti and Metcalf (1984) experimented with using a solar box cooker to help heat glass containers of river water for achieving disinfection. The technique of solar disinfection offers advantages in that it is inexpensive, easily understood, and easy to apply at either zero or very low cost. Solar disinfection is independent of energy sources other than sunlight, and is independent of the supply chains needed for chemical disinfectants. Solar disinfection is dependent upon access to PET bottles, although those are available as discarded items virtually everywhere. Solar disinfection does involve a relatively high labor demand and also this technique requires a long treatment time. Treatment by SODIS may take 6 h in sunlight and 2 days if the sun is obscured by clouds. Two days may well be a prohibitive length of time for people to await a drink of water. Also, solar water disinfection can have limited aspirational appeal in that this is perceived to be a 'poor people's method'. A very serious detriment is that the water is warm following treatment and people do not find it refreshing to drink warm water. Another drawback to solar disinfection is that

Fig. 8.35 Solar pasteurization of water also is called solar disinfection and abbreviated SODIS. The upper image is titled "Indonesia-sodis-gross" by SODIS Eawag and is being used under the Creative Commons Attribution 3.0 Generic license. The lower image shows water in a clear plastic

the heated water has no residual disinfectant capacity, and unfortunately some components of the microbial population that survive thermal treatment can experience regrowth in the treated water.

There are thermal indicators which consist of small sealed tubes containing a low melting wax that can be inserted into the bottles and those indicators will show when the water has reached a safe temperature because the wax will have liquified (Safapour and Metcalf 1999). Safapour and Metcalf (1999) used a cardboard reflector to redirect sunlight onto water contained in a black painted jar, and inside the jar those researchers had placed the reusable water pasteurization indicators (WAPI), which were a clear polycarbonate tube partially filled with a soybean wax that melts at about 70 °C. It was suggested that if the WAPI tube wax melts and falls to the bottom of the tube, then it would indicate that pasteurization conditions had been reached.

Information from Keogh et al. (2017) has indicated that ideally SODIS treatment requires the water to have a turbidity of less than 30 NTU (nephelometric turbidity units), although turbidities of natural water can exceed 200 NTU. Those authors suggested that reducing the turbidity of water would assist in making certain solar disinfection is effective (Keogh et al. 2017). Another point to notice is that, if the water initially has a high turbidity, then biofilms as well as sludge layers can develop within those bottles that contain the treated water. Bacterial regrowth can occur in those sludge layers. Keogh et al. (2017) indicated that reducing turbidity prior to solar disinfection may reduce the potential regrowth of bacteria (Keogh et al. 2017). Keogh et al. (2017) suggested improving microbial stability of highly turbid water by first using *Moringa* seed powder as a flocculation agent, subsequently decanting the cleared water, and then using SODIS to pasteurize the decanted water.

It has been mentioned several times in the literature that the solar disinfection process will receive benefit from the sun's ultraviolet light radiation, but it is not clear how much ultraviolet light actually would be transmitted into the water containers (Agrawal and Bhalwar 2009).

8.6.3.2.2 Boiling

Boiling water for a sufficient length of time will make the water safe to drink and the recommendations for duration of boiling range from 1 min to 5 min (Doerr and ECHO Staff 2005), Federal Emergency Management Agency (2006), United States Environmental Protection Agency (2017). The length of boiling time required is dependent upon the altitude at which the boiling occurs (United States Environmental Protection Agency 2017). Boiling is a simple and very effective way of killing all classes of microorganisms although the needed fuel can be expensive. Boiling water

Fig. 8.35 (continued) bottle versus a bottle whose exterior has been painted black to increase solar heating and usage of the image is courtesy of the author

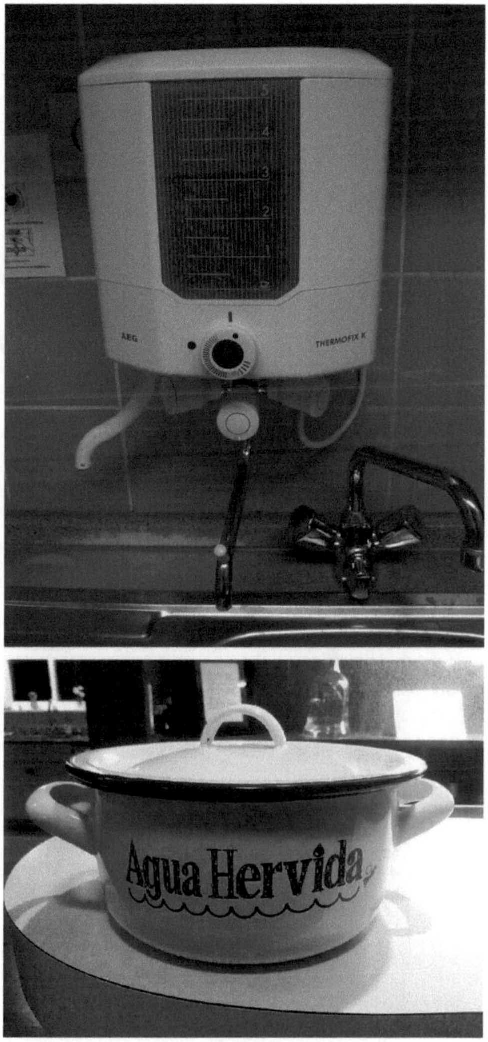

Fig. 8.36 The upper image shows an installed electric water boiling apparatus. The image is titled "Kochendwassergerät" *by Tetris L* and is being used under the Creative Commons Attribution 1.0 Generic license. The lower image shows a designated pot with a lid for boiling and then storing water, a translation of the lettering on this pot is 'Boiled Water'. Usage of the lower image is courtesy of the author

involves the possibility of scalding accidents due to the very hot water temperatures, and that is a particular hazard if small children are assigned responsibility for accomplishing the task of boiling the water. Boiled water easily can become recontaminated once it becomes cooled because the boiled water lacks any disinfectant capability (World Health Organization 2017) and for that reason boiled water should be stored carefully, kept covered, and preferably be kept in the same container that was used for the boiling. In at least Colombia, it is possible to buy pots marked "Aqua Hervida" [boiled water] made just for the purpose of boiling water and then storing the boiled water covered in the same container. There also are commercial kitchen appliances which will boil water automatically and then allow easy dispensing of the water (Fig. 8.36).

Fig. 8.37 These images show the components and assembly of a solar still designed for using heat from sunlight to distill water. Usage of these images is courtesy of the author

8.6.3.2.3 Distillation

Distillation will eliminate all microbial contaminants, salts, and turbidity. Effective distillation can be done using either solar energy or fuel as a source of heat (Skinner and Shaw 2019c; Federal Emergency Management Agency 2006). Figure 8.37 shows a solar distillation apparatus for which the design is based upon using a small collecting bucket placed inside of a larger bucket. The larger bucket will hold the water that is to be distilled. All of this is covered over with a sheet of clear plastic that is sealed to the outer sides of the large bucket using tape. A rock sits as a weight atop the center of the plastic, directly above the center of the collecting bucket, so that the condensed water will drip into the collecting bucket. A section of relatively small diameter plastic tubing, one end of which is inserted into the collecting bucket and the other end of which extends beyond the outside of the solar still, is used to obtain distilled water from the collecting bucket using either suction or siphoning. A section of large diameter plastic tubing with a funnel attached to one end of the tubing is used for supplying the outer bucket, that funnel is on the end of the tube which extends outside of the still. The outside of the large bucket can either be painted black or covered with black plastic to increase the effectiveness of having sunlight heat the water.

8.6.3.3 Combination Treatments that Are Designed to Both Remove Turbidity and Disinfect the Water

Combined flocculation and disinfection can be done using commercially produced products that are available as tablets and powders. The use of ferric sulfate as a coagulant with sodium dichloroisocyanurate as a disinfectant has been evaluated by Légaré-Julien et al. (2018) whose studies used commercially marketed double-layered tablets for which a 5 min manual stirring period is recommended, followed by 55 min of settling, and then cloth filtration. Allowing a settling period of twice that time was suggested for cold water temperatures (Légaré-Julien et al. 2018). The use of ferric sulfate as a coagulant in combination with calcium hypochlorite as a disinfectant has been evaluated by several researchers and that combination is available as a commercial powdered product called "PUR Water Purification Sachets" that are sold in sealed packets. The procedure used for that product is to add the contents of a sealed packet to an open container holding 10 L of water, stir for 5 min, let the solids settle to the bottom of the bucket and then filter the water through a cotton cloth into a second container. After that filtration, there is a waiting period of 20 min for the hypochlorite to further inactivate microorganisms (Centers for Disease Control and Prevention 2014a; Laurent 2005; United Nations Development Programme 2007). It has been reported that prevalence of diarrhea was approximately 40% lower in households which used that combination of ferric sulfate as a coagulant in combination with calcium hypochlorite as disinfectant (Chiller et al. 2006). Crump et al. (2004) compared the combination of ferric sulfate and calcium hypochlorite versus using either alum alone, sodium hypochlorite alone, and alum plus disinfection with sodium hypochlorite. The findings of Crump et al. (2004) were that: the combined flocculant-disinfectant product would be useful in settings where source waters are both highly microbially contaminated and turbid, and that all high turbidity samples treated with alum alone failed to reach the World Health Organizations bacteriologic potability standard.

8.7 The Water Recycling Unit Used on the International Space Station

When humans began sealing ourselves into living quarters that kept us isolated from the natural life support systems of the terrestrial surface on which we had evolved, it became necessary to develop artificial life support systems which could sustain us for at least a short while. Efforts at developing artificial life support systems began in 1896 with the Holland submarine "Plunger" and have progressed to the closed ecological systems that are being used onboard the International Space Station (Hurst et al. 1997).

Figure 8.38 shows the designs of several earth orbital "Space stations" as they have been imagined and of one which exists in our present reality. The lower left part of that image is titled "Space Wheel" by Syd Mead and is a futuristic suggestion from the 1950's of what an earth orbital station eventually might look like. The central part of that image is a drawing from the beginning of this century titled "Global Economy Saipan" by Andrew Au that hangs on my office wall, and for me it depicts construction of biomechanical orbital stations. The upper right part of that image is a photograph of the International Space Station.

The prospect of prolonged residence in space with either minimal or no resupply from earth has presented an enormous technical challenge. One of the most technically challenging tasks has been the successful adaptation of water treatment technology for use by crew members onboard the International Space Station, which represents a very isolated household. The water recycling unit on the International Space Station was designed with the goal of providing 10 L of drinkable water per day for each person in the station. All possible sources of water available to the International Space Station need to be collected and recycled. The only supplemental water is that which comes on resupply flights from earth. I was fortunate to have participated in testing the viral removal efficiency for that recycling unit (Roman and Hurst 1998). Combining treatment technologies into a multistage process allows for increased assurance of health safety. My guess would be that successfully removing chemicals was more challenging than the demands required for microbial removal.

The sources of water used for testing the recycling unit came from facilities at the NASA (National Aeronautics and Space Administration) Marshall Space Flight Center in Huntsville, Alabama where the testing was conducted. Figure 8.39 shows some of those water sources. The upper left image shows the closed entry door to the testing room, which was a temporary building housed inside of a larger building. Much of the source water would be collected from peoples usage of that testing room. The person seen in the upper left image has walked across a tacky sheet on the floor to remove particulates from the bottoms of his shoes, and he is shown using a powered shoe brush required to eliminate particles from the sides of his shoes prior to his entering that room. Inside of that room was a shower shown in the upper right image, from which the drainage water was collected. The room also contained toilet facilities shown in the lower left image, from which the urine provided an additional source of water. The lower right image shows a laundry washing machine

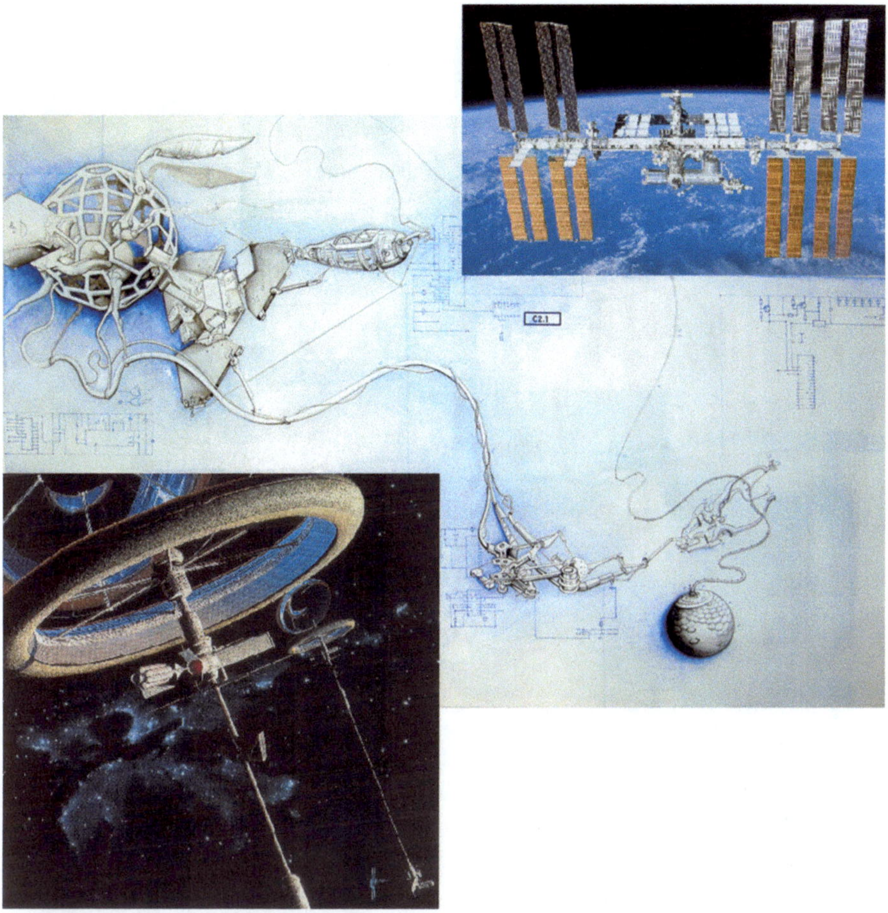

Fig. 8.38 This image shows the designs of earth orbital space stations and is a composite created by Christon J. Hurst, used with his permission. The upper right image is a photograph of the International Space Station, is titled "International Space Station after undocking of STS 132" by NASA/Crew of STS-132 and it is a public domain image. The central image is titled Macromeme: Global Economy Saipan by Andrew Au, imagined as depicting construction of biomechanical orbital stations and is used here with permission of the owner. The lower left image is titled "Space Wheel", Copyright© Syd Mead, Inc. www.sydmead.com and used with permission

which supplied spent water and a clothing dryer from which escaping humidity was collected. That laundry equipment was not in the testing room. Figure 8.40 shows the other areas inside of the testing room. The upper image shows a sink from which water used for hand washing, wet shaving and tooth brushing was collected, and a microwave oven which would have released some steam to also be recycled. The lower image shows exercise equipment plus shirts which had been worn during

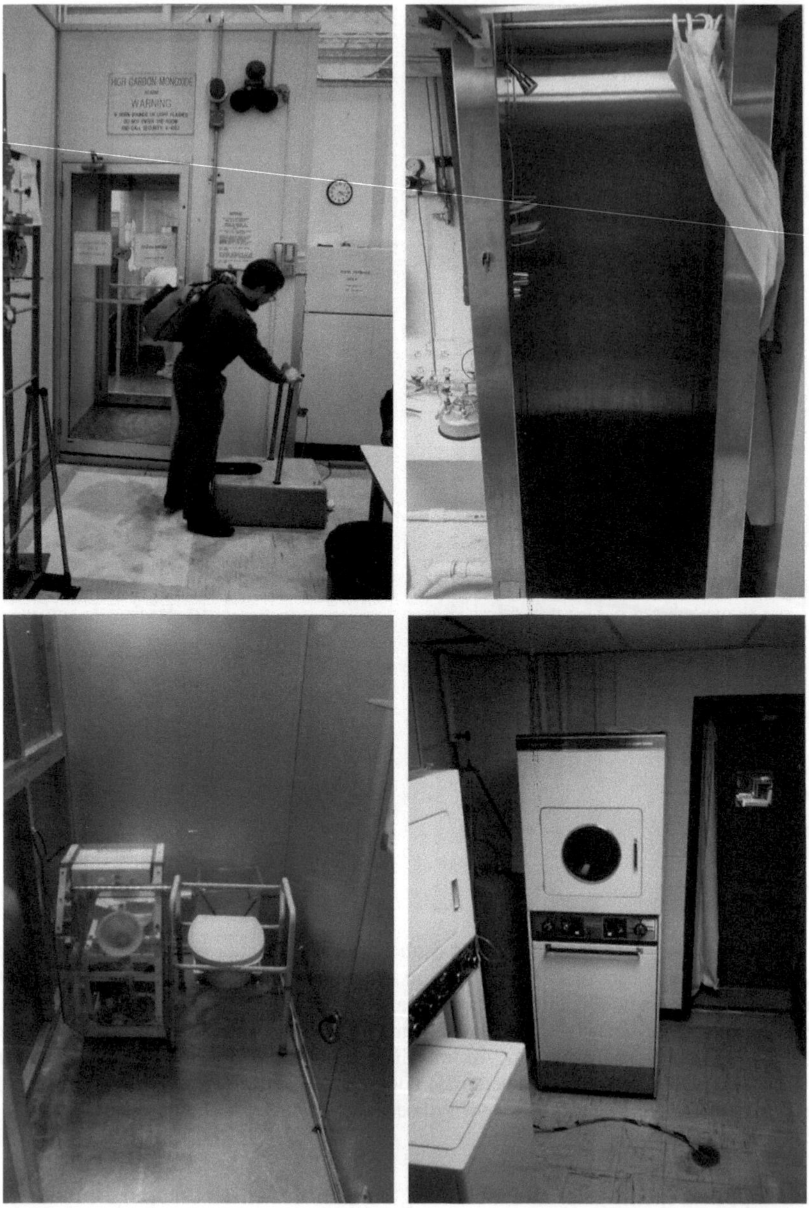

Fig. 8.39 These images show sources of water that were used for evaluating the water reclamation unit developed at NASA in Huntsville, Alabama for the International Space Station. Most of the water was collected from sources located within in a small temporary structure contained inside of a much larger building. Clockwise from the upper left these photographs show: a person using a powered shoe brush to clean their shoes before entering the temporary structure; a shower bath; a laundering machine and clothes dryer; and a toilet with urine and fecal collection separated. The laundering machine and dryer were housed separately. Usage of these images is courtesy of the author

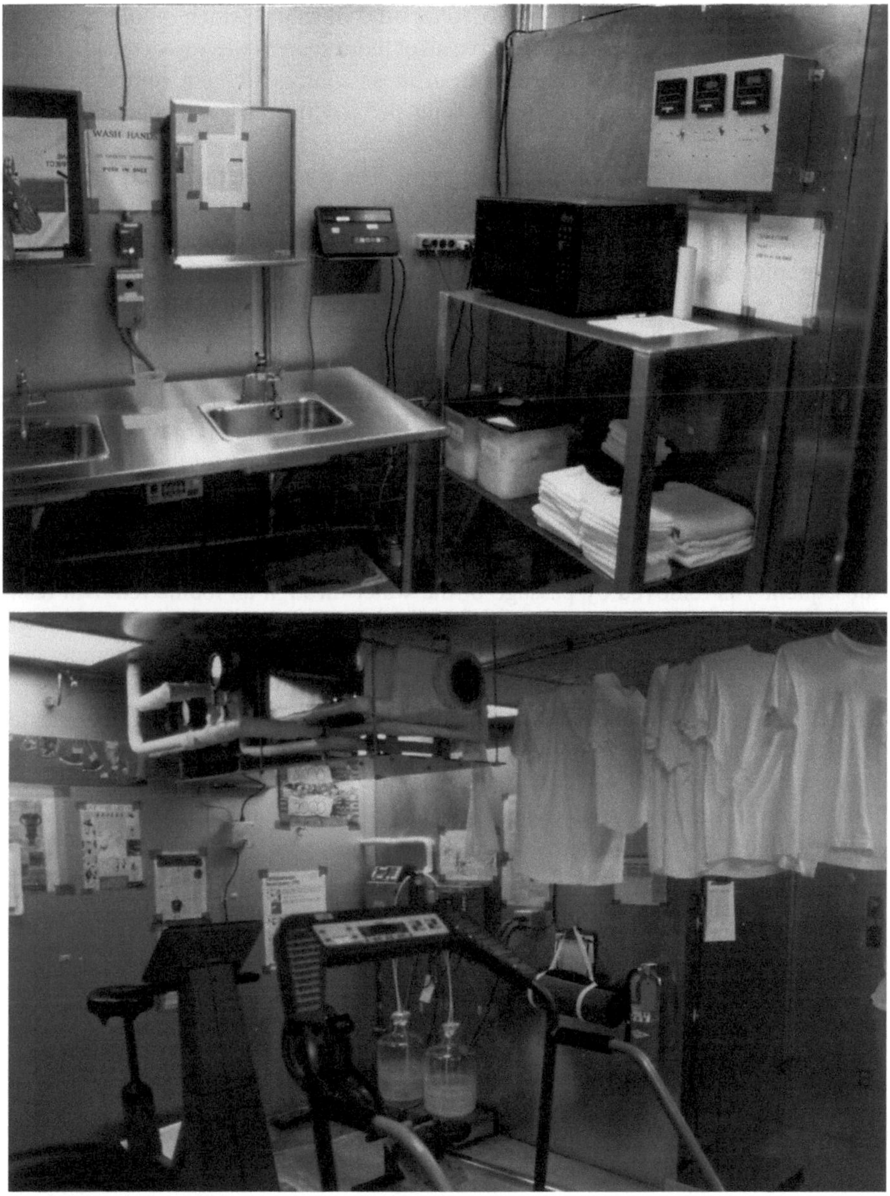

Fig. 8.40 These images show additional sources of water that were used for evaluating the water reclamation unit developed at NASA in Huntsville, Alabama for the International Space Station. The upper image shows a sink used for hand washing and dental hygiene plus a microwave oven for heating meals. The lower image shows exercise equipment and t-shirts hung to dry. In the bottom center area of the lower photograph condensate from a room dehumidifier can be seen collecting into glass storage containers. Usage of these images is courtesy of the author

exercising and then been hung up to collect the moisture which would evaporate from those shirts. The water collected from all of those sources was registered by weight. When the quantity of water collected from those sources was not sufficient, it was supplemented with 'ersatz' water which contained a defined list and concentrations of compounds.

Figure 8.41 shows the water recycling unit. The wastewater recycled by this unit included a combination of water containing human metabolic waste from humidity condensate, showering, hand washing, urine distillate, tooth cleaning, and facial wet shaving plus water from laundering. A laboratory prepared solution also was processed which represented the equivalent of water from fuel cells and a mixture analogous to animal humidity condensate plus equipment off-gassing contaminants. Processing of the water begins with the upper left image in which can be seen two large stainless steel pressure cans that hold the water mixture to be treated. The upper left image also shows a thin vertical white cartridge filter that removes particulates from the water. The step of particulate filtration is followed by the water passing through resin columns packed in stainless steel tubes which are called Unibeds, shown in the upper right image. The Unibeds include ion exchange resin, organic adsorption resin, inorganic adsorption resin, and iodine release resin. The iodine subsequently is removed by passage of the water through an iodine removal resin. Catalytic oxidation is the next stage of treatment, shown by equipment in the lower right image, and that treatment consists of oxygenation, high temperature catalyzation, and de-gassing. The water next passes through an ion exchange column that releases iodine. Finally, the product water is stored in containers behind the white panels that are shown on the bottom of a rack in the lower left image and the water will be filtered yet once more prior to use. That recycling system was repackaged to fit into a double wide rack for mounting in the International Space Station. Figure 8.42 shows a display which demonstrates what a double rack looks like from the front and how that rack unit fits into the curved sides of the space station. The water recycling processor shown in Fig. 8.41 needed to be reconfigured to fit into a double rack.

Fig. 8.41 These images show the International Space Station Water Processor. Clockwise from the upper left these images show the sequence of water processing: the stainless steel tanks which store wastewater that is to be processed and visible in the lower left quadrant of the metal rack is a particulate filter with 0.5 micron porosity; stainless steel tubes called unibeds which contain ion exchange and adsorbent resins; a catalytic oxidation reactor plus a process control monitor; and two enclosed storage tanks on the bottom half of a rack for containing the processed water. Usage of these images is courtesy of the author

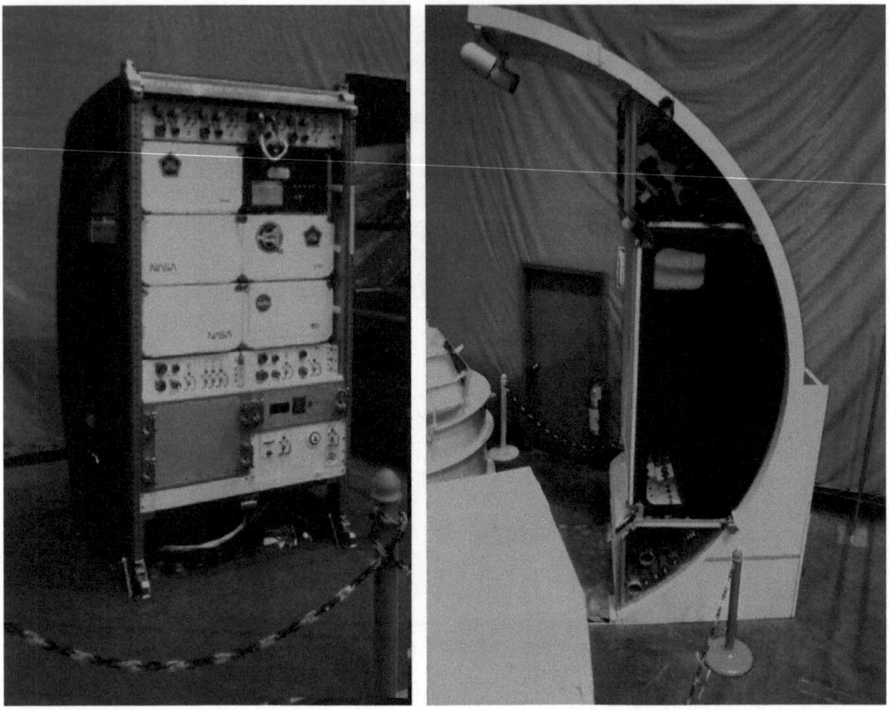

Fig. 8.42 These images demonstrate what a double width equipment rack looks like for the International Space Station and how those racks fit against the interior curved sides of the space station. The water recycling unit was reconfigured to fit into one double width equipment rack. Usage of these images is courtesy of the author

Compliance with Ethical Standards

Conflict of Interest Christon J. Hurst declares that he has no conflict of interest.

Ethical Approval This article does not contain any studies with human participants or animals.

References

Abiyu A, Yan D, Girma A et al (2018) Wastewater treatment potential of *Moringa stenopetala* over *Moringa olifera* as a natural coagulant, antimicrobial agent and heavy metal removals. Cogent Environ Sci 4:1433507. https://doi.org/10.1080/23311843.2018.1433507

Agrawal VK, Bhalwar R (2009) Household water purification: low-cost interventions. Med J Armed Forces India 65(3):260–263. https://doi.org/10.1016/S0377-1237(09)80019-1

Al Moosa ME, Khan MA, Alalami U et al (2015) Microbiological quality of drinking water from water dispenser machines. Int J Environ Sci Dev 6(9):710–713. https://doi.org/10.7763/IJESD.2015.V6.685

Alo MN, Anyim C, Elom M (2012) Coagulation and antimicrobial activities of *Moringa oleifera* seed storage at 3°C temperature in turbid water. Adv Appl Sci Res 3(2):887–894

Amenu D (2014) Microbiological quality of bottled water, international perspective. World J Nat Appl Sci 1(1):1–10

Bajszár G, Dekonenko A (2010) Stress-induced Hsp70 gene expression and inactivation of *Cryptosporidium parvum* oocysts by chlorine-based oxidants. Appl Environ Microbiol 76:1732–1739

Brikké F, Bredero M (2003) Linking technology choice with operation and maintenance in the context of community water supply and sanitation. A reference document for planners and project staff. World Health Organization and IRC Water and Sanitation Centre, Geneva. https://www.who.int/water_sanitation_health/hygiene/om/wsh9241562153.pdf. Accessed 10 Jan 2019

Bruni M, Spuhler D (2018a) Rapid sand filtration. Sustainable sanitation and water management toolbox. Norwegian University of Life Sciences and Sustainable Sanitation and Water Management, Willisau. https://sswm.info/sswm-university-course/module-6-disaster-situations-planning-and-preparedness/further-resources-0/rapid-sand-filtration. Accessed 2 Feb 2019

Bruni M, Spuhler D (2018b) Slow sand filtration. Sustainable sanitation and water management toolbox. Norwegian University of Life Sciences and Sustainable Sanitation and Water Management, Willisau. https://sswm.info/sswm-university-course/module-6-disaster-situations-planning-and-preparedness/further-resources-0/slow-sand-filtration. Accessed 2 Feb 2019

California Department of Public Health (2014) Water vending machine inspection procedure and operating requirements. State of California, Sacramento. https://www.cdph.ca.gov/Programs/CEH/DFDCS/CDPH%20Document%20Library/FDB/FoodSafetyProgram/Water/WVMIProcedure.pdf. Accessed 23 January 2019

Centers for Disease Control and Prevention (2009) Drinking water treatment methods for backcountry and travel use. Centers for Disease Control and Prevention, Atlanta. https://www.cdc.gov/healthywater/drinking/travel/backcountry_water_treatment.html. Accessed 3 Dec 2018

Centers for Disease Control and Prevention (2012) Ceramic filtration. Centers for Disease Control and Prevention, Atlanta. https://www.cdc.gov/safewater/ceramic-filtration.html. Accessed 5 Dec 2018

Centers for Disease Control and Prevention (2014a) Flocculant/disinfectant powder. Centers for Disease Control and Prevention, Atlanta. https://www.cdc.gov/safewater/flocculant-filtration.html. Accessed 5 Dec 2918

Centers for Disease Control and Prevention (2014b) Slow sand filtration. Centers for Disease Control and Prevention, Atlanta. https://www.cdc.gov/safewater/pdf/sand2011.pdf. Accessed 13 Dec 2018

Centers for Disease Control and Prevention (2017) Making water safe in an emergency. Centers for Disease Control and Prevention, Atlanta. https://www.cdc.gov/healthywater/emergency/drinking/making-water-safe.html. Accessed 4 Dec 2018

Chan S, Pullerits K, Riechelmann J et al (2018) Monitoring biofilm function in new and matured full-scale slow sand filters using flow cytometric histogram image comparison (CHIC). Water Res 138:27–36. https://doi.org/10.1016/j.watres.2018.03.032

Chiller TM, Mendoza CE, Lopez MB et al (2006) Reducing diarrhoea in Guatemalan children: randomized controlled trial of flocculant–disinfectant for drinking-water. Bull World Health Organ 84(1):28–35

Ciochetti DA, Metcalf RH (1984) Pasteurization of naturally contaminated water with solar energy. Appl Environ Microbiol 47:223–228

Clark PA, Pinedo CA, Fadus M et al (2012) Slow-sand water filter: design, implementation, accessibility and sustainability in developing countries. Med Sci Monit 18(7):RA105–RA117

Clasen T, Edmondson P (2006) Sodium dichloroisocyanurate (NaDCC) tablets as an alternative to sodium hypochlorite for the routine treatment of drinking water at the household level. Int J Hyg Environ Health 209(2):173–181. https://doi.org/10.1016/j.ijheh.2005.11.004

Colwell RR, Huq A, Islam MS, et al (2003) Reduction of cholera in Bangladeshi villages by simple filtration. Proc Natl Acad Sci USA 100:1051–1055. https://doi.org/10.1073/pnas.0237386100 (See instructions top of page 1053—and notes on first page about boiling)

Crump JA, Okoth GO, Slutsker L et al (2004) Effect of point-of-use disinfection, flocculation and combined flocculation–disinfection on drinking water quality in western Kenya. J Appl Microbiol 97:225–231. https://doi.org/10.1111/j.1365-2672.2004.02309.x

Daud MK, Nafees M, Ali S (2017) Drinking water quality status and contamination in Pakistan. Biomed Res Int 2017:7908183. https://doi.org/10.1155/2017/7908183

de Kwaadsteniet M, Dobrowsky PH, van Deventer A et al (2013) Domestic rainwater harvesting: microbial and chemical water quality and point-of-use treatment systems. Water Air Soil Pollut 224:1629. https://doi.org/10.1007/s11270-013-1629-7

Despins C, Farahbakhsh K, Leidl C (2009) Assessment of rainwater quality from rainwater harvesting systems in Ontario, Canada. J Water Supply Res Technol-Aqua 58(2):117–134. https://doi.org/10.2166/aqua.2009.013

Dodson LL, Bargach J (2015) Harvesting fresh water from fog in rural Morocco: research and impact. Dar Si Hmad's fogwater project in Aït Baamrane. Proc Eng 107:186–193. https://doi.org/10.1016/j.proeng.2015.06.073

Doerr B, ECHO Staff (2005) Technical note 52 moringa water treatment. ECHO Community, North Fort Myers. https://www.echocommunity.org/en/resources/2d26ef0f-7476-48f5-a2f5-3768c3714b71. Accessed 1 Jan 2019

Federación Nacional de Cafeteros de Colombia, Universidad del Valle Facultad de Ingeniería, Gobernación Departamental del Valle del Cauca (1988) Agua Potable para Comunidades en Zonas Cafeteras del Valle del Cauca, Programa de Potabilización de Agua para Comunidades de la Zona Cafetera del Valle del Cauca. Presentado al 8vo Congreso de Ingenieros de la Federación Nacional de Cafeteros, Manizales 15–17 Septiembre (1988). https://www.ircwash.org/sites/default/files/255.1-88AG-5951.pdf. Accessed 6 Jan 2019

Federal Emergency Management Agency (2006) Food and water in an emergency. FEMA 477, Federal Emergency Management Agency, Jessup. https://www.fema.gov/pdf/library/f&web.pdf. Accessed 11 Dec 2018

Fewtrell L, Kay D, Wyer M et al (1996) Survey of the microbiological quality of bottled waters (EPG l/9/78) Centre for Research into Environment and Health, the Environment Centre. University of Leeds, Leeds

Filtron (2018) Capacitación en uso y mantenimiento. Filtron SA, San Marcos. https://www.filtronnica.com/espa%C3%B1ol/capacitacion/. Accessed 13 Dec 2018

Florida Department of Agriculture and Consumer Services (2017) Water vending machine requirements. State of Florida, Tallahassee. https://www.freshfromflorida.com/content/download/70052/1633610/water_vending_machine_requirements.pdf. Accessed 23 Jan 2019

Fredrickson T (2017) Almost half of coin-operated water vending machines substandard. Bangkok Post 25 May 2017, Bangkok. https://www.bangkokpost.com/learning/advanced/1256079/almost-half-of-coin-operated-water-vending-machines-substandard. Accessed 6 Jan 2019

Galvis Castaño G, Latorre Montero J, Visscher JT (1999) Filtración en múltiples etapas: tecnología innovativa para el tratamiento de agua. Universidad del Valle; Cinara, Cali. https://www.ircwash.org/sites/default/files/255.9-99FI-17025.pdf. Accessed 31 Mar 2019

Ghaderpoori M, Dehghani MH, Fazlzadeh M (2009) Survey of microbial quality of drinking water in rural areas of Saqqez, Iran. Am Eurasian J Agric Environ Sci 5(5):627–632

Gutiérrez JP, van Halem D, Rietveld L (2017) Riverbank filtration for the treatment of highly turbid Colombian rivers. Drink Water Eng Sci 10:13–26. https://doi.org/10.5194/dwes-10-13-2017

Habtemariam S (2017) The African and Arabian Moringa species: chemistry, bioactivity and therapeutic applications. Elsevier, Amsterdam, pp 155–165. https://doi.org/10.1016/B978-0-08-102286-3.00011-7

Hashmi I, Sara Q, Farooq S (2012) Microbiological quality of drinking water in urban communities, Rawalpindi, Pakistan. Desalin Water Treat 41:240–248. https://doi.org/10.1080/19443994.2012.664721

Huq A, Yunus M, Sohel SS et al (2010) Simple sari cloth filtration of water is sustainable and continues to protect villagers from cholera in Matlab, Bangladesh. mBio 1(1):e00034–e00010. https://doi.org/10.1128/mBio.00034-10

Hurst CJ (2001) Disinfection of water: drinking water, recreational water and wastewater. In: Block SS (ed) Disinfection, sterilization and preservation, 5th edn. Lippincott Williams & Wilkins, Philadelphia, pp 1023–1047

Hurst CJ (2018) Understanding and estimating the risk of waterborne infectious disease associated with drinking water. In: Hurst CJ (ed) The connections between ecology and infectious disease, Advances in environmental microbiology, vol 5. Springer, Cham, pp 59–114. https://doi.org/10.1007/978-3-319-92373-4_3

Hurst CJ, Roman MC, Garland JL et al (1997) Microbiological aspects of space exploration. ASM News 63:611–615

Innovación para el Desarrollo y la Cooperación Sur-Sur (2018) Filtrón, ceramic filter for drinking water. http://www.ideassonline.org/public/pdf/br_28_60.pdf. Accessed 11 Dec 2018

International Network to Promote Household Water Treatment and Safe Storage (2007) Combating waterborne disease at the household level. World Health Organization, Geneva

Jeyakumar R, Parimalarenganayaki S, Elango L (2017) River bank filtration for natural treatment of water in India: a review. Int J Civil Eng Technol 8(8):1203–1212

Johnson GA (1916) The typhoid toll. J Am Water Works Assoc 3(2):249–326

Johnson R, Scherer T (2011) Filtration: sediment, activated carbon and mixed media (WQ1029 Revised). North Dakota State University Extension Service, Fargo. https://www.ag.ndsu.edu/publications/environment-natural-resources/filtration-sediment-activated-carbon-and-mixed-media. Accessed 14 Dec 2018

Johnson R, Scherer T (2013) Reverse osmosis (WQ1047). North Dakota State University Extension Service, Fargo. https://www.ag.ndsu.edu/pubs/h2oqual/watsys/wq1047.pdf. Accessed 31 Jan 2019

Karim MR, Shelly AB, Biswas M (2005) People perception and acceptance of rainwater harvesting in a coastal area in Bangladesh. In: 12th International rainwater catchment systems conference, New Delhi, Nov 2005

Keogh MB, Elmusharaf K, Borde P et al (2017) Evaluation of the natural coagulant *Moringa oleifera* as a pretreatment for SODIS in contaminated turbid water. Sol Energ 158:448–454. https://doi.org/10.1016/j.solener.2017.10.010

Laurent P (2005) Household drinking water systems and their impact on people with weakened immunity. Médecins Sans Frontières Holland, Public Health Department, Amsterdam. http://www.who.int/household_water/research/HWTS_impacts_on_weakened_immunity.pdf. Accessed 2 Dec 2018

Leclerc H, Moreau A (2002) Microbiological safety of natural mineral water. FEMS Microbiol Rev 26:207–222

Légaré-Julien F, Lemay O, Vallée-Godbout U et al (2018) Laboratory efficacy and disinfection by-product formation of a coagulant/disinfectant tablet for point-of-use water treatment. Water 10:1567. https://doi.org/10.3390/w10111567

Luzi S, Tobler M, Suter F et al (2016) SODIS manual guidance on solar water disinfection. Swiss Federal Institute of Aquatic Science and Technology, Dubendorf

Manitoba Conservation and Water Stewardship (2014) Drinking water and your health: water storage tanks (cisterns). Manitoba Government, Winnipeg. https://www.gov.mb.ca/waterstewardship/odw/public-info/fact_sheets/pdf/water_factsheet_cisterns.pdf. Accessed 28 Jan 2019.

Mendonça MHM, Roseno SAM, Cachoeira TRL et al (2017) Análise bacteriológica da água de consumo comercializada por caminhões-pipa. Rev Ambient Água 12(3):468–475. https://doi.org/10.4136/ambi-agua.1934

Meuli C, Wehrle K (2001) Spring catchment. Series of manuals on water supply, Vol 4. Swiss Centre for Development Cooperation in Technology and Management, Saint Gallen. http://skat.ch/book/spring-catchment/. Accessed 11 Jan 2019

Mwabi JK, Mamba BB, MNB M (2012) Removal of *Escherichia coli* and faecal coliforms from surface water and groundwater by household water treatment devices/systems: a sustainable solution for improving water quality in rural communities of the Southern African development community region. Int J Environ Res Public Health 9(1):139–170. https://doi.org/10.3390/ijerph9010139

Nagaraj V, Skillman L, Ho G et al (2017) Characterisation and comparison of bacterial communities on reverse osmosis membranes of a full-scale desalination plant by bacterial 16S rRNA gene metabarcoding. Biofilms Microbiomes 3:13. https://doi.org/10.1038/s41522-017-0021-6

Nair AT, Ahammed MM (2014) Biosand filtration: a sustainable option for household treatment of drinking water. In: Paper presented at international symposium on integrated water resources management (IWRM-2014) CWRDM, Kozhikode, Kerala, 19–21 Feb 2014. https://www.researchgate.net/publication/260865510. Accessed 5 Dec 2018

Naser AM, Higgins EM, Arman S et al (2018) Effect of groundwater iron on residual chlorine in water treated with sodium dichloroisocyanurate tablets in rural Bangladesh. Am J Trop Med Hyg 98(4):977–983. https://doi.org/10.4269/ajtmh.16-0954

New England Water Treatment Technology Assistance Center (2018) Optimizing performance of intermittent slow sand filters for microbial removals. New England Water Treatment Technology Assistance Center, University of New Hampshire, Durham. https://www.unh.edu/erg/sites/www.unh.edu.erg/files/media/WTTAC/issf_final_report.pdf. Accessed 12 Dec 2018

Oxfam (2012) Hygiene promotion for HWTS in emergencies. Oxfam, Oxford. https://www.pseau.org/outils/ouvrages/oxfam_uk_hygiene_promotion_for_hwts_in_emergencies_2012.pdf

Parr J, Shaw R (2019) Technical brief 49: Choosing an appropriate technology. Water, Engineering and Development Centre, Loughborough University, Leicestershire, pp 66–68. https://www.lboro.ac.uk/research/wedc/resources/pubs/guides/. Accessed 12 Jan 2019

Payment PL, Richardson J, Siemiatycki R et al (1991) A randomized trial to evaluate the risk of gastrointestinal disease due to consumption of drinking water meeting current microbiological standards. Am J Public Health 81:703–708

Payment PJ, Siemiatycki L, Richardson G et al (1997) A prospective epidemiological study of gastrointestinal health effects due to the consumption of drinking water. Int J Environ Health Res 7:5–31

PennState Extension (2016) Rainwater cisterns: design, construction, and treatment. Pennsylvania State University, University Park. https://extension.psu.edu/rainwater-cisterns-design-construction-and-treatment. Accessed 28 Jan 2019

Rahman S, Khan MTR, Akib S et al (2014) Sustainability of rainwater harvesting system in terms of water quality. Sci World J 18(2014):721357. https://doi.org/10.1155/2014/721357

Ramos P (2018) Manabí todavía purifica agua con filtros coloniales. El Telégrafo, Santiago de Guayaquil. https://www.eltelegrafo.com.ec/noticias/cultura/1/intercultural-filtros-agua-coloniales-manabi. Accessed 9 Dec 2018

Roman MC, Hurst CJ (1998) Assessment of the viral removal capability of the International Space Station water recovery and management system. Life Support Biosph Sci 5:31–34

Ruckart PZ, Ettinger AS, Hanna-Attisha M et al (2019) The flint water crisis: a coordinated public health emergency response and recovery initiative. J Public Health Manag Pract 25(Suppl 1):S84–S90. https://doi.org/10.1097/PHH.0000000000000871

Safapour N, Metcalf RH (1999) Enhancement of solar water pasteurization with reflectors. Appl Environ Microbiol 65:859–861

Saleh MA, Abdel-Rahman FH, Woodard BB et al (2008) Chemical, microbial and physical evaluation of commercial bottled waters in greater Houston area of Texas. J Environ Sci Health A 43:335–347. https://doi.org/10.1080/10934520701795400

Sánchez-Martín J, Beltrán-Heredia J, Peres JA (2012) Improvement of the flocculation process in water treatment by using *Moringa oleifera* seeds extract. Braz J Chem Eng 29(03):495–501

Skinner B, Shaw R (2019a) Technical brief 56: Buried and semi-submerged water tanks. Water, Engineering and Development Centre, Loughborough University, Leicestershire, pp 93–96. https://www.lboro.ac.uk/research/wedc/resources/pubs/guides/ Accessed 12 Jan 2019

Skinner B, Shaw R (2019b) Technical brief 58: Household water treatment 1. Water, Engineering and Development Centre, Loughborough University, Leicestershire, pp 93–96. https://www.lboro.ac.uk/research/wedc/resources/pubs/guides/ Accessed 12 Jan 2019

Skinner B, Shaw R (2019c) Technical brief 59: Household water treatment 2. Water, Engineering and Development Centre, Loughborough University, Leicestershire, pp 105–108. https://www.lboro.ac.uk/research/wedc/resources/pubs/guides/ Accessed 12 Jan 2019

State of Queensland (Queensland Health) (2015) Water carriers: a guide for local government on general food safety and hygiene requirements. State of Queensland, Brisbane

Sule IO, Agbabiaka TO, Saliu BK et al (2014) Comparative bacteriological assessment of input and output water from tanker trucks. Ilorin J Sci 1(1):171–178

Tellen V, Nkeng G, Dentel S (2010) Improved filtration technology for pathogen reduction in rural water supplies. Water 2(2):285–306. https://doi.org/10.3390/w2020285

Texas Commission on Environmental Quality (2018) Water supply division and regional areas public water system water hauler guidance. State of Texas, Dallas. https://www.tceq.texas.gov/assets/public/permitting/watersupply/pdw/Water%20Hauler%20Guidance.pdf. Accessed 1 Dec 2018

Totaro M, Casini B, Valentini P et al (2018) Assessing natural mineral water microbiology quality in the absence of cultivable pathogen bacteria. J Water Health 16(3):425–434. https://doi.org/10.2166/wh.2018.183

Tunggolou J, Payus C (2017) Application of *Moringa oleifera* plant as water purifier for drinking water purposes. J Environ Sci Technol 10(5):268–275. https://doi.org/10.3923/jest.2017.268.275

United Nations Development Programme (2007) Case study Procter & Gamble: providing safe drinking water to the poor. United Nations Development Programme, New York. http://growinginclusivemarkets.org/media/cases/Developing%20Countries_P&G_2008.pdf. Accessed 10 Dec 2018

United States Agency for International Development, National Demonstration Water Project, Institute for Rural Water, National Environmental Health Association (1982) Water for the world: technical notes: human resources, rural water supply. United States Agency for International Development, Washington. https://www.ircwash.org/sites/default/files/201-7288.pdf

United States Environmental Protection Agency (2017) Emergency disinfection of drinking water. United States Environmental Protection Agency, Washington. https://www.epa.gov/ground-water-and-drinking-water/emergency-disinfection-drinking-water. Accessed 3 Dec 2018

United States Government Publishing Office (2018) Processing and bottling of bottled drinking water. Title 21 chapter I subchapter B Part 129. In: Electronic code of federal regulations, United States Government, Washington. https://www.ecfr.gov/cgi-bin/text-idx?SID=0c417253f5a12d9cc7d0d83a94655ab7&mc=true&node=pt21.2.129&rgn=div5. Accessed 1 Dec 2018

van der Laan H, van Halem D, Smeets PWMH et al (2014) Bacteria and virus removal effectiveness of ceramic pot filters with different silver applications in a long term experiment. Water Res 51:47–54. https://doi.org/10.1016/j.watres.2013.11.010

van Halem D, Heijman SGJ, Soppe AIA et al (2007) Ceramic silver-impregnated pot filters for household drinking water treatment in developing countries: material characterization and performance study. Water Sci Technol Water Supply 7(5-6):9–17. https://doi.org/10.2166/ws.2007.142

World Health Organization (2011a) Evaluating household water treatment options: health-based targets and microbiological performance specifications. World Health Organization, Geneva

World Health Organization (2011b) Safe drinking-water from desalination WHO/HSE/WSH/11.03. World Health Organization, Geneva. https://www.who.int/water_sanitation_health/publications/2011/desalination_guidance_en.pdf. Accessed 14 Jan 2019

World Health Organization (2015) Boil water technical brief WHO/FWC/WSH/15.02. World Health Organization, Geneva. http://apps.who.int/iris/handle/10665/155821. Accessed 5 Dec 2018

World Health Organization (2017) Guidelines for drinking-water quality, 4th edn, incorporating the 1st addendum. World Health Organization, Geneva. https://www.who.int/water_sanitation_health/publications/drinking-water-quality-guidelines-4-including-1st-addendum/en/. Accessed 13 Jan 2019

World Health Organization (2018a) Silver as a drinking-water disinfectant. World Health Organization, Geneva. http://www.who.int/water_sanitation_health/publications/silver-02032018.pdf?ua=1. Accessed 5 Dec 2018

World Health Organization (2018b) Fact sheet on drinking water. World Health Organization, Geneva. https://www.who.int/en/news-room/fact-sheets/detail/drinking-water. Accessed 13 Jan 2019

Wrigley T (2007) Microbial counts and pesticide concentrations in drinking water after alum flocculation of channel feed water at the household level, in Vinh Long Province, Vietnam. J Water Health 5(1):171–178

Yang Y-A, Chong A, Song J (2018) Why is eradicating typhoid fever so challenging: implications for vaccine and therapeutic design. Vaccines 6:45. https://doi.org/10.3390/vaccines6030045

Yari AR, Mohammadi MJ, Geravandi S et al (2018) Assessment of microbial quality of household water output from desalination systems by the heterotrophic plate count method. J Water Health 16(6):930–937. https://doi.org/10.2166/wh.2018.082

Zahran S, McElmurry SP, Sadler RC (2017) Four phases of the flint water crisis: evidence from blood lead levels in children. Environ Res 157:160–172. https://doi.org/10.1016/j.envres.2017.05.028

Zereffa EA, Bekalo TB (2017) Clay ceramic filter for water treatment. Mater Sci Appl Chem 34:69–74. https://doi.org/10.1515/msac-2017-0011

Chapter 9
Microbiome of Drinking Water Distribution Systems

Laurence Mathieu, Tony Paris, and Jean-Claude Block

Abstract Drinking water distribution systems appear as vulnerable engineered systems constantly sown and colonized by microorganisms representing several hundreds of species. Such microorganisms originate firstly from finished potable drinking water, secondly from growth in the distribution systems in the bulk water and on the surface of pipes and reservoirs as biofilms, and thirdly from accidental or continuously low introduction of pathogens.

This biomass is a source of technological problems (microbiologically influenced corrosion, red water, odor, and taste of tap water) caused by some specific microbes. Investigation and control of the microbial population community and its behavior in drinking water distribution systems should be the cornerstone of efforts to protect distribution system integrity, water quality, and public health. This silent contamination could also be responsible for endemic gastrointestinal illnesses attributable to tap water meeting current standards.

As shown throughout this literature review (mainly limited to papers published from 1990 to 2014), distribution systems must be considered as bioreactors. First, we present bugs systematically found in drinking water distribution systems all over the world (bacteria, viruses, yeasts, fungi, protozoa, microcrustaceans, rotifers, and oligochaete worms). Then, we analyze and discuss several items related to biofilms grown under conditions relevant to drinking water environments including mechanisms of biofilm formation, structure, cohesiveness, biodiversity, and pathogen

L. Mathieu (✉)
EPHE-PSL Research University, LCPME, Nancy, France

Laboratoire de Chimie Physique et Microbiologie pour les Matériaux et l'Environnement, CNRS-Lorraine University, Villers-lès-Nancy, France
e-mail: laurence.mathieu@univ-lorraine.fr

T. Paris
Eurofins – IPL Est, Nancy, France

J.-C. Block
Laboratoire de Chimie Physique et Microbiologie pour les Matériaux et l'Environnement, CNRS-Lorraine University, Villers-lès-Nancy, France

C. J. Hurst (ed.), *The Structure and Function of Aquatic Microbial Communities*, Advances in Environmental Microbiology 7,
https://doi.org/10.1007/978-3-030-16775-2_9

reservoirs. Finally, the chapter concludes with a review of some of the parameters governing biofilm accumulation in drinking water distribution systems.

9.1 Introduction

Drinking water distribution systems have been in use for several centuries, but the first modern ones meant for transporting drinking water in every part of a city were built at the beginning of the eighteenth century. Such public drinking water supplies have a substantial impact on public health as it gives access to safe, potable drinking water and de facto prevents the epidemiological transmission of waterborne pathogens. Most drinking waters are transported (after more or less complex treatments depending on the quality of the resource) to the consumers thanks to underground distribution networks, which represent huge and costly, engineered systems consisting of a combination of reservoirs and pipes.

Distribution systems are challenging habitats for micro- and macroorganisms. However, bacteria, viruses, yeasts, fungi, protozoa, microcrustaceans, rotifers, and oligochaete worms occur systematically all over the world in reservoirs and pipes that transport drinking water through municipal water distribution systems. Those types of organisms also exit in private areas (i.e., after the water meter) representing household supply infrastructures where high times of retention, small diameter pipes with high area/volume ratios, and generally higher temperatures favor microorganism growth that resultantly exceeds guideline values for microbiological parameters (Völker et al. 2010; Proctor and Hammes 2015). Indeed, many waterborne microorganisms become autochthonous in distribution systems despite the relatively low concentrations of organic and inorganic nutrients needed for their carbon and energy requirements and addition of chemical disinfectants (generally oxidants such as chlorine or chloramine).

Live organisms (saprophytic and pathogenic) are everywhere even in the very recently built drinking water distribution systems. They originate firstly from finished potable drinking water, secondly from growth in the distribution systems in the bulk water and on the surface of pipes and reservoirs as biofilms, and thirdly from accidental or continuously low introduction of pathogens. In the first instance, distribution systems are constantly seeded with indigenous bacteria escaping from or produced in drinking water treatment plants (especially from sandfilters or biofilters) (Servais et al. 1991; Pinto et al. 2012; Bichai et al. 2014; Lautenschlager et al. 2014). This biomass shapes the downstream microbial water and biofilm community structure (Prévost et al. 1998; Velten et al. 2011; Pinto et al. 2012; Wang et al. 2013b). Second, the growth of both saprophytic bacteria, some of which being of sanitary interest (e.g., *Legionella*), and microinvertebrates occurs in the bulk water as well as within biofilms and also loose deposits, all of which contributing to increasing biomass throughout the network. Third, there occur accidental rare massive intrusions of pathogens variously caused by ineffective treatments, treatment breakthroughs, or leakages associated with negative transient pressure

(Besner et al. 2011), and all of which can be the cause of major waterborne outbreaks (Hrudey and Hrudey 2004; Craun 2012; Ashbolt 2015). There additionally are low continuous intrusions of fecal pathogens (e.g., rotavirus, *Campylobacter jejuni*) and fecal indicators (Westrell et al. 2003; Miles et al. 2009). This silent contamination could be responsible for endemic gastrointestinal illness rates as described by Payment et al. (1997), who reported that 14–40% of gastrointestinal illnesses were attributable to tap water meeting current standards and that the water distribution systems appeared partly responsible for these illnesses.

Drinking water distribution systems thus appear as vulnerable engineered systems constantly sown and colonized by microorganisms representing several hundreds of species. Walls of pipes, reservoirs, taps, valves, seals, and fitting accessories appear as the critical key areas for survival and growth of many of these microorganisms in the form of biofilms. This sessile biomass is also the source of technological problems (microbiologically influenced corrosion, red water, odor, and taste of tap water). Investigation and control of the microbial population community and its behavior in drinking water distribution systems should be the cornerstone of efforts to protect distribution system integrity, water quality, and public health.

In this chapter, after a brief description of distribution systems that must be considered as bioreactors, the most relevant scientific publications (mainly based on those published from 1990 to 2014) about microorganisms and microinvertebrates in drinking waters will be presented in an overview. Then, we will analyze and discuss several items related to biofilms grown under conditions relevant to drinking water environments including mechanisms of formation, structure and cohesiveness, biodiversity, and pathogen reservoirs. Finally, the chapter will conclude with a review of some of the parameters governing biofilm accumulation in drinking water distribution systems.

9.2 Drinking Water Distribution Systems: Multiphasic Unsteady-State Bioreactors

Drinking water distribution networks represent huge systems combining reservoirs and pipes (as an example, public networks in France equal to approximately 900,000 km in total; the length of private networks could be equivalent or greater). Pipes of 2 m to 2 cm in diameter are made of very different materials depending on the year of laying in the past century (Malm et al. 2012). Materials used in networks are highly diverse: lined or unlined cast iron, ductile iron, galvanized stainless steel (GS), stainless steel (SS), linings (asphaltic, cement, epoxy), polyvinyl chloride (PVC), plasticized polyvinyl chloride (PVCp), high-density polyethylene (HDPE), polyethylene cross-linked with aluminum foil between both layers (PEx), polypropylene (PP), ethylene propylene diene monomer (EPDM), and copper especially in private houses. All these materials are colonized by microorganisms, whatever the topographical and physicochemical characteristics of their surface.

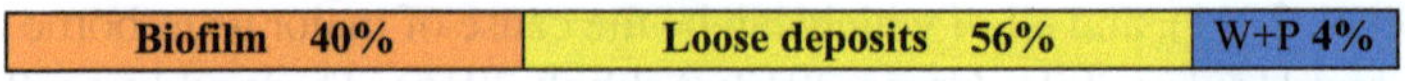

Fig. 9.1 Comparison of viable biomass in percentage (%) (as inferred from ATP results) of different phases within 1 m-long water pipes (PVC, 110 mm in diameter): from left to right pipe wall biofilm, loose deposits, and water + particles (W+P) (adapted from Liu et al. 2014a)

9.2.1 Where Is Most of the Biomass Located in the Network?

Drinking water distribution systems have been engineered for transporting water but work de facto as slow bioreactors producing biomass in different phases of the system: in the bulk water, on particles in suspension, on pipe walls in biofilm, and in loose deposits which represent material easily removed from the network by water flushing. The distribution of the biomass between these phases can differ greatly from one network to another, but it is generally admitted that most of the biomass is not found in the bulk water but either attached to surfaces (biofilm) or accumulated into loose deposits (also called deposits, soft deposits, sediments, or hydraulically mobile sediments). Indeed, a mass balance carried out by Kjellerup et al. (2006) showed that 77% of the bacteria were located at surfaces and 23% in the bulk water. In another work, it was shown that soft deposits easily recovered from the pipes by water flushing represent the first reservoir of bacteria (in combination with other bugs), ahead of biofilm, water, and suspended particles (Fig. 9.1). Interestingly, in the study of Liu et al. (2014a), bacteria from the loose deposit phase (dominated by *Sphingomonas* spp.) were clearly different from the bacteria of the bulk water (dominated by *Polaromonas* spp.).

In addition, Gauthier et al. (1999a) counted up to 10^9 CFU/g of dry deposits and Zacheus et al. (2001) up to 10^9 bacterial cells and 10^4 fungi/g. Van Lieverloo et al. (2012) found that aquatic sow bugs (Asellidae) and oligochaete worms, both known to have caused rare though embarrassing consumer complaints, represented 98% of the biomass in deposits flushed from mains. Loret and Greub (2010) underlined that the concentration of free-living amoebae (FLA) was especially high in sediments, which constitute ecological niches where they can feed bacteria. Last, Lipponen et al. (2002) reported high numbers of nitrifying bacteria and nitrification potentials in the sediments.

Exposure of consumers' drinking water to particle-associated bacteria appears low (25–30 cells attached on a single particle leading to 1–3.5 $\times$ 10^3 cells/mL) (Stoodley et al. 2001; Liu et al. 2013a). However, the presence of such bacterial aggregates should not be underestimated as they clearly affect QMRA (quantitative microbial risk assessment) results especially in the presence of a threshold in the dose-response relationship (Maul 2014).

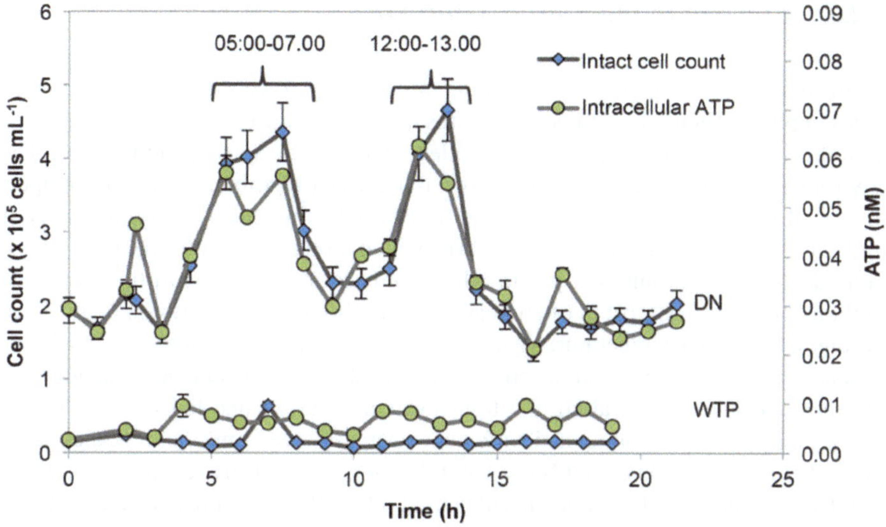

Fig. 9.2 Diurnal changes in bacterial parameters (cells and ATP) of water treatment plant (WTP) and distribution system (DN) at one single point of use on the network. The figure is by Nescerecka et al. (2014) and used under the Creative Commons Attribution 2.0 Generic license

9.2.2 Water Distribution Systems Work as Unsteady-State Bioreactors

Steady state in distribution systems could be defined by a constant concentration of all elements (bacteria, organic matter, chlorine). It cannot be reached in distribution systems because the water flow determined by consumer use is highly variable (high during the day; low at night in large pipes or even stagnant in house plumbing) with some peaks of consumption in the early morning, around noon, and later in the afternoon (Digiano and Zhang 2004). Flow rates vary from 0 (water stagnation) to 1.5 m/s (during peak hours) leading to variable water residence times from several days (up to 1 month at worst due to bad hydraulic conditions) to a few hours, respectively. Water stagnation is known to seriously affect bacterial water quality. As an example, overnight stagnation at household temperatures results in a significant increase in bacterial cell concentration in the first samples flushed from the taps (Lautenschlager et al. 2010).

As a result the microbial characteristics of distributed waters highly vary on a daily basis due to bacterial growth especially during stagnation episodes and biomass shearing from biofilm and mobilization of loose deposits during high flow rate events (Fig. 9.2).

9.2.3 *Flowing to the Tap: A Long Way*

From treatment to tap, finished drinking waters may travel several days through miles of pipes. Then, many microorganisms take advantage of this long distance and period of time to grow (inaccurately called regrowth by some practitioners). The environmental conditions found in the drinking water distribution systems support microbial life. Indeed, the so-called oligotrophic drinking waters carry hundreds of μg/L of assimilable organic matter (AOM), which represent feast for bacteria whose individual mass is around 0.2 pg (in other words, AOM can support the growth of millions of bacteria/L). pH is often near the neutrality (ranging between 6 and 8, on average), and water temperature typically varies in cold waters from 4 °C to 25 °C depending on the season and the geographical location. As a result bacterial numbers increase considerably all along the transport chain (up to one log in some cases) being favored by concomitant disinfectant decrease (see some field studies, e.g., Prévost et al. 1998; Power and Nagy 1999; Niquette et al. 2001; Kormas et al. 2010; Bal Krishna et al. 2013; Lautenschlager et al. 2013; Nescerecka et al. 2014). Interestingly, the bacterial community structure may also change along the way. For example, Lautenschlager et al. (2013) reported that during transport through the network, the proportion of *Proteobacteria* phylum varied with an increase in *Betaproteobacteria* at the end of the section studied in the network where hydraulic transport time was greater than 48 h (Fig. 9.3).

In some systems, analysis of the bacterial community and bacterial counts suggest that temporal variation patterns are stronger than spatial patterns. As an example, Pinto et al. (2014) showed that the patterns in spatial dynamics were weaker than those observed for the temporal trends, which exhibited seasonal cycling correlating with temperature and source patterns and also demonstrated reproducibility on an annual time scale. McCoy and Van Briesen (2014) analyzing the variance of *Alpha-*, *Beta-, and Gammaproteobacteria* and bacterial concentrations in different sites of a distribution system suggested that temporal patterns are stronger than spatial patterns.

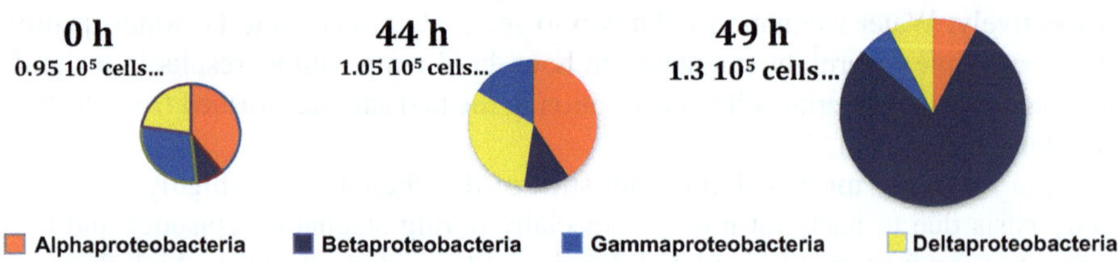

Fig. 9.3 Pyrosequencing analysis of drinking water sampled on taps located at different distances from the plant defined by the transport time of the water from 0 to 49 h for the most distant point (adapted from Lautenschlager et al. 2013)

9.3 Bacteria and Bugs in Drinking Waters

The numerous and biologically diverse microorganisms and macroorganisms within distribution systems form complex ecosystems.

9.3.1 A Few Million Bacteria per Liter of Drinking Water

What the consumers and water authorities know on a routine analysis basis about microbial characteristics of drinking water is very limited and exclusively based on mandatory controls and, to make it short, on analysis of fecal indicators such as *Escherichia coli* along with assessment of the heterotrophic plate count (HPC). HPC method has been used for decades but is time-consuming (72-h incubation at 20 °C) and provides very different information depending on the physiological state of the bacteria, the availability of nutrients, and the disinfectant concentration. Moreover, the use of culture-dependent methods has long been recognized to underestimate the number of live bacteria (LeChevallier et al. 1980; Martiny et al. 2005). Thus, HPC numbers found in the literature vary from 0 CFU/mL (chlorinated waters just at the intake of the distribution system) to 10,000 CFU/mL in distribution systems (Mathieu et al. 1995; Zacheus et al. 2001; Chen et al. 2013; Vaz-Moreira et al. 2013).

In truth, potable waters (defined by zero cultivable *Escherichia coli*/100 mL and zero cultivable enterococci/100 mL) carry millions of bacterial cells per liter. Throughout many distribution systems all over the world, most of the bacterial cell numbers as counted either by microscopy after staining with specific fluorochromes or using fast flow cytometry method (FCM) are around 10^8 cells/L (with variations ranging from 10^7 to 5×10^8 cells/L). Many of those cells are indirectly defined as active due do their high nucleic acid content (HNA bacteria) (Phe et al. 2005; Berney et al. 2008; Ramseier et al. 2011).

The concentrations of microorganisms could even be higher than indicated by the cited data because some very small-sized bacteria could have been missed by these analytical methods. According to Rinta-Kanto et al. (2004) and Wang et al. (2007), ultramicrobacteria passing through 0.02 or 0.01 μm porosity filters represent 0.1% of all cells in drinking water. Cultured ultramicrocells isolated from drinking waters and identified by their 16S ribosomal DNA sequences were closely related to members of the *Proteobacteria*, *Actinobacteria*, and *Firmicutes* (Silbaq 2009).

Estimation of the viable bacterial biomass in water by assessment of ATP concentration (a culture-independent method) appears successful, quite easy, and rapid (Berney et al. 2008; Hammes et al. 2008, 2010a). Good correlation was observed between ATP concentration not with the total number of cells (Liu et al. 2013b), but with the number of cells with undamaged membranes (Hammes and Egli 2010; Hammes et al. 2010a). Undamaged bacteria in tap waters could represent half of the total number of cells (Kahlisch et al. 2012). To sum up such the promising ATP technique could substitute for the currently mandatory HPC method.

9.3.2 *Bacterial Diversity in Drinking Waters: Dominance of Proteobacteria*

Like all natural aquatic environments, water distribution systems are home to millions of bacteria/L and hundreds of species of microorganisms. Additionally, each combination of water resource and treatment plant produces drinking water with a unique bacterial composition.

Different methods have been used to study the structure of bacterial communities in distribution systems. Culture-dependent methods have long been, and still are, used for isolating specific groups or species, often those of either pathogenic or metabolic interest such as *Actinobacteria*, *Actinomycetes*, *Aeromonas*, *Helicobacter*, *Legionella*, *Mycobacteria*, and nitrifiers depending on the technical or health-related issues to be addressed. More recently, less than 10 years ago, our understanding of the microbial ecology of distribution systems greatly improved thanks to molecular biology investigation, fingerprint determination, and sequencing-based approaches. These techniques are still being upgraded, and next-generation techniques will provide less expensive and time-consuming methodologies as well as new software, tools for analyzing gene sequences, and new data banks (Douterelo et al. 2014a; Liu et al. 2016).

As shown in Table 9.1, bacteria phyla are very diverse in drinking water (11 phyla and surely more than 100 species). Interestingly, the 17 publications reported in Table 9.1 show a large dominance of *Proteobacteria* whatever the geographical location, water resource origin, or season. However, each water distribution system has a unique bacterial composition (especially at the level of genus and species).

9.3.3 *Other Bugs in Drinking Water*

Drinking water distribution systems harbor a relatively wide-ranging ecosystem including fungi, amoebae, and microinvertebrates, which form with heterotrophic bacteria a trophic food chain.

9.3.3.1 Fungi

Fungi are diverse in drinking water and composed of many genera and species (Hageskal et al. 2009). *Aspergillus*, *Fusarium*, *Trichoderma*, *Mucor*, *Rhizopus*, etc. have been found sometimes in high concentrations (1–2200 CFU/mL) in drinking water distribution systems (Hageskal et al. 2006; Pereira et al. 2009, 2010; Poitelon et al. 2009a; Siqueira et al. 2011; Oliveira et al. 2013; Van der Wielen and Van der Kooij 2013; Al-gabr et al. 2014). Some opportunistically pathogenic species (e.g., *Fusarium oxysporum* and *F. dimerum*) which are capable of causing life-threatening

Table 9.1 Some examples of phyla and genera identified in drinking water by culture-independent techniques

Authors	Water sampling	Methods	Phyla/class and genera (dominants)
Douterelo et al. (2013)	Pilot made of recirculating loops (HDPE) fed with chlorinated drinking water + HDPE coupons	16S rRNA gene PCR followed by pyrosequencing	Dominance of *Proteobacteria* (*Alpha-* > *Beta*-class) followed by *Firmicutes* (*Clostridia* and *Bacilli* classes) At the genus level: *Methylocystis*, *Methylocella*, *Sphingopyxis*, *Polaromonas*, *Hydrogenophaga*, *Hyphomicrobium*, *Clostridium*, *Bosea*, *Nevskia*, *Mesorhizobium*
Douterelo et al. (2014b)	Two sections of a DWDS in the Northwest of England composed of HDPE or cast iron and fed with treated surface water	16S rRNA gene PCR followed by conventional cloning and sequencing	Dominance of *Proteobacteria* (mostly *Alpha-* > *Delta-* > *Gamma*-class) > *Firmicutes* (*Clostridia*, *Bacilli*) > *Actinobacteria* > *Spirochaeta* At the genus level: *Spirochaeta*, *Methylobacterium*, *Clostridium*, *Desulfobacterium*, *Flavobacterium*, *Lysinibacillus*, *Pseudomonas*, *Flavobacterium*, *Clostridium*, *Bacillus* Note: Differences in the structure of the bacterial community between samples from polyethylene and cast iron pipes
Eichler et al. (2006)	DWDS of the City of Braunschweig (Germany) fed with treated surface water + chlorine	16S rRNA and 16S rRNA gene-based single-strand confirmation polymorphism (SSCP) fingerprint	*Bacteroides* ≫ *Proteobacteria* (*Beta-* > *Alpha*-class) > *Actinobacteria* Note: DNA-based and RNA-based fingerprints differed in their major taxonomic groups

(continued)

Table 9.1 (continued)

Authors	Water sampling	Methods	Phyla/class and genera (dominants)
Henne et al. (2012, 2013)	DWDS of the City of Braunschweig (Germany) fed with treated surface water + chlorine	16S rRNA and 16S rRNA gene-based single-strand confirmation polymorphism (SSCP) fingerprint	*Proteobacteria* (*Alpha-* + *Beta*-class) > *Actinobacteria*, *Bacteroidetes*, *Planctomycetes* Note: *Gammaproteobacteria* proportion increases in RNA-based fingerprints
Holinger et al. (2014)	17 DWDS in cities along Arkansas and Mississippi river basins (USA)	16S rRNA gene PCR followed by pyrosequencing	*Proteobacteria* > *Cyanobacteria* (including chloroplasts) > *Actinobacteria* (with *Mycobacteria*) ≫ *Firmicutes*, *Bacteroidetes*
Kahlisch et al. (2012)	DWDS—City of Braunschweig (Germany) fed with treated surface water + chlorine	16S rRNA and 16S rRNA gene-based single-strand confirmation polymorphism (SSCP) fingerprint	16S rRNA genes dominated by *Betaproteobacteria*, *Bacteroidetes*, and *Actinobacteria* 16S rRNA dominated by *Alpha-*, *Beta-*, *Gammaproteobacteria*, *Bacteroides*, and *Cyanobacteria*
Li et al. (2010)	Four end points (red water events) of the Beijing (China) DWDS fed with treated surface water + chloramination	16S rRNA gene-based followed by conventional cloning and sequencing	Dominance of *Proteobacteria* (*Alpha-*, *Beta-*, *Gamma-*, *Delta-*, *Epsilon*-class), *Bacteroidetes*, *Actinobacteria* At the genus level: *Caulobacter*, *Brevundimonas*, *Sphingomonas*, *Rhodobacter*, *Afipia*, *Hyphomicrobium*, *Rhodocyclus*, *Ferribacterium*, *Propionivibrio*, *Gallionella*, *Comamonadaceae*, *Legionella*, *Pseudomonas*, *Nevskia*, *Sulfuricurvum*, *Flavobacteriaceae*, *Sphingobacterium*, *Actinomyces*, *Arthrobacter*, *Microbacterium*

(continued)

Table 9.1 (continued)

Authors	Water sampling	Methods	Phyla/class and genera (dominants)
Lin et al. (2014)	Reservoir tanks above DWDS (Hubei province, China), fed with treated surface waters from branches of Yangtze River + chlorine dioxide	16S rRNA gene PCR followed by pyrosequencing	*Proteobacteria* (85%) ≫ *Bacteroidetes* (14%) ≫ *Actinobacteria* (1%) At the genus level: *Enhydrobacter*, *Acinetobacter*, *Rheinheimera*, *Delftia*, *Acidovorax*, *Comamonas*, *Blastomonas*, *Brevundimonas*, *Porphyrobacter*, *Sphingobacterium*, *Chryseobacterium*, *Flavobacterium*, *Nubsella*
Liu et al. (2014b)	PVC pipe sections cut from a DWDS located in the northern part of the Netherlands and fed with unchlorinated drinking water	16S rRNA gene PCR followed by pyrosequencing	Dominance of *Proteobacteria* (*Beta-* > *Alpha-* > *Delta-*, *Gamma*-class), ≫ *Firmicutes* > *Actinobacteria*, *Chloroflexi*, *Bacteroidetes*, *Nitrospirae*, *Acidobacteria* At the genus level: Dominance of *Polaromonas* (69%), *Sphingomonas* (13%), *Acidovorax* (5%), *Janthinobacterium* (4%)
Martiny et al. (2005)	Loop-shaped reactor equipped with stainless steel coupons. Aerated filtered and groundwater (unchlorinated)	16S rRNA gene-based followed by conventional cloning and sequencing	Dominance of *Nitrospirae* ≫ *Proteobacteria* (*Alpha-* > *Beta-* = *Gamma-* = *Delta*-class > *Epsilon*-class) > *Acidobacteria* > *Planctomycetes*, *Bacteroidetes* Note: Richness showed a strong seasonal trend
Pinto et al. (2014)	Full-scale DWDS of Ann Arbor (Michigan, USA) carrying a combination of treated surface and groundwater	16S rRNA gene PCR followed by pyrosequencing	Dominance of *Proteobacteria* (*Beta-* > *Alpha-* > *Gamma*-class), *Betaproteobacteria* (especially *Acidovorax* and *Georgfuchsia*) dominated in summer when *Alphaproteobacteria* (especially *Brevundimonas*) dominated in winter Note: Richness showed a strong seasonal trend

(continued)

Table 9.1 (continued)

Authors	Water sampling	Methods	Phyla/class and genera (dominants)
Poitelon et al. (2009a, b)	Three DWDS of Paris (France) fed with treated surface water + chlorine Water sampled at the outlet of the plants	V6 ribosomal sequence tag (SARST-V6) method	Dominance of *Proteobacteria* (*Beta-* $\geq$ *Gamma-* > *Alpha- and Delta*-class) $\gg$ *Acidobacteria*, *Planctomycetes*, *Chlamydiae*, *Nitrospirae*
Prest et al. (2014)	Full-scale DWDS of Kralingen (Rotterdam area, Netherlands) fed with treated surface water + chlorine dioxide	16S rRNA gene PCR followed by pyrosequencing	*Proteobacteria* (*Beta*-class mainly) > *Planctomycetes* > *Bacteroides* > *Chloroflexi*, *Acidobacteria*, *Firmicutes*, *Actinobacteria* Note: Combined pyrosequencing with flow cytometry data
Revetta et al. (2010)	Cincinnati DWDS (Ohio, USA)—treated surface water Water sampled at the tap	16S rRNA (RT PCR) followed by conventional cloning and sequencing	*Proteobacteria* > *Cyanobacteria*, > *Actinobacteria*, *Bacteroidetes*, *Planctomycetes*, *Firmicutes*, *Actinobacteria* *Note:* Dominance of unclassified bacteria (>50%)
Vaz-Moreira et al. (2013)	Full-scale DWDS fed with chlorinated-treated water (Portugal)	16S rRNA gene-denaturing gradient gel electrophoresis (DGGE) profiling	*Proteobacteria* (*Alpha-* = *Beta-* $\gg$ *Gamma*-class) >>> *Actinobacteria* > *Bacteroides* > *Cyanobacteria* > *Planctomycetes* > *Aquificae*, *Acidobacteria*, *Firmicutes*
Wang et al. (2014a)	DWDS Pinellas county (Florida, USA). Annual chlorine burn performed during late summer to mitigate nitrification problems in chloraminated DWDS	qPCR + amplicon sequencing of the 16S rRNA gene	*Proteobacteria* dominate (26.8–99.7%) followed by *Bacteroides*, *Actinobacteria*, *Acidobacteria*, *Firmicutes*, and *Verrucomicrobia* At the class level: *Alpha-*, *Beta-*, *Gammaproteobacteria (48%) followed by Betaproteobacteria*

(continued)

Table 9.1 (continued)

Authors	Water sampling	Methods	Phyla/class and genera (dominants)
			(16.9%) and *Alphaproteobacteria (12.7%)* Note: Burn stages shift dominance to *Bacteroidetes*, *Firmicutes*, etc.
Williams et al. (2004)	Cincinnati DWDS (Ohio, USA)—treated surface water—ductile iron pipe loops + chlorine or chloramine	16S rRNA gene-based followed by conventional cloning and sequencing	*Alphaproteobacteria* dominate in both chlorinated and chloraminated DWDS. Other phyla detected *Cyanobacteria*, *Firmicutes*, *Betaproteobacteria*, *Gammaproteobacteria* At the genus level: *Sphingomonas*, *Brevundimonas*, *Caulobacter*, *Sphingobium*, *Acidovorax*, *Pseudomonas*, *Xanthomonas*, *Legionella*, *Nitrospira*, *Mycobacterium*, etc.

DWDS drinking water distribution system, *HDPE* high-density polyethylene, *PVC* polyvinyl chloride

infections in immunocompromised patients should be controlled in healthcare environments (Sautour et al. 2012), which means understanding the potential contribution of drinking water as a source of infectious disease within healthcare facilities. The disinfection of water systems contaminated with fungi appeared difficult as many strains are definitively unaffected by chloramine (Ma et al. 2015).

9.3.3.2 Free-Living Amoebae and Small Eukaryotes

Free-living amoebae (FLA) (*Naegleria*, *Acanthamoeba* spp., *Vermamoeba vermiformis*, etc.) are widespread in environmental waters and distribution systems [see the review by Loret and Greub (2010) and Thomas and Ashbolt (2011)]. While many FLA are ubiquitous and harmless to humans, several genera are pathogenic, particularly *Acanthamoeba* (keratitis, encephalitis) and *Naegleria* (encephalitis). Amoebae were found in 79% of household water samples from Ohio (USA) particularly in showerheads (52%) and kitchen sprayers (50%) (Stockman et al.

2011). Their prevalence and concentration correlated with the level of organic matter and the season (Valster et al. 2009; Marciano-Cabral et al. 2010). Wang et al. (2012a, c) found the occurrence of *Vermamoeba vermiformis* to be greater than that of *Acanthamoeba* spp., but the opposite was also reported in other distribution systems (Valster et al. 2009; Lin et al. 2014). Additionally, chlorinated disinfectant efficiency (bleach, chlorine dioxide, chloramines) is of course dose-dependent, and cysts are highly more resistant to disinfection than trophozoites (Dupuy et al. 2011, 2014).

FLA can act as hosts for amoeba-resisting bacteria, which represent a major concern in terms of public health as they can gain virulence by intracellular growth in amoebae and chlorine cannot easily disinfect intracellular pathogens. FLA diversity and intra-amoebal composition show significant differences among drinking water distribution systems. Many bacterial genera can be found in amoebae, including *Stenotrophomonas*, *Bradyrhizobium*, *Sphingomonas*, *Pseudoxanthomonas*, *Acidovorax*, *Legionella*, and *Chlamydiae* (Corsaro and Venditti 2010; Delafont et al. 2013, 2014).

Besides amoebae, populations of other small eukaryotes can be detected in drinking waters, such as dinoflagellates, ciliates, and metazoan (Poitelon et al. 2009a; Otterholt and Charnock 2011). Reported cases of diseases traceable to free-living indigenous eukaryotes in drinking waters seem however very rare.

9.3.3.3 Invertebrates

Invertebrates such as worms and crustacea are likely to be found in drinking water distribution systems. A large survey of invertebrates in drinking water was carried out in the Netherlands and showed systematic occurrence but large variations in their abundance (Van Lieverloo et al. 2012). Aquatic sow bugs (Asellidae, 1–12 mm up to 1500 per m^3) and oligochaete worms (Oligochaeta, 1–100 mm, up to 9900 per m^3) were found associated with loose deposits. Smaller crustaceans (0.5–2 mm) dominated in water from mains (Cladocera and Copepoda up to 14,000 per m^3). Common invertebrates in tap water were Rotifera (<1 mm) and nematode worms (Nematoda, <2 mm). A parallel study done in Denmark showed a nationwide occurrence of *Asellus aquaticus* (<2 mm) (Christensen et al. 2011). Annelida (worms), Platyhelminthes, and mollusks (snails) were also found (Fig. 9.4). In all the systems studied, the microbial quality of drinking water was high, and some biotic relationships were shown with corrosion, organic matter, and biofilm (Christensen et al. 2011; Van Lieverloo et al. 2012). Apart from the esthetic problems and some consumer complaints caused by larger animals like *A. aquaticus*, no hygienic problems seem to be associated with such bugs.

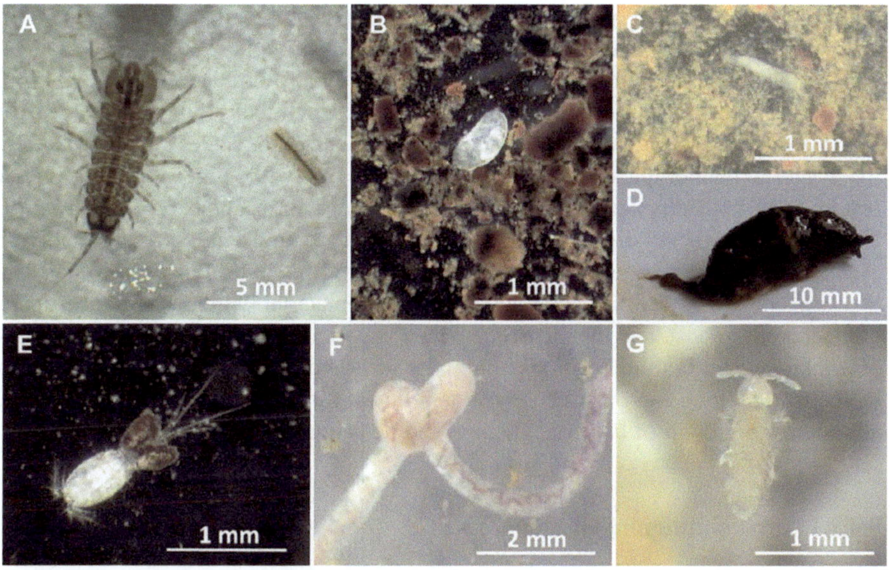

Fig. 9.4 Microinvertebrates sampled from a non-chlorinated drinking water supply. (**a**) Adult and juvenile *Asellus aquaticus* (Malacostraca), (**b**) seed shrimp (Ostracoda), (**c**) flatworm (Turbellaria), (**d**) land slug from a clean water tank, (**e**) *Cyclops* sp. (Maxillopoda), (**f**) *Tubifex* sp. (Clitellata), and (**g**) springtail (Entognatha). Reprinted from publication by Christensen et al. (2011), with permission from Elsevier

9.4 Drinking Water Biofilm Development

Drinking water biofilms result from a dynamic process of assembly of microorganisms, on surfaces, which is controlled by both physical and biological mechanisms. The outcome is a constantly renewed organo-mineral layer on the pipe surface, densely populated with many bacteria and bugs. Its high cohesiveness and biodiversity limit its easy cleaning by mechanical and chemical methods and prevent its complete eradication.

9.4.1 Stages of Biofilm Formation Under Conditions Relevant to Drinking Water Environments

Colonization of surfaces by bacteria begins through a series of recruitment processes that are well identified but not so easily depicted and understood in the case of drinking water. We now know that the typical models of disease-associated pathogenic bacterial biofilms (i.e., *Pseudomonas* or staphylococci), which have been illustrated at the molecular level (regulation pathways, signaling, genetics, etc.), must have parallels in drinking water systems. But equivalent detailed knowledge is missing in the field of drinking water biofilms, which to date have been more characterized by their structural, mechanical, and phenotypic aspects.

Several stages can be distinguished in the bacterial biofilm development process. The very first ones are clearly governed by physicochemical processes (convective transport, diffusion, and reversible adhesion), and then other stages are controlled by live bacteria (phenotypic differentiation, growth). Finally, the process of detachment results from both physical shear forces and cell migration under bacterial control. Understanding these different mechanisms better represents a real challenge especially for efforts at limiting development of biofilms in drinking water distribution systems (e.g., by selecting an appropriate material which should be inert, non-adherent, robust, and inexpensive!). Here, we briefly describe the different stages of drinking water biofilm development on the surface of new pipes newly installed in the network, which can be summarized as follows: formation of a conditioning film, adhesion and growth of bacteria, grazing, and detachment.

9.4.1.1 Preconditioning of Surfaces

Whatever the primary mechanism for biofilm formation may be, it begins with the attachment of a set of organic polymers recruited from the bulk water and forming patchy spots upon the substrate onto or besides which bacteria will land. This so-called conditioning film is formed within a few minutes or hours, essentially as soon as a material is exposed to natural waters, i.e., waters containing minerals and polymerlike organic molecules. This preliminary step in biofilm development is clearly under-explored, and only a few studies have attempted to characterize this tiny organic deposit. Some of these polymers have been described as gel colloidal particles consisting of acidic polysaccharides (approximately 0.2 μg/L of xanthan gum equivalent) (Van Nevel et al. 2012), but glycoproteins could represent the dominant polymers in conditioning films formed in drinking water (Francius et al. 2017). These first events certainly modify the characteristics of the surface, making it appropriate (or inappropriate) for further colonization by pioneer microorganisms (Dang and Lowell 2000).

9.4.1.2 Adhesion of Pioneer Bacteria

Bacterial adhesion in drinking water distribution systems is controlled by the physicochemical characteristics of both the cells and the pipe surfaces. On the one hand, negative charges (Walker et al. 2005; Ojeda et al. 2008) are distributed on the material surface and in the bio-interphase (i.e., the volume formed by the cell envelopes) of bacteria described as hydrodynamic permeable soft colloids (Gaboriaud et al. 2008; Gosselin et al. 2011). On the other hand, hydrophobic polymers in the bacterial envelopes (e.g., as mycolic acids in *Mycobacteria* envelopes) confer to the cells some hydrophobic properties, which will play a key role in "locking" the bacteria on the substratum more efficiently than can hydrophilic cells

(Van Loosdrecht et al. 1987; Tatchou-Nyamsi-König et al. 2008). As a result, depending on the surface characteristics of the pipe material, the polymers recruited for the conditioning film, and the hydrodynamic shear stress that governs the mass transfer at the wall, at least 10^3 bacterial cells/cm^2, will adhere onto the surface in a few hours.

9.4.1.3 Growth of Attached Drinking Water Bacteria

Development of a biofilm on the surface results from both continuous deposition of cells from the bulk water and bacterial growth. Attachment of bacteria to the pipe surface via electrostatic and hydrophobic interactions is not harmful for most bacteria. Live bacteria stay metabolically active and multiply, and some of them produce exopolymers (EPS), which will reinforce the adhesion of cells and cohesion of the growing multilayer biofilm (Percival et al. 1998; Flemming and Wingender 2010). The average apparent growth rate (μ) of drinking water biofilms varies from 0.041 per day to 0.2 per day (Block et al. 1993; Boe-Hansen et al. 2002; Manuel et al. 2007; Park et al. 2012; Pedersen 1990). This dynamic process is seemingly limited to microorganisms in the upper layer of the biofilm due to diffusion limitation phenomena in the deeper layers. Thus, biofilm activity potential drops rapidly within the first few days of biofilm development (Kalmbach et al. 1997). Additionally, there is a shift in the bacterial community structure along with an increase in cell surface area coverage, bacterial richness, and overall community diversity (Kalmbach et al. 1997; Martiny et al. 2003). For example, Douterelo et al. (2014c) reported that pioneer species (i.e., *Pseudomonas* spp. and *Janthinobacterium* spp.), which form the initial drinking water biofilm on polymeric material, were partly replaced by a multispecies biofilm within 3 weeks. Under drinking water conditions, it seems that cell coverage of the substratum is relatively slow (due to the fact that nutrients are limited, bacteria produce not only cells but also EPS, and the structure is continuously subjected to grazing and hydraulic shear stress—see below). Predicting how long it takes for biofilm to cover 80% of a new pipe surface by using data from Paris et al. (2007) gives a time of 900 days, which is in agreement with direct observations from Martiny et al. (2003).

9.4.1.4 Protozoa Grazing "in and on" Drinking Water Biofilms

Free-living amoebae use the biomass of biofilms as food. During the first period of drinking water biofilm formation, grazing favors large clumps by single cell or small aggregate consumption. Grazing may be very effective and was estimated equally to a consumption rate of $1–2 \times 10^4$ cells/cm^2 h (Sibille et al. 1998; Paris et al. 2007). It impacts biofilm construction by removing 80% of the attached cells in the first weeks of flowing drinking water into the systems (Paris et al. 2007; Declerck et al. 2009) or slowing down the average apparent biofilm generation time by up to 47 days (Pedersen 1990). As shown by Derlon et al. (2012), predation had a significant influence on both the total amount and the structure of biofilm developed on ultrafiltration membranes.

9.4.1.5 Drinking Water Biofilm Detachment

Detachment of bacterial cells or clumps from biofilms is of paramount importance to contamination of drinking water. It explains that the diversity of the bacteria flowing in a network is always a combination (in an unknown ratio) of cells originating from the treatment plant and those detached from the biofilm. Stoodley et al. (2001) using a drinking water lab reactor showed detachment of single cells and aggregates, which could contain from 10 to 1.7×10^3 cells (small aggregates were numerous). Such a detachment rate estimated by using the mathematical relations proposed by Van Der Wende et al. (1989) is de facto equal to the growth rate. Values of k_{shear} given in the literature are of the same order of magnitude, i.e., 0.039 per hour (Bois et al. 1997).

9.4.2 Biofilm Structure and Cohesiveness

The five stages briefly described above lead to the slow formation of mature biofilms, which are de facto very heterogeneous. The term heterogeneity refers to the extent of the nonuniform distribution of any selected constituent in any of the compartments of the biofilm system, such as the distribution of the biomass, the microscale chemical gradient, the microorganism species and their spatial organization, as well as the genetic expression and physiological state of the bacteria (Stewart and Franklin 2008; Lewandowski and Boltz 2011).

Drinking water biofilms formed under turbulent conditions are thin systems (from a few micrometers to more than 100 μm) composed of 10^5–10^7 cells/cm^2 in patchy distribution. This limited number of attached cells is clearly the result of both nutrient diffusion limitations, which prohibit the growth of many members of the consortium, and continuous detachment and erosion of the biofilm canopy. Under laminar flow or stagnation conditions, pipe surfaces are also partly covered with bigger soft, brown deposits composed of cells, organic matter, and iron oxides (a few millimeters high). Indeed, drinking water biofilms are far from being continuous films but rather dispersed populations of single cells and aggregates partly covering the surface area (Fig. 9.5), with sometimes the formation of streamer-like aggregates (up to 200 μm in length for around 1×10^6 bacterial cells).

The cohesion of biofilms is related to an exopolymer matrix (EPS) (Flemming 2016), whose composition in biofilms grown under drinking water conditions is not very well documented due to sensitivity limitations of most analytical techniques and interferences of metals trapped in the biofilms. Bacteria can produce different appendages such as curli fibrils (0.1–10 μm with a width of 4–12 nm). The fibrils consist mainly of CsgA protein, which is assembled in a fibrillar tertiary structure

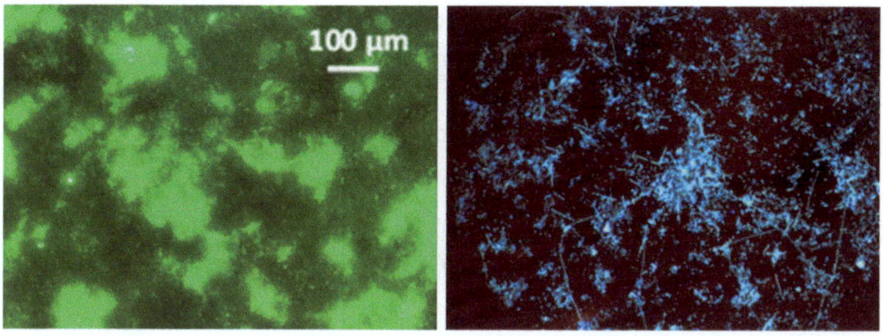

Fig. 9.5 Examples of biofilms on HDPE coupons colonized by 3-month-old drinking water biofilms without chlorination and under turbulent flow conditions at 20 °C. Microscopic observations after in situ staining with Sybr-II (left, magnitude × 200) and DAPI (right, magnitude × 1000)

(Larsen et al. 2007). The presence of adhesins with similar amyloid-like structure was also described for some *Gammaproteobacteria*. The tendency to produce amyloid adhesins seemed more pronounced in oligotrophic environments such as drinking water distribution systems (Larsen et al. 2007). In brief, by analogy with laboratory experiments or studies on thick environmental biofilms, it can be assumed that EPS are composed of polymers and heteropolymers such as polysaccharides, proteins, glycoproteins, lipids, and eDNA (Flemming and Wingender 2010).

The entanglement of the EPS matrix brings cohesion and viscoelasticity to the biofilm by forming a net, whose mesh size could vary from 5 nm for the most cohesive clusters (i.e., those with the lowest detachment shear stress applied with atomic force microscopy) to 11 nm for the least cohesive clusters (Mathieu et al. 2014; Abe et al. 2012). Entanglement is related to the nature and quantity of the EPS, and the degree of reticulation of the matrix depends on cation bridging and hydrophobic "glue." By referring both to the radius of gyration exhibited by polysaccharides in the range of 10–100 nm and the long polysaccharides detected in drinking water biofilms with lengths ranging from 1 to 10 μm (Abe et al. 2011), we can conclude that the macromolecules within the biofilm curl up, thus leading to higher entanglement. Mathieu et al. (2014) determined that each macromolecule could probably generate from 10 to 10^3 self-contact points and a 1 μm^3 biofilm could contain up to 10^4 macromolecules. They deduced that the maximum concentration of contact points due to the macromolecules of the network ranges from 10^4 to 10^7 per μm^3.

Such a huge network of exopolymers firmly attached to cells and to the substratum makes cleaning of water pipes very difficult. The cohesiveness of small biofilm aggregates is very high (i.e., their detachment requires high hydrodynamic strength). Flushing the pipes by rapid circulation of water (generating in this example a hydrodynamic shear stress of about 10 Pa) removes only large clusters and leaves on the surface small aggregates of bacteria which will start to multiply. Water flushing drinking water pipes appears effective in loose deposit mobilization and shearing of fragile scales but is quite ineffective in the removal of viscoelastic structures such as biofilms. Effective cleaning procedures should break up the matrix and change the elastic properties of bacterial biofilms. Only treatments such as oxidation induce changes in the mechanical properties of drinking water biofilms that destabilize the biofilm cohesiveness by creating breaks in the matrix structure (Mathieu et al. 2014; Tachikawa and Yamanaka 2014).

9.4.3 Bacterial Diversity in Biofilms

Understanding the microbial ecology in distribution systems is still a challenging task. Without a detailed inventory of the microorganisms growing in distribution systems, water utilities are forced, in a sense, to "fly blind" when making treatment decisions (Ingerson-Mahar and Reid 2012). The common questions arising when trying to study microorganisms in drinking water distribution systems irrespective of their life style are what types of microorganisms are present, how abundant they are, how their activities shape the environment or influence other organisms, and how the environment influences the structure and the function of the microorganisms present. Different methods have been used in an attempt to address some of these questions from culture-dependent to culture-independent techniques. The latter (genetic fingerprints, metagenomics, pyrosequencing, and so on) (see the review by Douterelo et al. 2014a) cast new light and provide exciting information. Still, the role of most of the bacteria detected in drinking water biofilms remains uncertain and the synergistic interactions unknown, just like the effect of protozoan grazing and the survival of intracellular bacteria. However, these new analytical methods in just a few years have profoundly changed our understanding of drinking water biofilms.

First, we do know that the biodiversity in biofilms is high, as a result of environmental parameters driven from both the bulk water (nutrients, electron acceptors, temperature, ions, pH) and the substratum (nutrients or toxic leaching from the material, physicochemical properties) and because of the heterogeneity of the organo-mineral deposits, which creates a mosaic of microenvironments. Second, the persistence of most of the attached bacteria and the development of a metastable ecosystem generate richness and diversity largely different from that of planktonic bacteria (Martiny et al. 2005; Henne et al. 2012; Douterelo et al. 2013; Liu et al. 2014a, b; Lürhig et al. 2015).

Bacterial biofilm communities have been for the most part explored in a descriptive way over the last decade. As shown in Table 9.2, three classes of *Proteobacteria* (α-, β-, γ-) are systematically present and predominate, and only their relative abundance varies. Besides *Proteobacteria*, other more or less dominant phyla are also detected such as *Firmicutes*, *Verrucomicrobia*, *Planctomycetes*, and *Bacteroidetes*. It could also be suspected that in specific situations (e.g., corroded materials, phosphate or disinfectant treatment, electron donors) some distinctive populations settle in (Norton and LeChevallier 2000; Martiny et al. 2005). At the class and genus levels, the inhabiting bacterial flora of drinking water biofilms clearly shows an extraordinary richness but does not allow us to draw, for the moment, any consensual scheme of colonization or any understanding and sense

Table 9.2 Some examples of community composition identified in drinking water distribution system biofilms by culture-independent techniques

Authors	Biofilm sampling	Methods	Dominant phyla and classes (major genera)	Comments
Chen et al. (2013)	Unlined cast iron pipe sections from a treated surface water from China—pipe tubercles scraped and grounded	16S rRNA gene-based PCR-DGGE followed re-amplification and conventional cloning and sequencing of the major DGGE bands	*Firmicutes*, *Actinobacteria*, and *Proteobacteria* (*Alpha-*, *Beta-*, *Gamma*-classes) At the genus level: *Rhizobium*, *Pseudomonas*, *Lactococcus*, *Brevundimonas*, *Rheinheimera*, *Arthrobacter*, *Bacillus*, *Herbaspirillum*	The major bacterial members in iron tubercle community were identical to those in the stagnant tap water *Alpha-*, *Beta-*, and *Gammaproteobacteria* shared comparable dominance within tubercles and stagnant tap water
Douterelo et al. (2013)	Pilot made of recirculating loops of high-density polyethylene pipes equipped with HDPE coupons and fed with chlorinated drinking water	16S rRNA gene PCR followed by pyrosequencing	*Proteobacteria* are dominating (*Gamma-* > *Beta-* > *Alpha*-classes) followed by *Firmicutes* (*Clostridia* and *Bacilli* classes) At the genus level: *Pseudomonas*, *Zooglea*, *Janthinobacterium*, and *Sphingomonas*	Bacterial community composition differed between biofilms and bulk water samples and before and after flushing The bacterial composition and community structure of biofilms changed between the three different hydraulic regimes Species richness and diversity tended to be higher at highly varied flow
Douterelo et al. (2014a, b)	Pilot made of recirculating loops (HDPE) + HDPE coupons Chlorinated drinking water	16S rRNA gene PCR followed by conventional cloning and sequencing	Dominance of *Proteobacteria* (*Beta-* > *Gamma-* > *Alpha*-classes) At the genus level: *Pseudomonas*, *Janthinobacterium*, *Methylophilus*, *Stenotrophomonas*, *Undibacterium*, *Dechloromonas*, *Acidovorax*, *Bacteroidetes*,	Sequencing analysis of the clone libraries showed changes in the biofilm bacterial community composition from day 7 to day 28 Gradual increase in species richness and diversity over time during the 28 days of biofilm accumulation Identify bacterial groups involved in initial attachment to

(continued)

Table 9.2 (continued)

Authors	Biofilm sampling	Methods	Dominant phyla and classes (major genera)	Comments
			Flavobacterium, *Curvibacter*, *Porphyrobacter*, *Sphingomonas*, *Nevskia*, *Sphingopyxis*, *Novosphingobium*, *Methylobacterium*, *Acinetobacter*	HDPE pipes and subsequent adhesion
Feazel et al. (2009)	Showerheads from nine cities in the USA	16S rRNA gene PCR followed by conventional cloning and sequencing	*Actinobacteria*, *Proteobacteria*, *Firmicutes* At the genus level: *Mycobacterium gordonae*, *Mycobacterium avium*, *Pseudomonas*, *Sphingomonas*, *Staphylococcus*, *Streptococcus*, *Burkholderia*, *Neisseria*, *Acinetobacter*, *Legionella*	Showerhead biofilms enriched in *Mycobacteria*, 100-fold above the water content
Henne et al. (2012)	Different sampling locations of the Braunschweig (G) DWDS fed with treated surface water + chlorine	16S rRNA and 16S rRNA gene-based single-strand confirmation polymorphism (SSCP) fingerprint	Dominance of *Proteobacteria* (*Alpha*- ≫ *Gamma*- > *Beta*-classes) > *Chlamydiales* >>> *Acidobacteria*, *Planctomycetes*, *Firmicutes*, *Actinobacteria*, *Chloroflexi*, *Bacteroidetes*, *Nitrospira*	Large differences in biofilm fingerprints, indicating a high spatial variability in DWDS Reduced richness compared to bulk water Higher relative abundances of single phylotypes in biofilms Relatedness of RNA-DNA within the biofilm fingerprints
Hong et al. (2010)	Two water meters (WMa and WMb) from Urbana Champaign DWDS (Illinois, USA)	16S rRNA gene PCR followed by pyrosequencing	Dominance of *Proteobacteria* (*Beta*- > *Alpha*- > *Gamma*-classes) ≫ *Firmicutes*, *Deinococcus-Thermus*, *Bacteroidetes*, *Actinobacteria*	The two water meters had different bacterial community compositions Presence of few methanotrophs belonging to the

			At genus level: *Acidovorax*, *Sphingomonas*, *Methylophilus*, *Lysobacter*, *Porphyrobacter*, *Bosea*, *Methylobacterium*, *Methylocystis*, *Bradyrhizobium*, *Sphingopyxis*	*Methylococcaceae* (*Gammaproteobacteria*) due to methane in the groundwater resource
Keinänen-Toivola et al. (2006)	Polycarbonate slides in annular reactor connected to Cincinnati DWDS (Ohio, USA). Fluoridated and chlorinated drinking water	16S rRNA and 16S rRNA gene-based followed by conventional cloning and sequencing	*Gammaproteobacteria* (*Nevskia ramosa*) and *Mycobacterium*	Sequences corresponded to several genera containing pathogenic species Presence of iron-oxidizing, nitrite-oxidizing genus
Kelly et al. (2014)	Ductile iron pipe sections from DWDS, Pinellas county (Florida, USA) fed with chlorinated drinking waters	Tag pyrosequencing of bacterial 16S rRNA genes of the extracted DNA	*Proteobacteria* (*Gamma*-, *Alpha*-classes), *Actinobacteria* At the genus level: *Methylomonas* (41% of the sequences), *Acinetobacter*, *Mycobacterium*, *Pseudomonas*, *Methylobacterium*	High seasonal variability Good relationship between diversity and bacterial abundance
Lin et al. (2014)	Reservoir tanks of the DWDS in Hubei province (China), fed with treated surface waters from branches of Yangtze River + chlorine dioxide Biofilm removed by scraping	16S rRNA gene PCR followed by pyrosequencing	Preponderance of *Proteobacteria* (44.6%) (*Alpha*- ≫ *Beta*- > *Gamma*-classes) ≫ *Gemmatimonadetes* > *Chloroflexi* > *Bacteroidetes* > *Nitrospirae* At the genus level: *Sphingobium*, *Porphyrobacter*, *Hyphomicrobium*, *Brevundimonas*, *Phenylobacterium*, *Nitrosomonadaceae*, *Hydrogenophaga*, *Mycobacterium*, *Clostridium*, *Planctomycetes*, *Nitrospira*	Differences in the bacterial community composition between biofilm and water samples Biofilms more diverse microbial communities than water

(continued)

Table 9.2 (continued)

Authors	Biofilm sampling	Methods	Dominant phyla and classes (major genera)	Comments
Liu et al. (2013a)	Particle-associated bacteria sampled at the entry of three unchlorinated DWDS	16S rRNA gene PCR followed by pyrosequencing	Dominance of *Proteobacteria* (*Beta*- and *Gamma*-class ≫ *Alpha* and *Delta*-class) > *Nitrospirae* > *Planctomycetes*, *Cyanobacteria*, *Euryarchaeota*, *Acidobacteria* > *Actinobacteria*, *Bacteroidetes*, *Crenarchaeota*, *Chloroflexi*, *Gemmatimonadetes* At the genus level: *Legionella*, *Nitrospira*, *Gallionella*, *Planctomycetes*, *Caulobacter*, *Hyphomicrobium*, *Aquabacterium*, *Comamonadaceae*, *Nitrosomonas*, *Crenothrix*, *Caldilinea*, *Rhodopirellula*, *Rhodospirillaceae*, *Leptolyngbya*	
Liu et al. (2014a)	PVC pipe sections cut from a DWDS located in the northern part of the Netherlands and fed with unchlorinated drinking water	16S rRNA gene PCR followed by pyrosequencing	Dominance of *Proteobacteria* (*Alpha*- ≫ *Beta*-, *Gamma*-class), >> *Actinobacteria* (3%), *Chloroflexi* (2%), *Bacteroidetes* (2%), *Nitrospirae* (1%), *Firmicutes* (1%), and *Acidobacteria* (1%) At the genus level: Dominance of *Sphingomonas* ≫ *Janthinobacterium*, *Pseudomonas*	Stable bacterial communities in bulk water, pipe wall biofilm, and suspended solids throughout the distribution system Bulk water bacteria (dominated by *Polaromonas* spp.) were clearly different from the biofilm bacteria (dominated by *Sphingomonas* spp.)

Liu et al. (2014b)	PVC and cast iron faucets located in a DWDS of Hubei province, China	16S rRNA gene PCR followed by pyrosequencing	Dominance of *Proteobacteria* (*Alpha*- ≫ *Beta*-, *Gamma*-class), *Actinobacteria* At the genus level: *Hyphomicrobium*, *Aquabacterium*, *Acinetobacter*, *Limnobacter*, *Mycobacterium*, *Nevskia*, *Porphyrobacter*, *Pseudomonas*, *Rhodobacter*	Difference in bacterial composition of the PVC biofilms (predominance of *Hyphomicrobium*) and cast iron biofilms (corrosion-associated bacteria) Bacterial communities in the bulk water and biofilm samples were significantly different
Martiny et al. (2005)	Loop-shaped reactor equipped with stainless steel coupons and fed with aerated filtered and unchlorinated groundwater	16S rRNA gene-based followed by conventional cloning and sequencing	Dominance of *Nitrospirae* ≫ *Proteobacteria* (*Gamma*- and *Alpha*-class > *Beta*- and *Delta*-class) > *Acidobacteria* > *Planctomycetes*, *Firmicutes*, *Bacteroidetes*	Attached and planktonic communities form separate clusters Increase in the biofilm richness over long period of time
Revetta et al. (2013)	Glass and PVC coupons from semi-closed pipe loop system or annual reactors fed with monochlorinated drinking water from Cincinnati DWDS (Ohio, USA)—treated surface water	16S rRNA gene-based followed by conventional cloning and sequencing	*Proteobacteria* (*Gamma*-, *Beta*-, *Alpha*-class), *Actinobacteria*, *Bacteroidetes*, *Nitrospirae*, *Firmicutes* At the genus level: Early stages of biofilm were dominated by *Serratia*, *Cloacibacterium*, *Diaphorobacter*, and *Pseudomonas* Older biofilms dominated by *Mycobacterium*, *Flavobacterium*, *Phenylobacterium*, *Acidovorax*	No significant difference in community structures between materials used Decrease in *Gammaproteobacteria* and increase in *Actinobacteria* with biofilm aging Increasing diversity with sampling time Phylogenetic structure of biofilm communities varied with biofilm aging

(continued)

Table 9.2 (continued)

Authors	Biofilm sampling	Methods	Dominant phyla and classes (major genera)	Comments
Schmeisser et al. (2003)	EPDM-coated valves in DWDS (northwest of Germany)	16S rRNA gene-based followed by conventional cloning and sequencing	Dominance of *Proteobacteria* (*Alpha-* = *Gamma-* > *Beta*-class) > *Actinobacteria* > *Firmicutes* At the family level: *Rhizobiales*, *Pseudomonas*, *Enterobacteriales*, *Burkholderiales*, *Caulobacterales*, *Actinomycetales*	
Williams et al. (2005)	Polycarbonate coupons in annular reactor connected to DWDS + chloramine or chlorine	16S rRNA gene-based followed by conventional cloning and sequencing	Dominance of *Proteobacteria* (*Alpha-*, *Beta-*, and *Gamma*-class), *Actinobacteria*, *Planctomycetes* At the genus level: – In the presence of chloramine: *Mycobacterium* and *Dechloromonas* – In the presence of chlorine: *Porphyrobacter*, *Hyphomicrobium*, *Bosea*, *Sphingomonas*, *Bradyrhizobium*	Disinfectant practice changed the biofilm bacterial communities Reduction in species richness in the chloraminated biofilm

DWDS drinking water distribution system, *EPDM* ethylene-propylene-diene monomer, *PVC* polyvinyl chloride

of the interactions between phyla and between genera. Indeed, each drinking water biofilm population has a unique bacterial population structure. Metabolomics could be an interesting way to document the biofilm activity by analyzing the chemical compounds (metabolites) that are either consumed or excreted by the microorganisms as a result of their metabolic activity. These metabolites can be used to infer the type of microorganisms present in the biofilm, as well as the nature of their activity (Beale et al. 2013).

9.5 Accumulation of Pathogens and Microorganisms of Public Health Interest in Drinking Water Biofilms

Whatever their structure and thickness, biofilms shape a new soft hydrated surface on the bare pipe material providing new specific properties. Thus, the hydrophilic-hydrophobic balance as well as the topography of the surface of the naked material may be greatly changed. The colonization of the drinking water biofilm by waterborne pathogens requires at least two things: (1) effective collisions with and physicochemical interactions between the biofilm surface and the pathogens and (2) the capacity for these microorganisms to adapt to this new environment despite being in competition for nutrients and space with the bacteria already attached.

First, predicting effective adhesion of pathogens onto biofilms is rather difficult due to the lack of information on the surface properties of microorganisms including waterborne pathogens despite growing efforts in this field (Gaboriaud et al. 2008; Ojeda et al. 2008). Additionally, characterization of biofilm is also limited and complex. Janjaroen et al. (2013) showed, for example, that the roughness of biofilms increases with age, which for us could be explained by higher accumulation of hydrophobic material within the EPS matrix. This makes sense as the authors found more *Escherichia coli* cells attached to "old" biofilms. Shen et al. (2015) also reported that *L. pneumophila* adhesion was enhanced by biofilm roughness. Indeed, as reported by Van Loosdrecht et al. (1987) and Tatchou-Nyamsi-König et al. (2008), hydrophobic bacteria do stick more efficiently despite their surface-negative charges, which determine a repulsive barrier. However, even few variations in envelope polymeric composition may greatly change the surface properties of bacteria as shown for *Legionella* (Gosselin et al. 2011) and the sorption of *Escherichia coli* shown to be growth phase dependent (Walker et al. 2005). The same was noted for nanoparticle-like bacteriophages GA and Qβ for which small surface property variations significantly affect the rate at which adsorption occurs (Fig. 9.6).

Second, many microorganisms survive for a relatively long period of time within distribution systems and partly colonize drinking water biofilms despite a competitive environment. For instance, some waterborne opportunistic pathogens occurring in premise plumbing systems, including strains of *Legionella* spp., non-tuberculosis *Mycobacteria*, *Pseudomonas*, fungi, and *Acanthamoeba*, are easily detected in

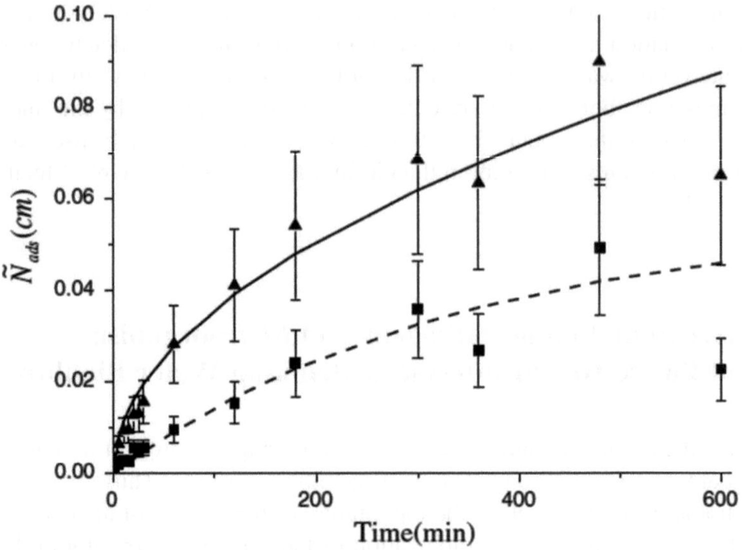

Fig. 9.6 Modeling of the adsorption of bacteriophages GA (filled triangle and full line) and Qβ (filled square and dashed line) under static conditions as a function of time. The full and dashed lines are, respectively, representative of the equations $y = (1.25 \times 10^{-5}t)^{0.5}$ and $y = 0.0552\exp(2.95 \times 10^{-3}t)$ with t being time expressed in min (Hébrant et al. 2014).

biofilms where they multiply. It means that both the pre-existing autochthonous biofilm population and antagonistic competition do not prevent biofilm colonization by allochthonous species. From this point of view, pre-colonization of new pipes with healthy microbial consortia capable of repelling opportunistic pathogens is still a faraway "silver bullet" solution (Wang et al. 2013a). In some cases, better survival of heterotrophic pathogens could even be suspected in the biofilm, as it would represent a necrotrophic organic matter resource (Temmerman et al. 2006; Valster et al. 2011) and a reactive barrier to disinfectant diffusion (see Sect. 9.6). Intracellular persistence or growth of pathogenic microbes within amoebae and other eukaryotic invertebrates in drinking water (Bichai et al. 2008) could explain the capacity of some microorganisms to colonize effectively drinking water biofilms (Garcia et al. 2013; Ovrutsky et al. 2013).

Biofilms can act as either temporary or long-term reservoirs and habitats for bacterial pathogens. As summarized in Fig. 9.7, some of them do find in biofilms almost systematically a favorable ecological environment (*Legionella*, *Mycobacteria*, *and Pseudomonas aeruginosa*), while for others (depending on temperature, availability of organic matter, and chlorine), occupancy is transient (*E. coli*, *Klebsiella*, *Salmonella*, *Campylobacter*, and *Helicobacter*). Last, viruses, *Giardia*, and *Cryptosporidium* oocysts (i.e., bugs unable to multiply outside hosts or under drinking water environmental conditions) will likely have a reduced persistence and wash out more or less rapidly (Helmi et al. 2008).

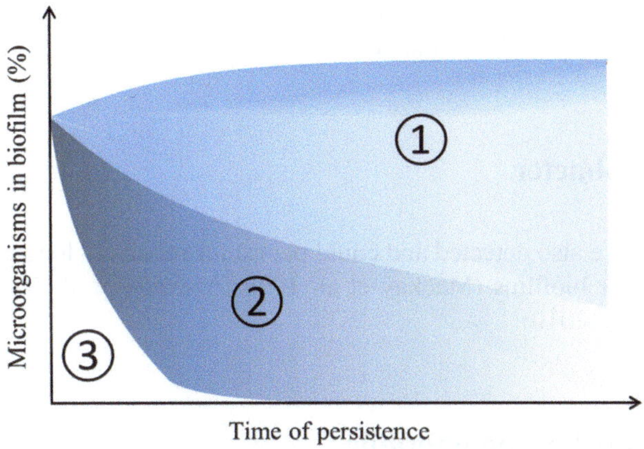

Fig. 9.7 Behavior of indigenous bacteria, waterborne pathogens, and microorganisms of sanitary interest in drinking water biofilms: ① zone with microbial growth rate higher or just slightly lower than the dilution rate: indigenous saprophytic bacteria (*Legionella*, *Mycobacteria*, *P. aeruginosa*, free-living amoebae); ② zone of transient colonization: *E. coli*, *Salmonella*, *Campylobacter*, and *Helicobacter*; and ③ zone of fast washout of microorganisms unable to grow in drinking water biofilms: virus, *Cryptosporidium*, *Giardia* (adapted from Batté et al. 2003)

Many of these microorganisms can also be internalized (e.g., *Pseudomonas*, *Mycobacteria*, *Candida* yeasts) or even multiply (e.g., *Legionella*) inside free-living amoebae which may be found in relatively high density (up to 10^3 FLA/cm^2 of biofilm) (Sibille et al. 1998).

9.5.1 Escherichia coli

Escherichia coli have rarely been detected in drinking water biofilms (Singh et al. 2003; Juhna et al. 2007; Feazel et al. 2009), which makes sense as best efforts are made to inactivate this fecal indicator in drinking waters. However, recent publications, for the most part, did not describe the community structure identified in the phylum *Gammaproteobacteria* at the genus level, and either noncultivable or dead *E. coli* could have been present (Schmeisser et al. 2003). Their growth was demonstrated (Camper et al. 1991; Szewzyk et al. 1994; Fass et al. 1996; Li et al. 2006) and allowed only transient stays in biofilms. Variations between the results from these rare assay reports could be explained by factors such as strain influence, local electrochemical gradients, and organic matter availability. Indeed, as shown in planktonic culture by Vital et al. (2010), growth of *Escherichia coli* and its effective competition with autochthonous bacterial communities depend on the concentration of organic matter. Their adhesion is also dependent on the ionic environment as *E. coli* adhesion rates increased with ionic strength on new PVC surfaces and very young biofilms, but seems independent of the solution chemistry for older biofilms.

This suggests that the physical structure of biofilms could play a role in facilitating the adhesion of *E. coli* cells (Janjaroen et al. 2013).

9.5.2 Helicobacter

Helicobacter were also detected and could persist for relatively long periods of time in drinking water biofilms (Mackay et al. 1998; Azevedo et al. 2006; Giao et al. 2008; Linke et al. 2010).

9.5.3 Legionella pneumophila

Legionella pneumophila (the most studied pathogenic species—causative agent of Legionnaires' disease) and **nonpathogenic species *Legionella*** have been found in water supplies at temperatures below 18 °C (Rodriguez-Martinez et al. 2015). They were recovered from 16% to 19% of samples collected in cold water systems of within-building distribution systems (Völker et al. 2010; Donohue et al. 2014) and detected in drinking water biofilms with molecular analytical tools (Williams et al. 2005; Wang et al. 2012a; Buse et al. 2014; Lin et al. 2014). The lowest concentration and the greatest diversity of *Legionella* were observed in the water supply with low organic matter concentration and high protozoan richness, and biofilms were found to harbor up to 100 *Legionella*/cm^2 (Wullings et al. 2011). The role of intracellular replication in amoebae appears essential for *Legionella* proliferation. As an example, Declerck et al. (2009) showed that biofilm-associated *Legionella* only increased after intracellular replication in *Acanthamoeba castellanii.* Many experimental systems built to test *Legionella* behavior in drinking water biofilms confirm these observations, i.e., that *Legionella* exhibit effective adhesion, long persistence and growth despite autochthonous communities, and chlorination cycles (Rogers et al. 1994; Van Der Kooij et al. 2005; Vervaeren et al. 2006; Langmark et al. 2007; Lehtola et al. 2007; Cooper et al. 2008; Hindre et al. 2008; Moritz et al. 2010; Stewart et al. 2012).

9.5.4 Mycobacteria

Environmental *Mycobacteria* (*M. avium*, *M. gordonae*, *M. intracellulare*, *M. lentiflavum*, *M. tiuscia*, etc.) were very frequently reported in distribution system biofilms at densities from 100 to 5×10^3 per cm^2 and can be considered as common inhabitants of public distribution systems (Falkinham et al. 2001; Schmeisser et al. 2003; September et al. 2004; Torvinen et al. 2004; Vaerewijck et al. 2005; Williams et al. 2005; Keinänen-Toivola et al. 2006; Feazel et al. 2009;

Wang et al. 2012a, c; Kelly et al. 2014). Slow but significant mycobacterial growth was expected or demonstrated in drinking water distribution systems and biofilms: (1) the number of *Mycobacteria* found in biofilms was higher than in the bulk water (Feazel et al. 2009), (2) growth was observed in laboratory assays under conditions relevant to drinking waters (Steed and Falkinham 2006), and (3) the highest numbers of *Mycobacteria* were found at the distal sites of the distribution systems (Falkinham et al. 2001; Torvinen et al. 2004).

9.5.5 Pseudomonas

Pseudomonas spp. occurrence in drinking water distribution systems was reported to be fairly high from 2% (Völker et al. 2010) to 13% (Wang et al. 2012a) in cold water systems of within-building distribution systems. They are frequently identified by culture-independent molecular techniques in drinking water biofilms (Schmeisser et al. 2003; Feazel et al. 2009; Chen et al. 2013; Douterelo et al. 2013, 2014b; Kelly et al. 2014; Lin et al. 2014; Liu et al. 2014a).

Pseudomonas aeruginosa, which is responsible for opportunistic infections, has become a major cause of nosocomial infections worldwide especially in intensive care units and neonatal units (Walker et al. 2014). The bacteria adopt a sedentary community lifestyle by forming a biofilm through various adhesive systems (Giraud et al. 2010) and are able to colonize water system biofilms (taps, u-bends, etc.) especially when some biodegradable organics are leached from the pipe material (Moritz et al. 2010).

9.6 Factors Controlling Biofilm Accumulation in Drinking Water Distribution Systems

Biomass variations in drinking water distribution systems occur due to the combined influence of several parameters (eight parameters and their effects are listed in Table 9.3), which themselves vary with water resources, treatments, seasons, and drinking water system construction materials (Langmark et al. 2007; Pinto et al. 2012; Henne et al. 2013; Wang et al. 2012a, b, 2013b, 2014a). These multiparametric environmental effects are not easily quantifiable at the level of species, and only few models have attempted to predict biomass variations (total number of cells or HPC). Standard correlation analysis techniques have indicated that biomass variations in drinking water are due to the combined influence of organic matter, chlorine, and temperature (Niquette et al. 2001; Zhang and Digiano 2002).

Among all the parameters listed in Table 9.3, some have been well studied as they represent operational parameters that practitioners either may control (i.e.,

Table 9.3 Some parameters (beneficial and detrimental effects) governing biomass accumulation in drinking water distribution systems and biofilms

Parameters	Major effects	References
Biodegradable or assimilable organic carbon (BDOC or AOC)	Substrate saturation constants (Ks) are shown to be as low as a few µg/L In the bulk water, consumption of 10 µg/L organic carbon would result in the formation of as many as 10^4–10^5 cells/mL (the yield factor should be less in biofilm as some of the matter is used for EPS production) For limiting growth in drinking water distribution systems, AOC should be limited to 50 µg/L or BDOC to 100 µg/L	Van Der Kooij and Hijnen (1988); LeChevallier et al. (1991); Servais et al. (1995); Van Der Kooij et al. (1995); Sibille et al. (1997); Niquette et al. (2001); Hammes et al. (2010b); Sack et al. (2014)
Disinfectants: chlorine or chloramine	Drinking water biofilms are shown to be systematically more difficult to inactivate than planktonic biomass due to the strong depletion of chlorine at the surface and within the biofilm (consumption by EPS) Monochloramine penetration within biofilm is better than that of chlorine (in spite of similar diffusion coefficients). This penetration does not necessarily result in immediate bacterial viability loss In drinking water distribution systems, chlorine (>1 mg/L Cl_2) is unable to eradicate biofilms	Chen and Stewart (1996); Gauthier et al. (1999b); Lu et al. (1999); Hallam et al. (2001); Lehtola et al. (2005); Steed and Falkinham III 2006; Szabo et al. (2007); Morrow et al. (2008); Berry et al. (2009); Lee et al. (2011); Hwang et al. (2012); Wang et al. (2012b, c, 2014a); Xue et al. (2012); Xue and Seo (2013); Mathieu et al. (2014)
Di-oxygen	Coliforms adhere quite well in oxygen-depleted environments Di-oxygen appears to be a major driver for eukaryote persistence	Farkas et al. (2013); Wang et al. (2014b)
Hydrodynamics	Accumulation of biomass on the surface is more rapid under turbulent conditions than under laminar flow Increasing shear stress resulted in lower cell number of cells per unit surface area of PVC and stainless steel material Water stagnation favors growth in bulk water (the total cell balance biofilm + water was constant). Growth rate under stagnation conditions is as high as 0.22 per hour	Lautenschlager et al. (2010); Manuel et al. (2010); Simoes et al. (2010)

(continued)

Table 9.3 (continued)

Parameters	Major effects	References
Phosphorus	Phosphorus can originate from water resource and water treatment (anticorrosion as phosphate) and also be released from some pipe materials Microbially available phosphorus (MAP) can show a wide concentration range in drinking waters (0.3–30 μg/L) Phosphorus can be sometimes the limiting factor for bacterial growth	Miettinen et al. (1997); Sathasivan and Ohgaki (1999); Appenzeller et al. (2001, 2002); Lehtola et al. (2002); Morton et al. (2005); Polanska et al. (2005); Gouider et al. (2009); Li et al. (2010); Jiang et al. (2011); Inkinen et al. (2014)
Temperature	A slight temperature increase (from 15 °C to 35 °C) induces a higher number of HPC in biofilms High temperature reduces biomass and biodiversity in biofilms (hot water systems at 60 °C or more) Temperatures >35 °C are required to control the occurrence of *Mycobacterium avium* in pipelines	Bagh et al. (2004); Norton et al. (2004); Silhan et al. (2006); Inkinen et al. (2014)
Pipe materials	Some elastomeric surfaces have more abundant biofilms due to organic leaching than other materials (except corroded iron surfaces). Leaching can be as high as 0.15–30 μg TOC/cm^2 day and 0.06 to 30 μg AOC/cm^2 day The effect of pipe construction material is measurable on prokaryote biomass not on eukaryotes Copper limits biofilm accumulation during the first months of pipe usage. In long-term biofilms grown on copper, the number of cells can be the same, but diversity and physiology may be affected by toxic Cu ions No differences in biofilm accumulation on stainless steel of different grades were found	Rogers et al. (1994); Kerr et al. (1999); Lu et al. (1999); Kalmbach et al. (2000); Niquette et al. (2000); Kilb et al. (2003); Schwartz et al. (2003); Lehtola et al. (2004, 2005); Gagnon et al. (2005); Moritz et al. (2010); Pavissich et al. (2010); Yu et al. (2010); Allion et al. (2011); Bucheli-Witschel et al. (2012); Wang et al. (2012c, 2014a); Buse et al. (2014); Douterelo et al. (2014b, c); Liu et al. (2014a)

(continued)

Table 9.3 (continued)

Parameters	Major effects	References
Corrosion of iron-bearing materials	Corrosion and corrosion products as well as iron (Fe II and Fe III) increase bacterial survival and cultivability and stimulate the rate of biofilm development Corrosion of iron or copper pipes induces a rapid consumption of chlorine at the pipe surface and a low efficacy of the disinfectant Growth of some species (*Legionella pneumophila*) can be inhibited under iron-rich conditions	Norton and LeChevallier (2000); Appenzeller et al. (2001, 2005); Butterfield et al. (2002); Grandjean et al. (2005, 2006); Lehtola et al. (2005); Szabo et al. (2007, 2012); Hindre et al. (2008); Wang et al. (2012b); Szabo and Minamyer (2014)

AOC assimilable organic carbon, *BDOC* biodegradable dissolved organic carbon, *EPS* exopolymers, *TOC* total organic carbon

disinfectant type and concentration, concentration of biodegradable organic matter, nature of pipe materials) or that vary during cleaning procedures (i.e., water flushing). None of these procedures allow biofilm prevention nor its eradication. We further discuss two of these major parameters (i.e., disinfectants and nutrients) below.

9.6.1 Chlorine and Other Chlorinated Oxidants

Chlorine and other chlorinated oxidants have been widely used for more than a century. They both injure and inactivate bacterial cells by oxidation and substitution reactions with envelope polymers (as shown by increases in membrane permeability) and affect key intracellular polymers such as nucleic acids (Saby et al. 1997; Phe et al. 2005; Ramseier et al. 2011). They also injure all living microorganisms including protozoa (Mogoa et al. 2010). However, complete biofilm disinfection is never achieved as chlorine penetration is limited by a reaction-diffusion interaction with all the organic material accumulated on the pipe wall surface (De Beer et al. 1994; Chen and Stewart 1996; Lee et al. 2011), and, as a consequence, some species escape the treatment depending on the nature and concentration of the disinfectant.

First, it is important to understand that the dominant bacterial species surviving after either chlorination or chloramination are different from those which predominated before treatment. As an example, Ling and Liu (2013) showed by 16S rRNA gene pyrosequencing analysis that chloramination selected *Acinetobacter* and *Acidobacteria* as dominant populations, while natural biofilm development leads to the selection of members of *Nitrospira* and *Bacteroides*. Roeder et al. (2010) also observed a shift in the bacterial composition of biofilms (compared by DGGE) and low similarities in treated versus untreated biofilms (especially with

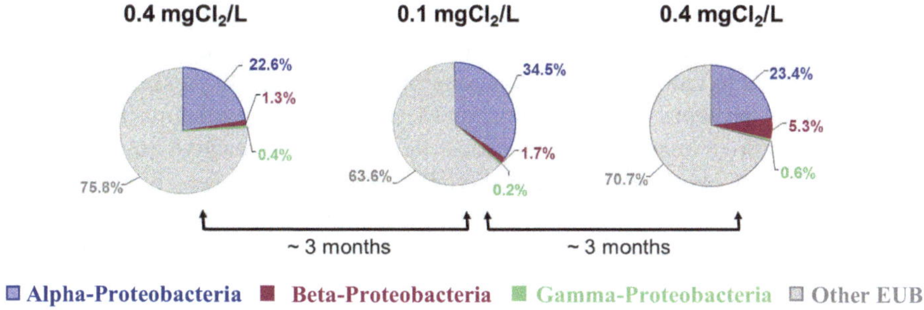

Fig. 9.8 Reversible shift in the *Proteobacteria* populations of drinking water biofilms when subjected to discontinuous chlorination (adapted from Mathieu et al. 2009)

chlorine dioxide). Changing disinfection regimes from chlorine to monochloramine in drinking water biofilms leads to population shifts, and the emergence of *Legionella* species in chlorinated biofilms and *Mycobacteria* in chloraminated ones (Santo Domingo et al. 2003; Pryor et al. 2004; Williams et al. 2005). Codony et al. (2005) showed that discontinuous chlorination reduced the susceptibility of biofilms to the disinfectant. Mathieu et al. (2009) reported that *Beta-* and *Gammaproteobacteria* survived better to increasing chlorination than did *Alphaproteobacteria* in drinking water biofilms. Interestingly, as soon as the selection pressure was over (by decreasing chlorine), the proportions of *Alpha-*, *Beta-*, and *Gammaproteobacteria* phyla returned in a few weeks to the initial equilibrium showing resilience of the bacterial community (Fig. 9.8).

Second, a physiological response of the bacteria remaining after disinfection could be noticed at several levels:

- On the one hand, either overproduction or synthesis of different EPS may be speculated from reports showing how discontinuous chloramination affects drinking water biofilm architecture (Milferstedt et al. 2013). The same is expected from the work of Schwering et al. (2014) who reported also novel characterization of morphologically distinct microcolonies after chlorination.
- On the other hand, resistance to chlorine is induced in many species after sublethal exposure. Knowing the key importance of chlorine for antimicrobial strategies, it is however surprising to observe that only limited knowledge is currently available regarding the ways in which bacteria sense and respond to reactive chlorine species. Most demonstrations of responses to either chlorine or chloramine have been largely done in planktonic cultures: according to Dukan and Touati (1996) *dps*, *katE*, and *katG* genes, as well as *oxyR* regulons, were involved in protective functions against chlorine. Du et al. (2015) showed that proteins involved in stress regulation and stress responses were among those upregulated under both starvation and chlorine or monochloramine disinfection. Glutathione appears as a key regulator in intracellular redox protection against chlorine (Chesney et al. 1996; Saby et al. 1999). *Escherichia coli* responded to monochloramine by activating not just one single antioxidant system but diverse

defense response including oxidative stress, DNA repair, and genes involved in attachment (fimbriae, curli) (Holder et al. 2013). Exposure of *Legionella pneumophila* to chlorine induced, among other things, the expression of cellular antioxidant proteins, stress proteins, and transcriptional regulators (Bodet et al. 2012). Although information about bacterial responses to reactive oxygen species (ROS) is vast, work addressing bacterial responses to reactive chlorine species has begun only recently. Transcriptomic and proteomic studies provide new insights on the bacteria defense strategies against these important antimicrobial compounds (see the review by Gray et al. 2013).

Most of these observations could be extrapolated to biofilm communities. Indeed, chlorine induced *soxS* (transcriptional activator of the superoxide response) to a greater extent in the dormant cells than in the active cells of the biofilm. In addition, chlorine-dependent induction of *soxS* was more prominent in aerobically grown cells (Kim et al. 2009). Comparative transcriptomic analysis revealed that planktonic *Escherichia coli* exposed to monochloramine shares a transcriptional fingerprint with cells grown under biofilm conditions that are known to decrease monochloramine sensitivity characterized by general metabolic inhibition, redox and oxidoreductase response, and cell envelope integrity response (Berry et al. 2010). In another situation (biofilm grown on Cu), the authors showed that UV irradiation intensified the *recA*-mediated SOS response (the main mechanism to repair DNA injuries and other damages) and suggested it could be responsible for differences in the taxonomic composition of biofilms (Jungfer et al. 2013).

To sum up, biofilms and the noncellular organic matter that accumulate on pipe walls have a high reducing potential, which leads to high chlorine consumption at the surface of the material as well as to chlorine having limited diffusion into the biofilm, thus reducing the exposure of biofilm-associated bacteria to chlorine and also to chloramine. As a consequence many biofilm bacteria do survive low disinfectant exposures and induce physiological protective reactions (e.g., oxidative stress resistance), which lead to higher resistance to further disinfection. As a result, continuous addition of disinfectants into drinking water distribution systems (0.1–0.5 mg/L Cl_2) limits growth of bacteria in the bulk water but cannot eradicate biofilms (Mathieu et al. 2009, 2014).

9.6.2 Organic Matter from Treated Water

Organic matter [measured either as dissolved organic carbon (DOC) or assimilable organic carbon (AOC) or biodegradable dissolved organic carbon (BDOC)] represents a key parameter for heterotrophic biomass growth in drinking water distribution systems (autotrophic bacteria develop much more slowly). Bacteria favor growth in biofilm when waterborne organic matter is limited to less than a few hundreds of μg/L, while the opposite situation is observed (preferred growth in the bulk water) when AOC > 500 μg/L (Tsai et al. 2004). Thus, threshold values have been proposed for AOC (<50 μg/L) and BDOC (<100 μg/L) to achieve a limitation of bacterial growth and result in the drinking water distribution system being defined as biostable.

The impact of more recalcitrant macromolecules has been explored on biofilm activity (Camper 2004; Rodrigues et al. 2010). Such macromolecules, i.e., humic substances or polysaccharides and proteins of phytoplanktonic and bacterial origin, cannot be assimilated directly due to their high molecular weight. Therefore, their biodegradation should occur first in the biofilms as extracellular enzymes are needed before their assimilation. Indeed, Sack et al. (2014) showed that biopolymers promote drinking water biofilm formation at microgram-per-liter levels. These polymers are bound to and hydrolyzed by cell-attached enzymes to be degraded into growth-promoting compounds (i.e., <700 Da). Furthermore, the biofilms grown with carbohydrates or proteins clearly differed in bacterial community composition. The protein-adapted biofilm communities were more diverse than those of the polysaccharide-adapted biofilm suggesting that proteins in distribution systems are more utilized and are therefore more likely to promote biofilm formation. The community structure also changed according to the nature of the polymers: *Cytophagia*, *Flavobacteriia*, *Gammaproteobacteria*, and *Sphingobacteriia* grew during polysaccharide addition, while *Alpha-*, *Beta-*, *Gammaproteobacteria*, *Cytophagia*, *Flavobacteriia, and Sphingobacteriia* grew during protein addition.

9.6.3 *Organic Matter Leached from Pipes*

Since the 1990s, plastic pipes have been recognized to support biofilm growth (Kilb et al. 2003; Schwartz et al. 2003; Moritz et al. 2010; Wang et al. 2012c). Biomass increase and population selection are material dependent (Rogers et al. 1994). As shown by Kalmbach et al. (2000), the bacterial community on soft PVC material differed significantly from those on other materials; the dominant species *Aquabacterium commune* was replaced by other *Betaproteobacteria* hybridizing to an amount of 66% with the *Aquabacterium citratiphilum*-specific probe beta 4. Such a selective effect could be related to plasticizers contained in soft PVC such as sebacate, azelate, and adipate, which are metabolizable by *Aquabacterium*. In another study, Liu et al. (2014b) showed that PVC biofilms were dominated by *Hyphomicrobium*-like phylotypes (66%), whereas the latter represented only 2–7% of the microbial community found on cast iron. Finally, by flushing pipes, Douterelo et al. (2014b) found that the highest species richness and diversity were in the samples of material mobilized from plastic pipes (as compared to cast iron) with a high relative abundance of *Alphaproteobacteria* (23%), *Clostridia* (10%), and *Actinobacteria* (10%) coming from the plastic pipe biofilm communities.

9.6.4 Other Nutrients Leached from Pipe Materials

Iron from corroded pipes (Norton and LeChevallier 2000; Grandjean et al. 2005; Wang et al. 2012b, 2015) and phosphorus from both some plastic materials and corroding cast iron (Lehtola et al. 2004; Morton et al. 2005) do contribute to the biofilm growth and diversity. Iron corrosion has been extensively studied, and many specialized species take advantage of this environment. As an example, Wang et al. (2012b, c) showed that iron-reducing bacteria *Shewanella*, iron-oxidizing bacteria *Sediminibacterium*, and sulfur-oxidizing bacteria *Limnobacter thiooxidans* strains promoted iron corrosion by synergistic interactions in the primary period (see also Tables 9.2 and 9.3).

To sum up, nutrients (quantity and quality) govern biofilm dynamics and speciation. Due to high affinity of most bacteria for nutrients on the order of a few μg/L (Van Der Kooij and Hijnen 1988; Van Der Kooij et al. 1995), it appears illusory to eradicate heterotrophic biofilms from drinking water distribution systems. Reducing organic matter thanks to the best available technology and limiting corrosion are the best ways to limit biofilm activity and diversity.

9.7 Conclusion

Drinking water distribution systems have been engineered for transporting water but work de facto as slow bioreactors producing biomass in different phases of the system (in the bulk water, on particles in suspension, in loose deposits, on pipe walls in biofilm). The microbial characteristics of distributed waters highly vary on a daily basis due to bacterial growth especially during stagnation episodes and biomass shearing from biofilm and mobilization of loose deposits during high flow rate events.

Potable waters (defined by zero cultivable *Escherichia coli*/100 mL and zero cultivable enterococci/100 mL) carry millions of bacterial cells per liter. Bacteria phyla are very diverse in drinking water (up to 48 phyla and in excess of 4000 unique operational taxonomical units, OTUs—Proctor and Hammes 2015), with a large dominance of *Proteobacteria* whatever the geographical location, water resource origin, or season. However, each water distribution system has a unique bacterial composition (especially at the level of genus and species). Moreover, drinking water distribution systems harbor a relatively wide-ranging ecosystem including fungi, free-living amoebae, small eukaryotes, and microinvertebrates, which form with heterotrophic bacteria a trophic food chain.

Besides drinking water biofilms formed under turbulent conditions are thin systems (from a few to more than 100 μm) composed of 10^5–10^7 cells/cm^2 patchy distributed. Three classes of *Proteobacteria* (α-, β-, γ-) are systematically present and predominate, and only their relative abundance varies in drinking water biofilms. Besides *Proteobacteria*, other more or less dominant phyla are also detected such as

Firmicutes, *Verrucomicrobia*, *Planctomycetes*, and *Bacteroidetes*. Biofilms can act as either temporary or long-term reservoirs and habitats for bacterial pathogens. Some of these pathogens do find in drinking water biofilms almost systematically a favorable ecological environment (*Legionella*, *Mycobacteria*, *and Pseudomonas aeruginosa*), while for others (depending on temperature, availability of organic matter, and chlorine), occupancy is transient (*E. coli*, *Klebsiella*, *Salmonella*, *Campylobacter*, and *Helicobacter*). Last, viruses, *Giardia*, and *Cryptosporidium* oocysts (i.e., bugs unable to multiply outside hosts or under drinking water environmental conditions) will likely have reduced persistence and washout more or less rapidly.

Microbiome variations in drinking water distribution systems occur due to the combined influence of several parameters, which themselves vary with water resources, treatments, seasons, and drinking water system construction materials. Only recently, less than 10 years ago, our understanding of the microbial ecology of distribution systems greatly improved thanks to molecular biology investigation, fingerprint determination, and sequencing-based approaches. These techniques are still being improved, and next-generation techniques will provide less expensive and time-consuming methodologies as well as new software, tools for analyzing the gene sequences, and new data banks. For the time being, understanding the microbial ecology in distribution systems is still a challenging task. Without a detailed inventory of the microorganisms growing in distribution systems and their interactions, water utilities are forced, in a sense, to "fly blind" when making treatment decisions and managing microbial contamination risks.

Compliance with Ethical Standards

Conflict of Interest Laurence Mathieu declares that she has no conflict of interest. Tony Paris declares that he has no conflict of interest. Jean-Claude Block declares that he has no conflict of interest.

Ethical Approval This article does not contain any studies with human participants or animals performed by any of the authors.

References

Abe Y, Polyakov P, Skali-Lami S et al (2011) Elasticity and physico-chemical properties during drinking water biofilm formation. Biofouling 27(7):739–750

Abe Y, Skali-Lami S, Block JC et al (2012) Cohesiveness and hydrodynamic properties of young drinking water biofilms. Water Res 46(4):1155–1166

Al-gabr HM, Zheng T, Yu X (2014) Occurrence and quantification of fungi and detection of mycotoxigenic fungi in drinking water in Xiamen City, China. Sci Total Environ 466–467:1103–1111

Allion A, Lassiaz S, Peguet L et al (2011) A long term study on biofilm development in drinking water distribution system: comparison of stainless steel grades with commonly used materials. Revue de Métallurgie 108:258–268

Appenzeller BM, Batte M, Mathieu L et al (2001) Effect of adding phosphate to drinking water on bacterial growth in slightly and highly corroded pipes. Water Res 35(4):1100–1105

Appenzeller BM, Duval YB, Thomas F et al (2002) Influence of phosphate on bacterial adhesion onto iron oxyhydroxide in drinking water. Environ Sci Technol 36(4):646–652

Appenzeller BM, Yanez C, Jorand F et al (2005) Advantage provided by iron for *Escherichia coli* growth and cultivability in drinking water. Appl Environ Microbiol 71(9):5621–5623

Ashbolt NJ (2015) Microbial contamination of drinking water and human health from community water systems. Curr Environ Health Rep 2:95–106

Azevedo NF, Pacheco AP, Keevil CW, Vieira MJ (2006) Adhesion of water stressed *Helicobacter pylori* to abiotic surfaces. J Appl Microbiol 101:718–724

Bagh LK, Albrechtsen HJ, Arvin E et al (2004) Distribution of bacteria in a domestic hot water system in a Danish apartment building. Water Res 38(1):225–235

Bal Krishna KC, Sathasivan A, Ginige MP (2013) Microbial community changes with decaying chloramine residuals in a lab-scale system. Water Res 47(13):4666–4679

Batté M, Appenzeller BRM, Grandjean D et al (2003) Biofilms in drinking water distribution systems. Rev Environ Sci Biotechnol 2:147–168

Beale DJ, Barratt R, Marlow DR et al (2013) Application of metabolomics to understanding biofilms in water distribution systems: a pilot study. Biofouling 29(3):283–294

Berney M, Vital M, Hulshoff I et al (2008) Rapid, cultivation-independent assessment of microbial viability in drinking water. Water Res 42(14):4010–4018

Berry D, Xi C, Raskin L (2009) Effect of growth conditions on inactivation of *Escherichia coli* with monochloramine. Environ Sci Technol 43(3):884–889

Berry D, Holder D, Xi C et al (2010) Comparative transcriptomics of the response of to the disinfectant monochloramine and to growth conditions inducing monochloramine resistance. Water Res 44(17):4924–4931

Besner MC, Prévost M, Regli S (2011) Assessing the public health risk of microbial intrusion events in distribution systems: conceptual model, available data, and challenges. Water Res 45 (3):961–979

Bichai F, Payment P, Barbeau B (2008) Protection of waterborne pathogens by higher organisms in drinking water: a review. Can J Microbiol 54:509–524

Bichai F, Dullemont Y, Hijnen W et al (2014) Predation and transport of persistent pathogens in GAC and slow sand filters: a threat to drinking water safety? Water Res 64:296–308

Block JC, Haudidier K, Paquin JL et al (1993) Biofilm accumulation in drinking water distribution systems. Biofouling 6:333–343

Bodet C, Sahr T, Dupuy M et al (2012) *Legionella pneumophila* transcriptional response to chlorine treatment. Water Res 46(3):808–816

Boe-Hansen R, Albrechtsen H-J, Arvin E, Jorgensen C (2002) Bulk water phase and biofilm growth in drinking water at low nutrient conditions. Water Res 36:4477–4486

Bois F, Fahmy T, Block JC et al (1997) Dynamic modeling of bacteria in a pilot drinking-water distribution system. Water Res 31(12):3146–3156

Bucheli-Witschel M, Kotzsch S, Darr S et al (2012) A new method to assess the influence of migration from polymeric materials on the biostability of drinking water. Water Res 46 (13):4246–4260

Buse HY, Lu J, Struewing IT et al (2014) Preferential colonization and release of *Legionella pneumophila* from mature drinking water biofilms grown on copper *versus* unplasticized polyvinylchloride coupons. Int J Hyg Environ Health 217(2–3):219–225

Butterfield PW, Camper AK, Biederman JA et al (2002) Minimizing biofilm in the presence of iron oxides and humic substances. Water Res 36(15):3898–3910

Camper AK (2004) Involvement of humic substances in regrowth. Int J Food Microbiol 92 (3):355–364

Camper K, McFeters GA, Characklis WG, Jones WL (1991) Growth kinetics of coliform bacteria under conditions relevant to drinking water distribution systems. Appl Environ Microbiol 57 (8):2233–2239

Chen X, Stewart PS (1996) Chlorine penetration into artificial biofilm is limited by a reaction-diffusion interaction. Environ Sci Technol 30:2078–2083

Chen L, Jia RB, Li L (2013) Bacterial community of iron tubercles from a drinking water distribution system and its occurrence in stagnant tap water. Environ Sci Process Impacts 15 (7):1332–1340

Chesney JA, Eaton JW, Mahoney JJ Jr (1996) Bacterial glutathione: a sacrificial defense against chlorine compounds. J Bacteriol 17(7):2131–2135

Christensen SC, Nissen E, Arvin E et al (2011) Distribution of *Asellus aquaticus* and microinvertebrates in a non-chlorinated drinking water supply system-effects of pipe material and sedimentation. Water Res 45(10):3215–3224

Codony F, Morato J, Mas J (2005) Role of discontinuous chlorination on microbial production by drinking water biofilms. Water Res 39(9):1896–1906

Cooper IR, White J, Mahenthiralingam E et al (2008) Long-term persistence of a single *Legionella pneumophila* strain possessing the mip gene in a municipal shower despite repeated cycles of chlorination. J Hosp Infect 70(2):154–159

Corsaro D, Venditti D (2010) Phylogenetic evidence for a new genotype of *Acanthamoeba* (*Amoebozoa, Acanthamoebida*). Parasitol Res 107(1):233–238

Craun GF (2012) The importance of waterborne disease outbreak surveillance in the United States. Ann Ist Super Sanita 48(4):447–459

Dang H, Lowell CR (2000) Bacterial primary colonization and early succession on surfaces in marine waters as determined by amplified rRNA gene restriction analysis and sequence analysis of 16S rRNA genes. Appl Environ Microbiol 66(2):467–475

De Beer D, Srinivasan R, Stewart PS (1994) Direct measurement of chlorine penetration into biofilms during disinfection. Appl Environ Microbiol 60(12):4339–4344

Declerck P, Behets J, Margineanu A et al (2009) Replication of *Legionella pneumophila* in biofilms of water distribution pipes. Microbiol Res 164(6):593–603

Delafont V, Brouke A, Bouchon D et al (2013) Microbiome of free-living amoebae isolated from drinking water. Water Res 47(19):6958–6965

Delafont V, Mougari F, Cambau E et al (2014) First evidence of amoebae-mycobacteria association in drinking water network. Environ Sci Technol 48(20):11872–11882

Derlon N, Peter-Varbanets M, Scheidegger A et al (2012) Predation influences the structure of biofilm developed on ultrafiltration membranes. Water Res 46(10):3323–3333

Digiano FA, Zhang W (2004) Uncertainty analysis in a mechanistic model of bacterial regrowth in distribution systems. Environ Sci Technol 38(22):5925–5931

Donohue MJ, O'Connell K, Vesper SJ et al (2014) Widespread molecular detection of *Legionella pneumophila* serogroup 1 in cold water taps across the United States. Environ Sci Technol 48 (6):3145–3152

Douterelo I, Sharpe RL, Boxall JB (2013) Influence of hydraulic regimes on bacterial community structure and composition in an experimental drinking water distribution system. Water Res 47 (2):503–516

Douterelo I, Boxall JB, Deines P et al (2014a) Methodological approaches for studying the microbial ecology of drinking water distribution systems. Water Res 65:134–156

Douterelo I, Husband S, Boxall JB (2014b) The bacteriological composition of biomass recovered by flushing an operational drinking water distribution system. Water Res 54:100–114

Douterelo I, Sharpe R, Boxall J (2014c) Bacterial community dynamics during the early stages of biofilm formation in a chlorinated experimental drinking water distribution system: implications for drinking water discolouration. J Appl Microbiol 117(1):286–301

Du Z, Nandakumar R, Nickerson KW, Li X (2015) Proteomic adaptations to starvation prepare *Escherichia coli* for disinfection tolerance. Water Res 69:110–119

Dukan S, Touati D (1996) Hypochlorous acid stress in *Escherichia coli*: resistance, DNA damage, and comparison with hydrogen peroxide stress. J Bacteriol 178(21):6145–6150

Dupuy M, Mazoua S, Berne F, Bodet C, Garrec N, Herbelin P, Ménard-Szczebara F, Oberti S, Rodier M-H, Soreau S, Wallet F, Héchard Y (2011) Efficiency of water disinfectants against *Legionella pneumophila* and *Acanthamoeba*. Water Res 45:1087–1094

Dupuy M, Berne F, Herbelin P, Binet M, Berthelot N, Rodier M-H, Soreau S, Héchard Y (2014) Sensitivity of free-living amoeba trophozoites and cysts to water disinfectants. Int J Hyg Environ Health 217:335–339

Eichler S, Christen R, Holtje C et al (2006) Composition and dynamics of bacterial communities of a drinking water supply system as assessed by RNA- and DNA-based 16S rRNA gene fingerprinting. Appl Environ Microbiol 72(3):1858–1872

Falkinham JO III, Norton CD, LeChevallier MW (2001) Factors influencing numbers of *Mycobacterium avium*, *Mycobacterium intracellulare*, and other Mycobacteria in drinking water distribution systems. Appl Environ Microbiol 67(3):1225–1231

Farkas A, Butiuc-Keul A, Ciataras D et al (2013) Microbiological contamination and resistance genes in biofilms occurring during the drinking water treatment process. Sci Total Environ 443:932–938

Fass S, Dincher ML, Reasoner DJ et al (1996) Fate of *Escherichia coli* experimentally injected in a drinking water distribution system. Water Res 30(9):2215–2221

Feazel LM, Baumgartner LK, Peterson KL et al (2009) Opportunistic pathogens enriched in showerhead biofilms. Proc Natl Acad Sci U S A 106(38):16393–16399

Flemming HC (2016) EPS – then and now. Microorganisms 4(41):2–18. https://doi.org/10.3390/microorganisms4040041

Flemming HC, Wingender J (2010) The biofilm matrix. Nat Rev Microbiol 8(9):623–633

Francius G, El Zein R, Mathieu L et al (2017) Nano-exploration of organic conditioning film formed on polymeric surfaces exposed to drinking water. Water Res 109:155–163

Gaboriaud F, Gee ML, Strugnell R et al (2008) Coupled electrostatic, hydrodynamic, and mechanical properties of bacterial interfaces in aqueous media. Langmuir 24(19):10988–10995

Gagnon GA, Rand JL, O'Leary KC et al (2005) Disinfectant efficacy of chlorite and chlorine dioxide in drinking water biofilms. Water Res 39(9):1809–1817

Garcia A, Goni P, Cieloszyk J et al (2013) Identification of free-living amoebae and amoeba-associated bacteria from reservoirs and water treatment plants by molecular techniques. Environ Sci Technol 47(7):3132–3140

Gauthier V, Gérard B, Portal J-M et al (1999a) Organic matter as loose deposits in a drinking water distribution system. Water Res 33(4):1014–1026

Gauthier V, Redercher S, Block JC (1999b) Chlorine inactivation of *Sphingomonas* cells attached to goethite particles in drinking water. Appl Environ Microbiol 65(1):355–357

Giao MS, Azevedo NF, Wilks SA et al (2008) Persistence of *Helicobacter pylori* in heterotrophic drinking-water biofilms. Appl Environ Microbiol 74(19):5898–5904

Giraud C, Bernard C, Ruer S et al (2010) Biological 'glue' and 'Velcro': molecular tools for adhesion and biofilm formation in the hairy and gluey bug *Pseudomonas aeruginosa*. Environ Microbiol Rep 2(3):343–358

Gosselin F, Duval JF, Simonet J et al (2011) Impact of the virulence-associated MAb3/1 epitope on the physicochemical surface properties of *Legionella pneumophila* sg1: an issue to explain infection potential? Colloids Surf B Biointerfaces 82(2):283–290

Gouider M, Bouzid J, Sayadi S et al (2009) Impact of orthophosphate addition on biofilm development in drinking water distribution systems. J Hazard Mater 167(1–3):1198–1202

Grandjean D, Fass S, Tozza D et al (2005) Coliform culturability in over- *versus* undersaturated drinking waters. Water Res 39(9):1878–1886

Grandjean D, Jorand F, Guilloteau H et al (2006) Iron uptake is essential for *Escherichia coli* survival in drinking water. Lett Appl Microbiol 43(1):111–117

Gray MJ, Wholey W-Y, Jakob U (2013) Bacterial responses to reactive chlorine species. Annu Rev Microbiol 67:141–160

Hageskal G, Knutsen AK, Gaustad P et al (2006) Diversity and significance of mold species in Norwegian drinking water. Appl Environ Microbiol 72(12):7586–7593

Hageskal G, Lima N, Skaar I (2009) The study of fungi in drinking water. Mycol Res 113:165–172

Hallam NB, West JR, Forster CF et al (2001) The potential for biofilm growth in water distribution systems. Water Res 35(17):4063–4071

Hammes F, Egli T (2010) Cytometric methods for measuring bacteria in water: advantages, pitfalls and applications. Anal Bioanal Chem 397(3):1083–1095

Hammes F, Berney M, Wang Y et al (2008) Flow-cytometric total bacterial cell counts as a descriptive microbiological parameter for drinking water treatment processes. Water Res 42 (1–2):269–277

Hammes F, Goldschmidt F, Vital M et al (2010a) Measurement and interpretation of microbial adenosine tri-phosphate (ATP) in aquatic environments. Water Res 44(13):3915–3923

Hammes F, Vital M, Egli T (2010b) Critical evaluation of the volumetric "bottle effect" on microbial batch growth. Appl Environ Microbiol 76(4):1278–1281

Hébrant M, Pelleïeux S, Mathieu L et al (2014) Distinct adsorption kinetics of Qß and GA bacteriophages on drinking water biofilms. Adsorption 20(5–6):823–828

Helmi K, Skraber S, Gantzer C et al (2008) Interactions of *Cryptosporidium parvum*, *Giardia lamblia*, vaccinal poliovirus type 1, and bacteriophages phiX174 and MS2 with a drinking water biofilm and a wastewater biofilm. Appl Environ Microbiol 74(7):2079–2088

Henne K, Kahlisch L, Brettar I et al (2012) Analysis of structure and composition of bacterial core communities in mature drinking water biofilms and bulk water of a citywide network in Germany. Appl Environ Microbiol 78(10):3530–3538

Henne K, Kahlisch L, Hofle MG et al (2013) Seasonal dynamics of bacterial community structure and composition in cold and hot drinking water derived from surface water reservoirs. Water Res 47(15):5614–5630

Hindre T, Bruggemann H, Buchrieser C et al (2008) Transcriptional profiling of *Legionella pneumophila* biofilm cells and the influence of iron on biofilm formation. Microbiology 154 (Pt 1):30–41

Holder D, Berry D, Dai D et al (2013) A dynamic and complex monochloramine stress response in *Escherichia coli* revealed by transcriptome analysis. Water Res 47(14):4978–4985

Holinger EP, Ross KA, Robertson CE et al (2014) Molecular analysis of point-of-use municipal drinking water microbiology. Water Res 49:225–235

Hong PY, Hwang C, Ling F et al (2010) Pyrosequencing analysis of bacterial biofilm communities in water meters of a drinking water distribution system. Appl Environ Microbiol 76 (16):5631–5635

Hrudey SE, Hrudey EJ (2004) Safe drinking water. Lessons from recent outbreaks in affluent nations. IWA Publishing, London

Hwang C, Ling F, Andersen GL et al (2012) Microbial community dynamics of an urban drinking water distribution system subjected to phases of chloramination and chlorination treatments. Appl Environ Microbiol 78(22):7856–7865

Ingerson-Mahar M, Reid A (2012) Microbes in pipes: the microbiology of water distribution systems. A report on an American Academy of Microbiology Colloquium, Boulder, CO. ASM, Washington, DC

Inkinen J, Kaunisto T, Pursiainen A et al (2014) Drinking water quality and formation of biofilms in an office building during its first year of operation, a full scale study. Water Res 49:83–91

Janjaroen D, Ling FQ, Monroy G et al (2013) Roles of ionic strength and biofilm roughness on adhesion kinetics of *Escherichia coli* onto groundwater biofilm grown on PVC surfaces. Water Res 47(7):2531–2542

Jiang D, Chen Y, Ni G (2011) Effect of total phosphorus (TP) and microbially available phosphorus (MAP) on bacterial regrowth in drinking water distribution system. Syst Eng Procedia 1:124–129

Juhna T, Birzniece D, Larsson S et al (2007) Detection of *Escherichia coli* in biofilms from pipe samples and coupons in drinking water distribution networks. Appl Environ Microbiol 73 (22):7456–7464

Jungfer C, Friedrich F, Varela Villarreal J et al (2013) Drinking water biofilms on copper and stainless steel exhibit specific molecular responses towards different disinfection regimes at waterworks. Biofouling 29(8):891–907

Kahlisch L, Henne K, Grobe L et al (2012) Assessing the viability of bacterial species in drinking water by combined cellular and molecular analyses. Microb Ecol 63(2):383–397

Kalmbach S, Manz W, Szewzyk U (1997) Isolation of new bacterial species from drinking water biofilms and proof of their in situ dominance with highly specific 16S rRNA probes. Appl Environ Microbiol 63(11):4164–4170

Kalmbach S, Manz W, Bendinger B et al (2000) *In situ* probing reveals *Aquabacterium commune* as a widespread and highly abundant bacterial species in drinking water biofilms. Water Res 34 (2):575–581

Keinänen-Toivola MM, Revetta RP, Santo Domingo JW (2006) Identification of active bacterial communities in a model drinking water biofilm system using 16S rRNA-based clone libraries. FEMS Microbiol Lett 257(2):182–188

Kelly JJ, Minalt N, Culotti A et al (2014) Temporal variations in the abundance and composition of biofilm communities colonizing drinking water distribution pipes. PLoS One 9(5):e98542

Kerr CJ, Osborn KS, Robson GD et al (1999) The relationship between pipe material and biofilm formation in a laboratory model system. J Appl Microbiol Symposium Suppl 85:29S–38S

Kilb B, Lange B, Shaule G et al (2003) Contamination of drinking water by coliforms from biofilms grown on rubber-coated valves. Int J Hyg Environ Health 206:563–573

Kim J, Park HJ, Lee JH et al (2009) Differential effect of chlorine on the oxidative stress generation in dormant and active cells within colony biofilm. Water Res 43(20):5252–5259

Kjellerup BV, Gudmonsson G, Sowers K et al (2006) Evaluation of analytical methods for determining the distribution of biofilm and active bacteria in a commercial heating system. Biofouling 22(3–4):145–151

Kormas KA, Neofitou C, Pachiadaki M et al (2010) Changes of the bacterial assemblages throughout an urban drinking water distribution system. Environ Monit Assess 165(1–4):27–38

Langmark J, Storey MV, Ashbolt NJ et al (2007) The effects of UV disinfection on distribution pipe biofilm growth and pathogen incidence within the greater Stockholm area, Sweden. Water Res 41(15):3327–3336

Larsen P, Nielsen JL, Dueholm MS et al (2007) Amyloid adhesins are abundant in natural biofilms. Environ Microbiol 9(12):3077–3090

Lautenschlager K, Boon N, Wang Y et al (2010) Overnight stagnation of drinking water in household taps induces microbial growth and changes in community composition. Water Res 44(17):4868–4877

Lautenschlager K, Hwang C, Liu WT et al (2013) A microbiology-based multi-parametric approach towards assessing biological stability in drinking water distribution networks. Water Res 47 (9):3015–3025

Lautenschlager K, Hwang C, Ling F et al (2014) Abundance and composition of indigenous bacterial communities in a multi-step biofiltration-based drinking water treatment plant. Water Res 62:40–52

LeChevallier MW, Seidler RJ, Evans TM (1980) Enumeration and characterization of standard plate count bacteria in chlorinated and raw water supplies. Appl Environ Microbiol 40 (5):922–930

LeChevallier MW, Schulz W, Lee RG (1991) Bacterial nutrients in drinking water. Appl Environ Microbiol 57(3):857–862

Lee WH, Wahman DG, Bishop PL et al (2011) Free chlorine and monochloramine application to nitrifying biofilm: comparison of biofilm penetration, activity, and viability. Environ Sci Technol 45(4):1412–1419

Lehtola MJ, Miettinen IT, Vartiainen T et al (2002) Changes in content of microbially available phosphorus, assimilable organic carbon and microbial growth potential during drinking water treatment processes. Water Res 36(15):3681–3690

Lehtola MJ, Miettinen IT, Keinanen MM et al (2004) Microbiology, chemistry and biofilm development in a pilot drinking water distribution system with copper and plastic pipes. Water Res 38(17):3769–3779

Lehtola MJ, Miettinen IT, Lampola T et al (2005) Pipeline materials modify the effectiveness of disinfectants in drinking water distribution systems. Water Res 39(10):1962–1971

Lehtola MJ, Torvinen E, Kusnetsov J et al (2007) Survival of *Mycobacterium avium*, *Legionella pneumophila*, *Escherichia coli*, and caliciviruses in drinking water-associated biofilms grown under high-shear turbulent flow. Appl Environ Microbiol 73(9):2854–2859

Lewandowski Z, Boltz JP (2011) Biofilms in water and water treatment. In: Treatise on water science. Elsevier, Chapter 4–15, pp 529–570. https://doi.org/10.1016/B978-0-444-53199-5.00095-6

Li J, McLellan S, Ogawa S (2006) Accumulation and fate of green fluorescent labeled *Escherichia coli* in laboratory-scale drinking water biofilters. Water Res 40(16):3023–3028

Li D, Li Z, Yu J et al (2010) Characterization of bacterial community structure in a drinking water distribution system during an occurrence of red water. Appl Environ Microbiol 76 (21):7171–7180

Lin W, Yu Z, Zhang H et al (2014) Diversity and dynamics of microbial communities at each step of treatment plant for potable water generation. Water Res 52:218–230

Ling F, Liu WT (2013) Impact of chloramination on the development of laboratory-grown biofilms fed with filter-pretreated groundwater. Microbes Environ 28(1):50–57

Linke S, Lenz J, Gemein S et al (2010) Detection of *Helicobacter pylori* in biofilms by real-time PCR. Int J Hyg Environ Health 213(3):176–182

Lipponen MT, Suutari MH, Martikainen PJ (2002) Occurrence of nitrifying bacteria and nitrification in Finnish drinking water distribution systems. Water Res 36(17):4319–4329

Liu G, Ling FQ, Magic-Knezev A et al (2013a) Quantification and identification of particle-associated bacteria in unchlorinated drinking water from three treatment plants by cultivation-independent methods. Water Res 47(10):3523–3533

Liu G, Van der Mark EJ, Verberk JQ et al (2013b) Flow cytometry total cell counts: a field study assessing microbiological water quality and growth in unchlorinated drinking water distribution systems. Biomed Res Int 2013:595872

Liu G, Bakker GL, Li S et al (2014a) Pyrosequencing reveals bacterial communities in unchlorinated drinking water distribution system: an integral study of bulk water, suspended solids, loose deposits, and pipe wall biofilm. Environ Sci Technol 48(10):5467–5476

Liu R, Zhu J, Yu Z et al (2014b) Molecular analysis of long-term biofilm formation on PVC and cast iron surfaces in drinking water distribution system. J Environ Sci (China) 26(4):865–874

Liu S, Gunawan C, Barraud N, Rice SA, Harry EJ, Amal R (2016) Understanding, monitoring, and controlling biofilm growth in drinking water distribution systems. Environ Sci Technol 50:8954–8976

Loret JF, Greub G (2010) Free-living amoebae: biological by-passes in water treatment. Int J Hyg Environ Health 213(3):167–175

Lu W, Kiéné L, Lévi Y (1999) Chlorine demand of biofilms in water distribution systems. Water Res 33(3):827–835

Lürhig K, Canbäck B, Paul CJ, Johansson T, Persson K, Radström P (2015) Bacterial community analysis of drinking water biofilms in southern Sweden. Microbes Environ 30(1):99–107

Ma X, Baron JL, Vikram A, Stout JE, Bibby K (2015) Fungal diversity and presence of potentially pathogenic fungi in a hospital hot water system treated with on-site monochloramine. Water Res 71:197–206

Mackay WG, Gribbon LT, Barer MR, Reid DC (1998) Biofilms in drinking water systems – a possible reservoir for *Helicobacter pylori*. Water Sci Technol 38(12):181–18S

Malm A, Ljunggren O, Bergstedt O et al (2012) Replacement predictions for drinking water networks through historical data. Water Res 46(7):2149–2158

Manuel CM, Nunes OC, Melo LF (2007) Dynamics of drinking water biofilm in flow/non-flow conditions. Water Res 41:551–562

Manuel CM, Nunes OC, Melo LF (2010) Unsteady state flow and stagnation in distribution systems affect the biological stability of drinking water. Biofouling 26(2):129–139

Marciano-Cabral F, Jamerson M, Kaneshiro ES (2010) Free-living amoebae, *Legionella* and *Mycobacterium* in tap water supplied by a municipal drinking water utility in the USA. J Water Health 8(1):71–82

Martiny AC, Jorgensen TM, Albrechtsen HJ et al (2003) Long-term succession of structure and diversity of a biofilm formed in a model drinking water distribution system. Appl Environ Microbiol 69(11):6899–6907

Martiny AC, Albrechtsen HJ, Arvin E et al (2005) Identification of bacteria in biofilm and bulk water samples from a nonchlorinated model drinking water distribution system: detection of a large nitrite-oxidizing population associated with Nitrospira spp. Appl Environ Microbiol 71 (12):8611–8617

Mathieu L, Block JC, Prévost M et al (1995) Biological stability of drinking-water in the city of Metz distribution system. J Water SRT-Aqua 44(5):230–239

Mathieu L, Bouteleux C, Fass S et al (2009) Reversible shift in the alpha-, beta- and gamma-proteobacteria populations of drinking water biofilms during discontinuous chlorination. Water Res 43(14):3375–3386

Mathieu L, Bertrand I, Abe Y et al (2014) Drinking water biofilm cohesiveness changes under chlorination or hydrodynamic stress. Water Res 55:175–184

Maul A (2014) Heterogeneity: a major factor influencing microbial exposure and risk assessment. Risk Anal 34(9):1606–1617

McCoy ST, Van Briesen JM (2014) Comparing spatial and temporal diversity of bacteria in a chlorinated drinking water distribution system. Environ Eng Sci 31(1):32–41

Miettinen IT, Vartiainen T, Martikainen PJ (1997) Phosphorus and bacterial growth in drinking water. Appl Environ Microbiol 63(8):3242–3245

Miles SL, Gerba CP, Pepper IL et al (2009) Point-of-use drinking water devices for assessing microbial contamination in finished water and distribution systems. Environ Sci Technol 43 (5):1425–1429

Milferstedt K, Santa-Catalina G, Godon JJ et al (2013) Disturbance frequency determines morphology and community development in multi-species biofilm at the landscape scale. PLoS One 8(11):e80692

Mogoa E, Bodet C, Legube B et al (2010) *Acanthamoeba castellanii*: cellular changes induced by chlorination. Exp Parasitol 126(1):97–102

Moritz MM, Flemming HC, Wingender J (2010) Integration of *Pseudomonas aeruginosa* and *Legionella pneumophila* in drinking water biofilms grown on domestic plumbing materials. Int J Hyg Environ Health 213(3):190–197

Morrow JB, Almeida JL, Fitzgerald LA et al (2008) Association and decontamination of *Bacillus* spores in a simulated drinking water system. Water Res 42(20):5011–5021

Morton SC, Zhang Y, Edwards MA (2005) Implications of nutrient release from iron metal for microbial regrowth in water distribution systems. Water Res 39(13):2883–2892

Nescerecka A, Rubulis J, Vital M et al (2014) Biological instability in a chlorinated drinking water distribution network. PLoS One 9(5):e96354

Niquette P, Servais P, Savoir R (2000) Impacts of pipe materials on densities of fixed bacterial biomass in a drinking water distribution system. Water Res 34(6):1952–1956

Niquette P, Servais P, Savoir R (2001) Bacterial dynamics in the drinking water distribution system of Brussels. Water Res 35(3):675–682

Norton CD, LeChevallier MW (2000) A pilot study of bacteriological population changes through potable water treatment and distribution. Appl Environ Microbiol 66(1):268–276

Norton CD, LeChevallier MW, Falkinham JO 3rd (2004) Survival of *Mycobacterium avium* in a model distribution system. Water Res 38(6):1457–1466

Ojeda JJ, Romero-Gonzalez ME, Bachmann RT et al (2008) Characterization of the cell surface and cell wall chemistry of drinking water bacteria by combining XPS, FTIR spectroscopy, modeling, and potentiometric titrations. Langmuir 24(8):4032–4040

Oliveira BR, Crespo MT, San Romao MV et al (2013) New insights concerning the occurrence of fungi in water sources and their potential pathogenicity. Water Res 47(16):6338–6347

Otterholt E, Charnock C (2011) Identification and phylogeny of the small eukaryote population of raw and drinking waters. Water Res 45(8):2527–2538

Ovrutsky AR, Chan ED, Kartalija M, Bai X, Jackson M, Gibbs S, Falkinham JO, Iseman MD, Reynolds PR, McDonnell G, Thomas V (2013) Cooccurrence of free-living amoebae and nontuberculous mycobacteria in hospital water networks, and preferential growth of *Mycobacterium avium* in *Acanthamoeba lenticulata*. Appl Environ Microbiol 79(10):3185–3192

Paris T, Skali-Lami S, Block JC (2007) Effect of wall shear rate on biofilm deposition and grazing in drinking water flow chambers. Biotechnol Bioeng 97(6):1550–1561

Park S-K, Choi J-H, Hu JY (2012) Assessing bacterial growth potential in a model distribution system receiving nanofiltration membrane treated water. Desalination 296:7–15

Pavissich JP, Vargas IT, Gonzalez B et al (2010) Culture dependent and independent analyses of bacterial communities involved in copper plumbing corrosion. J Appl Microbiol 109 (3):771–782

Payment P, Siemiatycki J, Richardson L et al (1997) A prospective epidemiological study of gastrointestinal health effects due to the consumption of drinking water. Int J Environ Health Res 7:5–31

Pedersen K (1990) Biofilm development on stainless steel and PVC surface in drinking water. Water Res 24(2):239–243

Percival SL, Knapp JS, Edyvean RGJ et al (1998) Biofilms, mains water and stainless steel. Water Res 32(7):2187–2201

Pereira VJ, Basilio MC, Fernandes D et al (2009) Occurrence of filamentous fungi and yeasts in three different drinking water sources. Water Res 43(15):3813–3819

Pereira VJ, Fernandes D, Carvalho G et al (2010) Assessment of the presence and dynamics of fungi in drinking water sources using cultural and molecular methods. Water Res 44(17):4850–4859

Phe MH, Dossot M, Guilloteau H et al (2005) Nucleic acid fluorochromes and flow cytometry prove useful in assessing the effect of chlorination on drinking water bacteria. Water Res 39 (15):3618–3628

Pinto AJ, Xi C, Raskin L (2012) Bacterial community structure in the drinking water microbiome is governed by filtration processes. Environ Sci Technol 46(16):8851–8859

Pinto AJ, Schroeder J, Lunn M et al (2014) Spatial-temporal survey and occupancy-abundance modeling to predict bacterial community dynamics in the drinking water microbiome. MBio 5 (3):e01135–e01114

Poitelon JB, Joyeux M, Welte B et al (2009a) Identification and phylogeny of eukaryotic 18S rDNA phylotypes detected in chlorinated finished drinking water samples from three Parisian surface water treatment plants. Lett Appl Microbiol 49(5):589–595

Poitelon JB, Joyeux M, Welte B et al (2009b) Assessment of phylogenetic diversity of bacterial microflora in drinking water using serial analysis of ribosomal sequence tags. Water Res 43 (17):4197–4206

Polanska M, Huysman K, Van Keer C (2005) Investigation of microbially available phosphorus (MAP) in flemish drinking water. Water Res 39(11):2267–2272

Power KN, Nagy LA (1999) Relationship between bacterial regrowth and some physical and chemical parameters within Sydney's drinking water distribution system. Water Res 33 (3):741–750

Prest EI, El-Chakhtoura J, Hammes F et al (2014) Combining flow cytometry and 16S rRNA gene pyrosequencing: a promising approach for drinking water monitoring and characterization. Water Res 63:179–189

Prévost M, Rompré A, Coallier J et al (1998) Suspended bacterial biomass and activity in full-scale drinking water distribution systems: impact of water treatment. Water Res 32(5):1393–1406

Proctor CR, Hammes F (2015) Drinking water microbiology – from measurement to management. Curr Opin Biotechnol 33:87–94

Pryor M, Springthorpe S, Riffard S, Brooks T, Huo Y, Davis G, Satter SA (2004) Investigation of opportunistic pathogens in municipal drinking water under different supply and treatment regimes. Water Sci Technol 50:83–90

Ramseier MK, von Gunten U, Freihofer P et al (2011) Kinetics of membrane damage to high (HNA) and low (LNA) nucleic acid bacterial clusters in drinking water by ozone, chlorine, chlorine dioxide, monochloramine, ferrate(VI), and permanganate. Water Res 45(3):1490–1500

Revetta RP, Pemberton A, Lamendella R et al (2010) Identification of bacterial populations in drinking water using 16S rRNA-based sequence analyses. Water Res 44(5):1353–1360

Revetta RP, Gomez-Alvarez V, Gerke TL et al (2013) Establishment and early succession of bacterial communities in monochloramine-treated drinking water biofilms. FEMS Microbiol Ecol 86(3):404–414

Rinta-Kanto JM, Lethola MJ, Vartiainen T et al (2004) Rapid enumeration of virus-like particles in drinking water samples using SYBR green I-staining. Water Res 38:2614–2618

Rodrigues AL, Pereira MA, Janknecht P, Broto AG, Nogueira R (2010) Biofilms formed on humic substances: response to flow conditions and carbon concentrations. Bioresour Technol 101:6888–6894

Rodriguez-Martinez S, Sharaby Y, Pecellin M, Brettar I, Höfle M, Halpern M (2015) Spatial distribution of *Legionella pneumophila* MLVA-genotypes in a drinking water system. Water Res 77:119–132

Roeder RS, Lenz J, Tarne P et al (2010) Long-term effects of disinfectants on the community composition of drinking water biofilms. Int J Hyg Environ Health 213(3):183–189

Rogers J, Dowsett AB, Dennis PJ et al (1994) Influence of plumbing materials on biofilm formation and growth of *Legionella pneumophila* in potable water Systems. Appl Environ Microbiol 60 (6):1842–1851

Saby S, Sibille I, Mathieu L et al (1997) Influence of water chlorination on the counting of bacteria with DAPI (4′,6-diamidino-2-phenylindole). Appl Environ Microbiol 63(4):1564–1569

Saby S, Leroy P, Block JC (1999) *Escherichia coli* resistance to chlorine and glutathione synthesis in response to oxygenation and starvation. Appl Environ Microbiol 65(12):5600–5603

Sack EL, van der Wielen PW, van der Kooij D (2014) Polysaccharides and proteins added to flowing drinking water at microgram-per liter levels promote the formation of biofilms predominated by bacteroidetes and proteobacteria. Appl Environ Microbiol 80(8):2360–2371

Santo Domingo JW, Meckes MC, Simpson JM, Sloss B, Reasoner DJ (2003) Molecular characterization of bacteria inhabiting a water distribution system simulator. Water Sci Technol 47:149–154

Sathasivan A, Ohgaki S (1999) Application of new bacterial regrowth potential method for water distribution system – a clear evidence of phosphorus limitation. Water Res 33(1):137–144

Sautour M, Edel-Hermann V, Steinberg C et al (2012) *Fusarium* species recovered from the water distribution system of a French university hospital. Int J Hyg Environ Health 215(3):286–292

Schmeisser C, Stockigt C, Raasch C et al (2003) Metagenome survey of biofilms in drinking-water networks. Appl Environ Microbiol 69(12):7298–7309

Schwartz T, Hoffmann S, Obst U (2003) Formation of natural biofilms during chlorine dioxide and u.v. disinfection in a public drinking water distribution system. J Appl Microbiol 95(3):591–601

Schwering M, Song J, Louie M, Turner RJ, Ceri H (2014) Multi-species biofilms defined from drinking water microorganisms provided increased protection against chlorine disinfection. Biofouling 29(8):917–928

September SM, Brozel VS, Venter SN (2004) Diversity of nontuberculoid *Mycobacterium* species in biofilms of urban and semiurban drinking water distribution systems. Appl Environ Microbiol 70(12):7571–7573

Servais P, Billen G, Ventresque C et al (1991) Microbial activity in GAC filters at the Choisy-le-Roi treatment plant. J Am Water Works Assoc 83(2):62–68

Servais P, Laurent P, Billen G et al (1995) Development of a model of BDOC and bacterial biomass fluctuations in distribution systems. Revue des Sciences de l'Eau 8:427–464

Shen Y, Monroy GL, Derlon N, Janjaroen D, Huang C, Morgenroth E, Boppart SA, Ashbolt NJ, Liu W-T, Nguyen TH (2015) Role of biofilm roughness and hydrodynamic conditions in *Legionella pneumophila* adhesion to and detachment from simulated drinking water biofilms. Environ Sci Technol 49:4274–4282

Sibille I, Mathieu L, Paquin JL et al (1997) Microbial characteristics of a distribution system fed with nanofiltered drinking water. Water Res 31(9):2318–2326

Sibille I, Sime-Ngando T, Mathieu L et al (1998) Protozoan bacterivory and *Escherichia coli* survival in drinking water distribution systems. Appl Environ Microbiol 64(1):197–202

Silbaq FS (2009) Viable ultramicrocells in drinking water. J Appl Microbiol 106:106–117

Silhan J, Corfitzen CB, Albrechtsen HJ (2006) Effect of temperature and pipe material on biofilm formation and survival of *Escherichia coli* in used drinking water pipes: a laboratory-based study. Water Sci Technol 54(3):49–56

Simoes LC, Simoes M, Vieira MJ (2010) Adhesion and biofilm formation on polystyrene by drinking water-isolated bacteria. Antonie Van Leeuwenhoek 98(3):317–329

Singh R, Stine OC, Smith DL et al (2003) Microbial diversity of biofilms in dental unit water systems. Appl Environ Microbiol 69(6):3412–3420

Siqueira VM, Oliveira HM, Santos C et al (2011) Filamentous fungi in drinking water, particularly in relation to biofilm formation. Int J Environ Res Public Health 8(2):456–469

Steed K, Falkinham JO III (2006) Effect of growth in biofilms on chlorine susceptibility of *Mycobacterium avium* and *Mycobacterium intracellulare*. Appl Environ Microbiol 72 (6):4007–4011

Stewart PS, Franklin MJ (2008) Physiological heterogeneity in biofilms. Nat Rev Microbiol 6 (3):199–210

Stewart CR, Muthye V, Cianciotto NP (2012) *Legionella pneumophila* persists within biofilms formed by *Klebsiella pneumoniae*, *Flavobacterium* sp., and *Pseudomonas fluorescens* under dynamic flow conditions. PLoS One 7(11):e50560

Stockman LJ, Wright CJ, Visvesvara GS et al (2011) Prevalence of *Acanthamoeba* spp. and other free-living amoebae in household water, Ohio, USA – 1990-1992. Parasitol Res 108 (3):621–627

Stoodley P, Wilson S, Hall-Stoodley L et al (2001) Growth and detachment of cell clusters from mature mixed-species biofilms. Appl Environ Microbiol 67(12):5608–5613

Szabo J, Minamyer S (2014) Decontamination of biological agents from drinking water infrastructure: a literature review and summary. Environ Int 72:124–128

Szabo JG, Rice EW, Bishop PL (2007) Persistence and decontamination of *Bacillus atrophaeus* subsp. *globigii* spores on corroded iron in a model drinking water system. Appl Environ Microbiol 73(8):2451–2457

Szabo JG, Muhammad N, Heckman L et al (2012) Germinant-enhanced decontamination of *Bacillus* spores adhered to iron and cement-mortar drinking water infrastructures. Appl Environ Microbiol 78(7):2449–2451

Szewzyk U, Manz W, Amann R et al (1994) Growth and *in situ* detection of a pathogenic *Escherichia coli* in biofilms of a heterotrophic water-bacterium by use of 16S- and 23S-rRNAdirected fluorescent oligonucleotide probes. FEMS Microbiol Ecol 13:169–176

Tachikawa M, Yamanaka K (2014) Synergistic disinfection and removal of biofilms by a sequential two-step treatment with ozone followed by hydrogen peroxide. Water Res 64:94–101

Tatchou-Nyamsi-König JA, Dague E, Mullet M et al (2008) Adhesion of *Campylobacter jejuni* and *Mycobacterium avium* onto polyethylene terephtalate (PET) used for bottled waters. Water Res 42(19):4751–4760

Temmerman R, Vervaeren H, Boon N, Verstraete W (2006) Necrotrophic growth of *Legionella pneumophila*. Appl Environ Microbiol 72:4323–4328

Thomas JM, Ashbolt NJ (2011) Do free-living amoebae in treated drinking water systems present an emerging health risk? Environ Sci Technol 45(3):860–869

Torvinen E, Suomalainen S, Lehtola MJ et al (2004) *Mycobacteria* in water and loose deposits of drinking water distribution systems in Finland. Appl Environ Microbiol 70(4):1973–1981

Tsai YP, Paib TY, Qiu JM (2004) The impacts of the AOC concentration on biofilm formation under higher shear force condition. J Biotechnol 111:155–167

Vaerewijck MJ, Huys G, Palomino JC et al (2005) Mycobacteria in drinking water distribution systems: ecology and significance for human health. FEMS Microbiol Rev 29(5):911–934

Valster RM, Wullings BA, Bakker G et al (2009) Free-living protozoa in two unchlorinated drinking water supplies, identified by phylogenic analysis of 18S rRNA gene sequences. Appl Environ Microbiol 75(14):4736–4746

Valster RM, Wullings BA, van den Berg R et al (2011) Relationships between free-living protozoa, cultivable *Legionella* spp., and water quality characteristics in three drinking water supplies in the Caribbean. Appl Environ Microbiol 77(20):7321–7328

van der Kooij D, Hijnen WA (1988) Nutritional versatility and growth kinetics of an *Aeromonas hydrophila* strain isolated from drinking water. Appl Environ Microbiol 54(11):2842–2851

van der Kooij D, Veenendaal HR, Baars-Lorist C et al (1995) Biofilm formation on surfaces of glass and teflon exposed to treated water. Water Res 29(7):1655–1662

van der Kooij D, Veenendaal HR, Scheffer WJ (2005) Biofilm formation and multiplication of *Legionella* in a model warm water system with pipes of copper, stainless steel and cross-linked polyethylene. Water Res 39(13):2789–2798

Van der Wende E, Characklis WG, Smith C (1989) Biofilms and bacterial drinking water quality. Water Res 23(10):1313–1322

van der Wielen PW, van der Kooij D (2013) Nontuberculous mycobacteria, fungi, and opportunistic pathogens in unchlorinated drinking water in The Netherlands. Appl Environ Microbiol 79 (3):825–834

van Lieverloo JH, Hoogenboezem W, Veenendaal G et al (2012) Variability of invertebrate abundance in drinking water distribution systems in the Netherlands in relation to biostability and sediment volumes. Water Res 46(16):4918–4932

Van Loosdrecht MCM, Lyklema J, Norde W et al (1987) Electrophoretic mobility and hydrophobicity as a measure to predict the initial steps of bacterial adhesion. Appl Environ Microbiol 53 (8):1898–1901

Van Nevel S, Hennebel T, De Beuf K et al (2012) Transparent exopolymer particle removal in different drinking water production centers. Water Res 46(11):3603–3611

Vaz-Moreira I, Egas C, Nunes OC et al (2013) Bacterial diversity from the source to the tap: a comparative study based on 16S rRNA gene-DGGE and culture-dependent methods. FEMS Microbiol Ecol 83(2):361–374

Velten S, Boller M, Koster O et al (2011) Development of biomass in a drinking water granular active carbon (GAC) filter. Water Res 45(19):6347–6354

Vervaeren H, Temmerman R, Devos L et al (2006) Introduction of a boost of *Legionella pneumophila* into a stagnant-water model by heat treatment. FEMS Microbiol Ecol 58 (3):583–592

Vital M, Stucki D, Egli T et al (2010) Evaluating the growth potential of pathogenic bacteria in water. Appl Environ Microbiol 76(19):6477–6484

Völker S, Schreiber C, Kistemann T (2010) Drinking water quality in household supply infrastructure – a survey of the current situation in Germany. Int J Hyg Environ Health 213(3):204–209

Walker SL, Hill JE, Redman JA et al (2005) Influence of growth phase on adhesion kinetics of *Escherichia coli* D21g. Appl Environ Microbiol 71(6):3093–3099

Walker JT, Jhutty A, Parks S et al (2014) Investigation of healthcare-acquired infections associated with *Pseudomonas aeruginosa* biofilms in taps in neonatal units in Northern Ireland. J Hosp Infect 86(1):16–23

Wang Y, Hammes F, Boon N et al (2007) Quantification of the filterability of freshwater bacteria through 0.45, 0.22, and 0.1 microm pore size filters and shape-dependent enrichment of filterable bacterial communities. Environ Sci Technol 41(20):7080–7086

Wang H, Edwards M, Falkinham JO III et al (2012a) Molecular survey of the occurrence of *Legionella* spp., *Mycobacterium* spp., *Pseudomonas aeruginosa*, and amoeba hosts in two chloraminated drinking water distribution systems. Appl Environ Microbiol 78(17):6285–6294

Wang H, Hu C, Hu X et al (2012b) Effects of disinfectant and biofilm on the corrosion of cast iron pipes in a reclaimed water distribution system. Water Res 46(4):1070–1078

Wang H, Masters S, Hong Y et al (2012c) Effect of disinfectant, water age, and pipe material on occurrence and persistence of *Legionella*, mycobacteria, *Pseudomonas aeruginosa*, and two amoebas. Environ Sci Technol 46(21):11566–11574

Wang H, Edwards MA, Falkinham JO III et al (2013a) Probiotic approach to pathogen control in premise plumbing systems? A review. Environ Sci Technol 47(18):10117–10128

Wang H, Pryor MA, Edwards MA et al (2013b) Effect of GAC pre-treatment and disinfectant on microbial community structure and opportunistic pathogen occurrence. Water Res 47 (15):5760–5772

Wang H, Masters S, Edwards MA et al (2014a) Effect of disinfectant, water age, and pipe materials on bacterial and eukaryotic community structure in drinking water biofilm. Environ Sci Technol 48(3):1426–1435

Wang H, Proctor CR, Edwards MA et al (2014b) Microbial community response to chlorine conversion in a chloraminated drinking water distribution system. Environ Sci Technol 48 (18):10624–10633

Wang H, Hu C, Li X (2015) Characterization of biofilm bacterial communities and cast iron corrosion in bench-scale reactors with chloraminated drinking water. Eng Fail Anal 57:423–433

Westrell T, Bergstedt O, Stenstrom TA et al (2003) A theoretical approach to assess microbial risks due to failures in drinking water systems. Int J Environ Health Res 13(2):181–197

Williams MM, Domingo JWS, Meckes MC et al (2004) Phylogenetic diversity of drinking water bacteria in a distribution system simulator. J Appl Microbiol 96(5):954–964

Williams MM, Santo Domingo JW, Meckes MC (2005) Population diversity in model potable water biofilms receiving chlorine or chloramine residual. Biofouling 21(5–6):279–288

Wullings BA, Bakker G, van der Kooij D (2011) Concentration and diversity of uncultured *Legionella* spp. in two unchlorinated drinking water supplies with different concentrations of natural organic matter. Appl Environ Microbiol 77(2):634–641

Xue Z, Seo Y (2013) Impact of chlorine disinfection on redistribution of cell clusters from biofilms. Environ Sci Technol 47(3):1365–1372

Xue Z, Sendamangalam VR, Gruden CL et al (2012) Multiple roles of extracellular polymeric substances on resistance of biofilm and detached clusters. Environ Sci Technol 46 (24):13212–13219

Yu J, Kim D, Lee T (2010) Microbial diversity in biofilms on water distribution pipes of different materials. Water Sci Technol 61(1):163–171

Zacheus OM, Lehtola MJ, Korhonen LK et al (2001) Soft deposits, the key site for microbial growth in drinking water distribution networks. Water Res 35(7):1757–1765

Zhang W, DiGiano FA (2002) Comparison of bacterial regrowth in distribution systems using free chlorine and chloramine: a statistical study of causative factors. Water Res 36(6):1469–1482

Chapter 10
Isolation and Cultivation of Bacteria

Martin W. Hahn, Ulrike Koll, and Johanna Schmidt

Abstract Isolation and cultivation of microorganisms enables in combination with cultivation-independent methods comprehensive research on ecology and function of microbes. The availability of cultures provides opportunities for ecophysiological experiments, enables high-quality genome research and represents a mandatory prerequisite for the description of new taxa of prokaryotes. Unfortunately, access to microorganism by cultivation methods is still quite limited, and the majority of the microbial diversity remains uncultured so far. This chapter discusses the potential reasons for this lack in cultivability and reviews advances in cultivation methods for prokaryotes. Detailed analysis of the media and methods used by taxonomists for isolation of bacterial strains used for description of >1000 new species revealed that the description of new taxa in the ranks of species and genera is currently not methodically limited. On the other hand, isolation of strains enabling the description of taxa representing higher ranks is obviously strongly limited by the currently applied methods. Consequently, different cultivation strategies are required according to the scientific goals of the respective research. While taxonomists interested in the description of new species affiliated with any genus or family can isolate new strains suitable for this task by using standard cultivation methods and media, ecologists interested in cultivation of model organisms appearing in natural systems with high cell numbers, as well as taxonomists interested in isolation of strains representing new higher ranking taxa like orders, classes and phyla, should rather use high-throughput non-standard methods in combination with optimized screening protocols.

M. W. Hahn (✉) · U. Koll · J. Schmidt
Research Department for Limnology, University of Innsbruck, Mondsee, Austria
e-mail: martin.hahn@uibk.ac.at

C. J. Hurst (ed.), *The Structure and Function of Aquatic Microbial Communities*, Advances in Environmental Microbiology 7,
https://doi.org/10.1007/978-3-030-16775-2_10

10.1 Introduction

Isolation and cultivation of bacteria is an easy task. Take an agar plate of a standard medium, for instance, Marine Agar (Zobell 1941) for marine or R2A Medium (Reasoner and Geldreich 1985) for freshwater or soil bacteria, and spread 100 μL liquid sample or soil suspension on the plate. You will find several colonies on the plate after incubation at room temperature for a few days. Differences in colony morphology and coloration usually indicate that different taxa are present on such plates. Most of the grown colonies could be easily purified and propagated as pure cultures. So, why is cultivation of bacteria and other microorganisms still a big issue in microbiology and related disciplines dealing with ecology and diversity of microorganisms? The answer is that those easy to cultivate bacteria represent only a minor fraction of the current bacterial diversity on earth (Rappe and Giovannoni 2003; Hugenholtz et al. 1998). The majority of bacteria present in the environment, frequently estimated as 99% or >99% of cells (Pham and Kim 2012; Amann et al. 1995), is assumed to be difficult to be cultivated or partially even not culturable at all. Importantly, easy to cultivate microorganisms are assumed to represent taxa growing in their natural habitats only with low abundance and contributing, due to their low numbers, probably only marginally to the metabolism of ecosystems (Staley and Konopka 1985; Pace et al. 1986). The majority of bacteria are resistant to standard cultivation approaches and thusly have escaped cultivation by microbiologists, a truth that was overlooked by microbiologists for about a century. Recent advances in cultivation techniques resulted in isolation and cultivation of many difficult to be cultured microorganisms representing abundant taxa in, for example, marine or freshwater systems. Investigation of recently cultured taxa enabled, for instance, new insights in ecology and diversity of microorganisms abundant under natural conditions (Giovannoni et al. 2005a, b; Hahn et al. 2012b). However, despite these successes, the vast majority of microbial diversity is still not represented by cultivated organisms, which limits research in many scientific disciplines. This includes, for instance, such important fields of research as the development of new antibiotics, which is required due to the emerging problem of antibiotic resistance in bacterial pathogens. The development of new antibiotics is steadily decreasing since a couple of decades (Braine 2011), which has several reasons where one reason is the limited access to antibiotic substances encoded by so far uncultured bacteria.

10.2 What Is Isolation and Cultivation?

Microorganisms were cultivated by humans as fermentation starters since ancient times. Traditional fermentation starters were used for production or conservation of various food (e.g. sourdough, soybeans, milk products) and beverage products (e.g. beer). The scientific concept of purposeful establishment and maintenance of cultures and especially of pure cultures of microorganisms was mainly developed in the second half of the eighteenth century. One of the leading scientists was Robert Koch (Brock 1999). In the 1880s, mainly Koch and coworkers developed the concept of *pure cultures*, i.e. clonal cultures derived from a single cell, and the methodology for establishment and maintenance of such cultures (Blevins and Bronze 2010). Further key innovations established by Robert Koch, Walther Hesse, Fanny Hesse and Julius Richard Petri were the introduction of agar for solidification of liquid media (e.g. broth) (Hesse 1992) and the development of glass-made Petri dishes (Petri 1887). This methodology based on dishes filled with sterile solidified media is still the basic microbiological technique for isolation, cultivation and maintenance of prokaryotes. Nowadays, this direct plating of samples from the environment (e.g. aquatic samples or soil suspensions) onto agar plates is circumscribed by the term *standard methods* (or "traditional methods" or "cultivation with standard media") for isolation and cultivation of prokaryotes (Vartoukian et al. 2010; Hazan et al. 2012; Stewart 2012). Microorganisms responding to such treatments with macroscopic growth on agar plates are termed *readily culturable microorganisms* (Kell et al. 1998; Barer and Harwood 1999; Handelsman 2004).

Cultivation of microorganisms always includes the process of *isolation* of particular microbes from other members of their community. In many cases, the physical separation of the targeted microbes is done in one or two subsequent steps, for instance, by serial dilution of samples, plating of the dilution on agar plates and subsequent transfer of single colonies to another plate. The dilution-to-extinction method (Button et al. 1993; Connon and Giovannoni 2002) also uses dilution for isolation of bacteria. Precision of the dilution procedure can be increased by using a MicroDrop microdispenser system (Bruns et al. 2003a; Gich et al. 2005). Other methods for physical separation, i.e. isolation, of cells are filtration through or onto filters with a certain pore size (Hahn et al. 2004) and fluorescence-activated cell sorting (FACS), which is frequently used for isolation of picocyanobacteria (e.g. *Prochlorococcus* and *Synechococcus*). The latter method makes use of the autofluorescent traits of these phototrophic bacteria (Moore et al. 1998; Crosbie et al. 2003). Laser microscopes (optical tweezers) were used for isolation of archaeal cells from mixed cultures (Huber et al. 1995, 2000).

Frequently, microbiologists are interested in isolation or detection of specific prokaryotes, e.g. strains of a particular species or genus (e.g. pathogenic bacteria) or strains with a specific physiology (e.g. nitrifiers). Such targeted cultivation is usually performed by using *selective cultivation media* in combination with preferably selective incubation conditions and may also include a specific pretreatment of samples (e.g. fractionation by filtration). A high number of media suitable for specific isolation and cultivation of various microorganisms as well as many media with a broader applicability were developed during the past 130 years. A very broad collection of recipes of microbiological cultivation media is provided by the Handbook of Microbiological Media (Atlas 2010). All available media are more or less selective for certain groups of organisms sharing specific physiological traits. Consequently, there is no unselective medium suitable for cultivation of all readily culturable microorganisms. When considering the broad spectrum of microbial physiologies ranging from chemoorganoheterotrophs (e.g. *E. coli* and yeasts) and photolithoautotrophs (e.g. *Cyanobacteria*) to chemolithoautotrophs (e.g. nitrifiers), it is obvious that a universal medium meeting the demands of all existing microorganisms is impossible to be created.

In many cases, targeted isolation of specific readily culturable microorganisms is hampered by either their low abundance (cell numbers) under natural conditions or by the lack of a highly specific microbiological medium or the lack of selective cultivation conditions. For such microorganisms, *enrichment cultures* are employed, which increase the relative abundance of the targeted microorganisms by using semi-selective conditions. Frequently, liquid media are used for such enrichments, and samples from enrichments are subsequently spread on agar plates providing suitable growth conditions. One example for such a protocol is enrichment of *Vibrio cholera* in alkaline peptone water (Lesmana et al. 1985) for detection of this pathogen in stool samples of patients. In other cases, enrichment of target organisms can be achieved by sample fractionation, e.g. by filtration using filters with certain pore sizes (Watanabe et al. 2009). Sample fractions penetrating filters, e.g. 0.2 μm filters [isolation of spirochaetes, *Polynucleobacter* or other bacteria (Hahn 2004; Johnson 1977)], or sample fractions retained by filters [e.g. isolation of filamentous cyanobacteria and other filamentous bacteria (Rippka 1988)] may be of interest. The success of enrichment treatments may be influenced by various factors including sample transport and cultivation conditions (Alam et al. 2006) and the presence of superior competitors or specific phage (Muniesa et al. 2005).

Successful cultivation of a microorganism includes subcultivation and potential maintenance of the culture. The first subcultivation is in many cases unsuccessful, and a strain initially growing on an agar plate or in a liquid culture is lost (Kenters et al. 2011). Some researchers do not distinguish between temporary cultivation without successful subcultivation (Zengler et al. 2002, 2005) and maintenance and sustainable cultivation; however, this difference is crucial when it comes to physiological or taxonomic investigations of cultivated microorganisms.

As illustrated in Fig. 10.1, standard methods for isolation and cultivation of prokaryotes employ various adjustment "screws" enabling an increase of specificity of the method for cultivation of the targeted organisms. In other cases, a low specificity is wanted in order to isolate new taxa. Interestingly, a large fraction of newly described species is still isolated and cultivated by using low-specificity standard methods (see below). An example is direct plating of water samples or soil suspensions on R2A Agar, which yielded 11% of all newly described species in the years 2009 and 2010. All so-called standard methods have in common that they employ solidified media with high nutrient concentrations and aim on cultivation of microorganisms as macroscopically visible colonies.

Prokaryotes previously not cultivated by using standard methods, e.g. all bacteria and archaea not ready to grow on standard agar plates with high nutrient concentration, are considered to be *difficult to cultivate prokaryotes*. This may include strains, which are basically able to grow under such conditions but for which suitable media were not developed yet. It is obvious that it is difficult to draw a line separating readily culturable and difficult to be cultured taxa. On the other hand, it is easier to

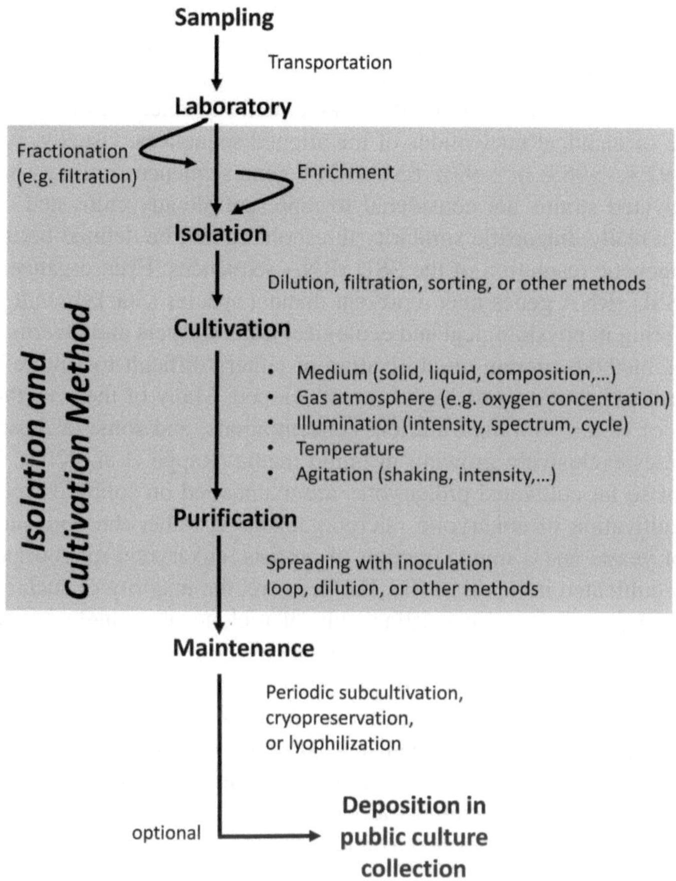

Fig. 10.1 General workflow of isolation and cultivation experiments. Steps highlighted by the grey area encompass what is usually meant by the term cultivation method

distinguish between previously cultivated and "uncultured" (or not yet cultured) taxa. This separation is usually done by performing comparisons of small subunit (SSU) rRNA gene sequences [also known as 16S rRNA (Prokaryotes) and 18S rRNA genes (Eukaryotes)] of organisms of interest with sequences deposited in public nucleic acid sequence databases (GenBank, EMBL-ENA, DDBJ). These databases contain very large collections of SSU rRNA gene sequences representing at least type strains of all prokaryotic species described so far, as well as large numbers of ribosomal sequences of other cultured and environmental sequences of uncultured microorganisms. Environmental sequences were retrieved by molecular methods from environmental DNA samples without cultivation of organisms (Amann et al. 1995). Such sequences are frequently labelled as "uncultured" or "environmental" or "cloned" sequence. Sequence comparisons are frequently performed by using the BLAST (Basic Local Alignment Search Tool) algorithm

(Altschul et al. 1990) provided, for instance, by NCBI (http://blast.ncbi.nlm.nih.gov/Blast.cgi). This algorithm searches databases for sequences similar to the query sequence (i.e. the SSU rRNA gene sequence of interest), aligns the similar sequences and calculates the sequence similarity (also termed sequence identity), which is the percentage of identical nucleotides of the aligned sequences. Usually, prokaryotes sharing >97%, >98% or >99% SSU rRNA gene sequence similarity with previously cultivated strains are considered to represent already cultivated organisms. However, a really diagnostic similarity threshold cannot be defined because of the low phylogenetic resolution of the SSU rRNA sequences. Even organisms sharing identical SSU rRNA genes may represent distinct species (Stackebrandt and Ebers 2006) differing in physiological and ecological traits (Jaspers and Overmann 2004).

Various methods aiming on cultivation of either "difficult to culture" or so far "uncultured" microorganisms have been developed. Many of these methods avoid utilization of media with high nutrient concentrations, and some of these methods yield cultures exclusively growing in liquid media (Rappe et al. 2002). While the majority of so far cultivated prokaryotes are maintained on solidified medium, the trend in cultivation of eukaryotic microorganisms is rather the opposite. Besides cultures of yeasts and a minor fraction of protists, eukaryotic microorganisms are frequently cultivated in liquid media. Furthermore, the majority of such cultures do not represent pure cultures consisting only of a clone of a single organism. Frequently such liquid cultures consist of a single eukaryote or a few eukaryotes mixed with a not characterized diversity of prokaryotes. Especially phagotrophic protists like ciliates and many other heterotrophic protists (protozoa) are maintained in cultures regularly fed with prey organisms like algae. A typical ciliate culture usually consists of the ciliate; its food, e.g. an algae; and an unknown number of diverse bacterial taxa. Such mixed cultures are termed xenic cultures, while pure cultures free of other organisms are termed axenic cultures.

10.3 Quantification of the Uncultured Fraction of Microbial Diversity

The scientific literature is full of quantitative statements on culturability of environmental bacteria (or microorganisms) which consider proportional estimation of the "not yet cultured" or "not culturable" fraction of bacterial (or microbial) diversity. Many papers state that of natural bacterial communities, only 1% or <1% of cells enumerated microscopically can be grown on standard media (i.e. represent readily culturable cells). Other papers state that only 1% or <1% of cells of such communities are culturable. Such statements usually do not mention methods considered for cultivation, for instance, standard or non-standard methods. In addition, some publications applied this 1% or <1% figure to the so far cultivated fraction of the global bacterial diversity. There are papers which use this 1% figure for bacterial cell numbers or diversity, and other papers use it for all microorganisms. The almost universal application of the (<) 1% figure seems to suggest that quantification of cultivation success and estimation of the fraction of so far uncultured taxa are both trivial issues but this is not the case. While efficiency of particular cultivation experiments can be measured, the latter issue can currently only be addressed by unfirm estimations and speculations.

The success of cultivation experiments can be quantified as cultivation efficiency in different ways by referring to parameters characterizing a particular sample used for a cultivation experiment. Those parameters are usually cell numbers, but taxa numbers or community fractions could also be used (Fig. 10.2). Nonquantitative measures of cultivation success could be presentation of new taxa not represented by previously described genera, families or even taxa of higher rank (Tamaki et al. 2011; Mori et al. 2009; Zhang et al. 2003).

The traditional measure of cultivation efficiency is the comparison of the number of cells present in a sample with the number of cultures obtained from the sample (Jannasch and Jones 1959). The number of cells in the sample is nowadays usually enumerated either by epifluorescence microscopy (Bruns et al. 2003b) or by flow cytometry (Button and Robertson 2001) of stained samples; however, application of these methods to samples from soil, sediments or faeces is difficult. The determination of the number of cultivated cells depends on the kind of cultivation methods used in the experiment. If solidified media are employed, the number of colonies (CFU, colony-forming units) is used as a measure of cultivated cells (Jannasch and Jones 1959). If exclusively liquid media are used, the number of cultivated cells is determined as most probable number (MPN) by dilution experiments (Bruns et al. 2002). The cultivation efficiency (CE_c) is expressed as percentage of total number of counted cells represented by cultures. In standard cultivation experiments, i.e. spreading of samples on agar plates, the cultivation efficiency equals the ratio of cultivable cells to the total number of countable cells expressed as a percentage {[(number of colonies obtained per sample volume)/(number of cells counted per sample volume)] × 100} (Fig. 10.2). The determined efficiencies usually depend on the kind of investigated sample and the kind of cultivation method used for the experiment (Jannasch and Jones 1959; Bruns et al. 2002; Janssen et al. 2002; Davis et al. 2005; Sait et al. 2006). Some review papers provided overviews on quantitative results of cultivation experiments by categorization of various types of environmental sample (Amann et al. 1995; Schleifer 2004; Puspita et al. 2012). The data presented by Amann and colleagues, as well as by Schleifer represent exclusively quantifications comparing microscopical counts with the number of colony-forming units (CFU) on agar plates, while the review by Puspita and colleagues also included most probable numbers obtained from cultivation experiments using liquid media. The reported cultivation efficiencies ranged from 0.0007 to 0.2% for desert soil (Connon et al. 2007) to 58% for human faeces (Wilson and Blitchington 1996). For some habitat types, for instance, for soil, the three reviews presented diverging figures. In the oldest review paper, a cultivation efficiency of 0.3% (Torsvik et al. 1990) is mentioned, while the latest review reported a range of 2.4–19% (Sait et al. 2002). This difference may have resulted from the more sophisticated cultivation methods employed by Sait and colleagues. However, due to the broad variety of

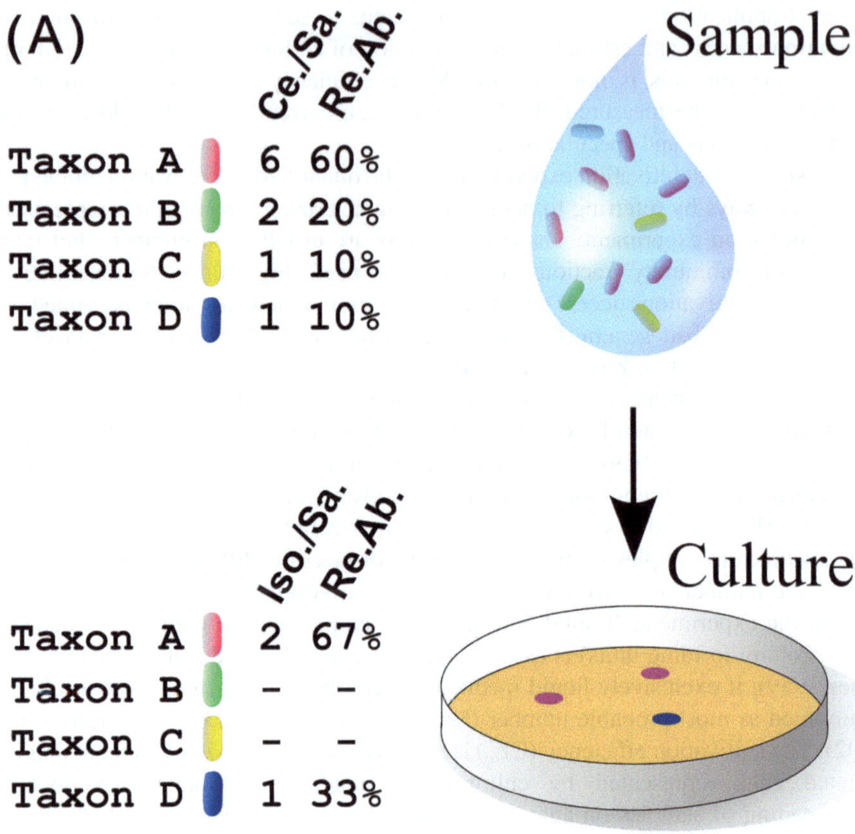

(B)

	Cultivation efficiency
Cell number	30%
Taxon number	50%
Community (taxa)	70%
Taxon A	33%
Taxon B,C	0%
Taxon D	100%

Fig. 10.2 Theoretical example on quantification of cultivation success. (**a**) A sample containing ten cells representing four taxa is used in a cultivation experiment. The experiment yields three cultures representing two taxa. (**b**) The three isolates represent 30% of cells contained in the sample (cultivation efficiency based on cell numbers, CE_C), 50% of taxa contained in the sample (CE_T) and 70% of the community contained in the sample (sum of relative abundance of cultured taxa in sample, i.e. community fraction, CE_{CF}). The specific cultivation efficiencies for taxon A, B, C and D are 33%, 0%, 0% and 100%, respectively. *Ce./Sa.* cells/sample, *Iso./Sa.* isolates/sample, *Re.Ab.* relative abundance

reported efficiencies and methodological differences between the reviewed studies, it is not reasonable to calculate an average cultivation efficiency. Importantly, a rough trend of increasing cultivation efficiencies with increasing trophic status of habitats or samples is obvious (Puspita et al. 2012). Rather oligotrophic environments like desert soil, fresh and seawater tend to result in lower efficiencies than more nutrient-rich (i.e. higher concentrations of organic carbon) environments like soil, sediments, activated sludge or faeces. On the other hand, due to difficulties in microscopic enumeration of cell numbers in complex samples like soil, sediment and activated sludge samples, cultivation efficiencies determined for those samples may suffer from underestimation of total cell numbers. However, it is possible that such potential overestimation of cultivation efficiency is at least partially compensated by another bias. Complex and heterogeneous samples may contain clumped bacterial cells (e.g. microcolonies) or microscopic objects carrying several attached bacterial cells (e.g. small soil particles) which could result in underestimation of cultivation efficiency if such cells are separately counted by microscopy but formed only a joint colony on an agar plate. Janssen and colleagues determined for untreated soil samples that the potential cultivation efficiency cannot exceed about 65% of present prokaryotic cells due to the presence of a high percentage of clumped cells (Janssen et al. 2002). Sonication of samples was demonstrated to reduce the resulting underestimation of cultivation efficiency (Janssen et al. 2002). Another factor potentially resulting in underestimation of cultivation efficiency is formation of microcolonies overlooked in macroscopical counting of CFUs (Jannasch and Jones 1959). In a pioneering study, Jannasch and Jones investigated the influence of different media and cultivation methods on the cultivation efficiency of marine bacterioplankton (Jannasch and Jones 1959). Both formation of microcolonies on filter membranes and formation of macrocolonies on standard agar plates were considered for the determination of cultivation efficiencies. Standard cultivation (macrocolonies) resulted in an average cultivation efficiency of about 0.7% of microscopic counts, which was probably a too high figure due to underestimation of total bacterial numbers caused by microscopic limitations at the time the study was performed. However, counting of microcolonies resulted in a 20–35 times higher efficiency (about 13–23%) compared to the standard cultivation. Other investigations confirmed these trends for other habitat systems (Ferrari et al. 2005, 2008). Furthermore, prolonged incubation of agar plates was demonstrated to increase visibility of small colonies formed by slowly growing bacteria (Davis et al. 2005, 2011). Besides consideration of macro- and microcolonies, quantitative cultivation experiments conducted by using liquid low nutrient media usually result in increased cultivation efficiencies (Bruns et al. 2002, 2003a, b; Köpke et al. 2005). It has to be noted that most of these mentioned cultivation efficiencies were only based on temporary cultivation without any proof for feasibility of sustained cultivation.

A second but more theoretical perspective on cultivation efficiency is coverage of microbial diversity present in a sample or system by cultivation. This measure of cultivation efficiency (CE_T) compares taxon or species richness of the sampled community with the taxon richness of the cultivated fraction of the community

(Fig. 10.2). If a community would consist in total of 1000 taxa and a cultivation experiment would result in cultivation of 200 of those taxa, CE_T would be 20%. This approach requires binning of organisms present in the sample and represented by cultures in taxa. Because the majority of prokaryotes dwelling in natural environments represent undescribed species (see below), classification into species categories makes no sense. An alternative method, which is used in various diversity studies, is classification in operational taxonomic units (OTU) defined by sequence similarity or phylogenetic position of organisms (Wang et al. 2007; Schloss and Handelsman 2005). Usually such OTU classifications are based on sequences of SSU rRNA genes (Schloss and Westcott 2011). In sequence similarity-based binning, sequences of cultured or uncultured organisms, respectively, are compared pairwise and binned according to their similarity values. Threshold values frequently used for OTU classification are 97% or 99% SSU rRNA gene sequence similarity (Thompson et al. 2005).

A third taxon-based measure of cultivation efficiency (CE_{CF}) refers to the fraction of the studied community represented by obtained cultures (Fig. 10.2). This measure includes the relative abundance of taxa in the community and weights abundant taxa more than rare taxa. This measure is of interest if the aim of the cultivation experiment is to provide model organisms representing important players of a community. The probably most impressive example of cultivation of strains representing a large fraction of a community is the isolation of SAR11 bacteria (Rappe et al. 2002). SAR11 represents a clade of *Alphaproteobacteria* comprising by average 20–30% of marine bacterioplankton in surface waters (Morris et al. 2002). Strains affiliated to this clade were cultivated by a high-throughput technique based on dilution-to-extinction and cultivation in nutrient-poor media (Rappe et al. 2002; Connon and Giovannoni 2002). The first obtained isolates were described as *Candidatus* Pelagibacter ubique (Rappe et al. 2002). Further investigation of the obtained isolates provided deep insights in their physiology, ecology and evolution (Giovannoni et al. 2005a, b; Vergin et al. 2007; Tripp et al. 2008). Knowledge gained by such studies even enabled the development of improved media better suited for cultivation of members of the SAR11 clade (Carini et al. 2013). Other examples are the cultivation of *Polynucleobacter* strains representing at least at the time of sampling about 60% of the bacterioplankton community in a freshwater pond (Hahn 2003; Hahn et al. 2005; Jezberova et al. 2010). The contribution of more than 50% of the total cell number of freshwater bacteria by *Polynucleobacter* bacteria is not a usual figure; however, a contribution of more than 10% by this taxon are typical values for many freshwater systems (Jezberova et al. 2010, 2012). By targeted isolation of *Limnohabitans* (Kasalicky et al. 2013) and *Polynucleobacter* (Hahn et al. 2004; Watanabe et al. 2009), it is possible to cover by average about 20% of freshwater bacterioplankton (Jezbera et al. 2012). Strains affiliated with the genera *Polynucleobacter* and *Limnohabitans* are known to grow on agar plates solidified with a standard agar concentration of 1.5% (w/v) (Hahn 2003; Hahn et al. 2009; Kasalicky et al. 2010, 2013), and at least *Polynucleobacter* strains can be isolated from lake or pond water by direct plating of samples on modified R2A (Watanabe et al. 2009, 2012) or NSY agar plates (Hahn et al., unpublished data). But this does not mean that cultivation experiments with direct plating of samples from habitats with >10% *Polynucleobacter* bacteria known to grow efficiently on such media consistently would result in cultivation efficiencies of >10% (Hahn et al., unpublished data).

To our knowledge, large-scale cultivation experiments aiming on cultivation of a maximum number of taxa from a single environmental sample by using a large number of media and methods (Fig. 10.3) have not been performed so far. Application of a large number of different cultivation methods would be required because of selectivity of all methods and media. Frequently, taxon-based cultivation studies compare phylogenetic positions of obtained isolates and their previously cultured closest relatives (with sequences available from public databases) by construction of phylogenetic trees with SSU rRNA sequences (Zengler et al. 2002) and claim cultivation success if at least some of the obtained cultures are only distantly related (e.g. <94% or <97% SSU rRNA similarity) to previously cultured strains (Zengler et al. 2002; Hahn et al. 2004; Janssen et al. 1997, 2002; Sait et al. 2002; Connon and Giovannoni 2002; Page et al. 2004; Cho and Giovannoni 2004). In a much more frequent kind of study aiming on characterization of the diversity of natural communities of microorganism, SSU rRNA sequences of uncultured organisms obtained by cultivation-independent methods are also compared with sequences of cultured organisms but with all previously cultured and sequenced strains independently of their origin (Britschgi and Giovannoni 1991; Mullins et al. 1995; Hugenholtz et al. 1998; Zwart et al. 2002; Newton et al. 2011; Janssen 2006). Each such study focusing on a natural system demonstrated that the vast majority of microorganisms (including eukaryotic microorganisms) is not represented by closely related cultured relatives. However, this kind of analysis only suggests that the vast majority of extant microorganisms was not cultured so far, but does not provide a reliable estimation of the percentage of yet uncultured diversity.

It can be concluded that the cell number-based cultivation efficiency (CE_c) of direct plating on standard bacteriological media is frequently <1% of microscopical cell counts but may exceed 1% if nutrient-rich environments like activated sludge are investigated (Amann et al. 1995). Isolation experiments using liquid low nutrient media tend to result in higher CE_c (Bruns et al. 2003a, b). However, presentation of a general figure for CE_c of naturally occurring prokaryotic communities is impossible. Due to the selectivity of media and methods, the determination of cultivability of members of a certain community would require a large number of cultivation experiments employing a broad range of different media (Fig. 10.3). Since different media may cultivate overlapping fractions of taxa present in the studied community, the cumulative cultivation efficiency of the tested media has to be determined as taxon-based efficiency CE_T. Studies based on taxon richness, i.e. the number of taxa present in the studied community, are faced with the problem of determination of the richness of the rare microbial biosphere (Sogin et al. 2006; Kunin et al. 2010). This problem could be circumvented by exclusion of rare taxa contributing, for instance, less than 0.1% of the community. The presence of rare taxa culturable only with a few specific media would also be a problem in a large-scale experiment aiming on the estimation of the fraction of "unculturable" taxa (compare Fig. 10.3). However,

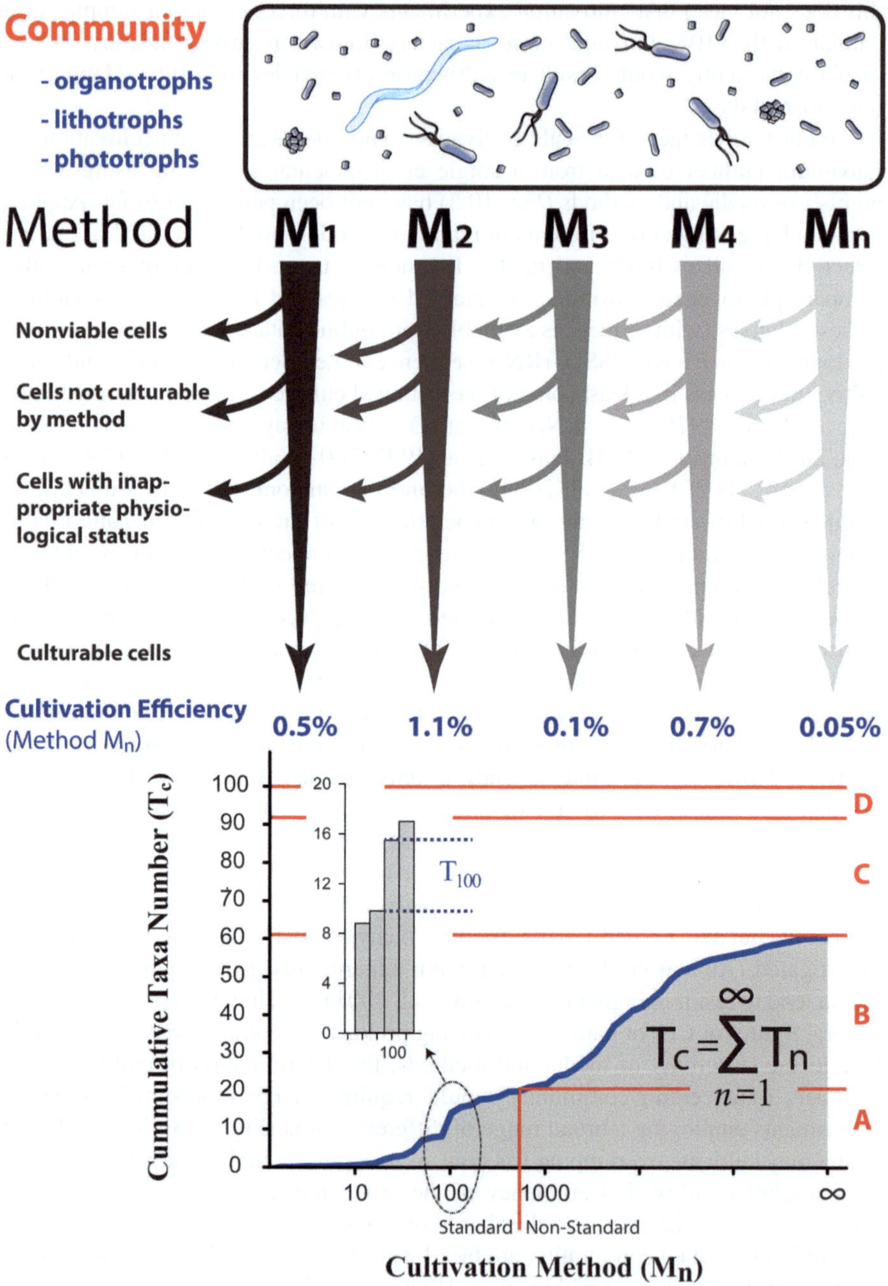

Fig. 10.3 Schematic illustration on hypothetical cumulative cultivation success resulting from an infinite number of cultivation experiments with different cultivation methods (M_1 to M_∞) applied to a natural community of prokaryotes in a particular habitat (e.g. water of a hot spring). Each method is able to cultivate a certain percentage of cells and taxa (e.g. strains sharing >99% SSU rRNA

due to the lack of appropriate experiments employing a large number of different media, even sound estimates of the fraction of basically cultivable taxa present, for instance, in marine prokaryotic communities are currently lacking. We can only speculate on the fraction of taxa resistant to cultivation by the currently available methods.

Obviously, the determination of CE_c for natural microbial communities by using one or a few standard media does not result in meaningful data. The majority of microbiologists acknowledge that cultivation methods are not suitable for determination of diversity of communities or abundance of particular taxa (Amann et al. 1995). This is similar in clinical microbiology; however, in contrast to environmental microbiology, clinical microbiologists developed a large number of cultivation-dependent assays for specific detection of the majority of bacterial pathogens of humans (Delmée et al. 2005; Becker et al. 2013; Gould et al. 2009). Such highly selective media and methods are largely lacking for the majority of readily culturable environmental prokaryotes.

A question of great scientific importance is about the global richness of microbial diversity, and the fraction of this diversity so far cultivated by microbiologists and described as species by taxonomists. Unfortunately, it is a highly difficult and currently probably impossible task to address these questions properly. First attempts to estimate the global number of operational taxonomic units defined by sequence similarities of SSU rRNA greater than 97% ($OTU_{97\%}$) were done (Curtis et al. 2002, 2006), but the vast diversity of prokaryotes and again the problem of accessing and quantification of the rare biosphere (Huse et al. 2010; Kunin et al. 2010; Sogin et al. 2006) limit the results of such efforts. Due to these reasons, and due to limitations in transforming OTU numbers into species numbers (Stackebrandt and Ebers 2006), the currently available estimations on the global number of undescribed prokaryotic species are based on vague estimations and assumptions, which results in a quite large range of estimates. While the current (March 2015) number of validly described species of prokaryotes is about 13,000 (www.bacterio.net), the estimates of prokaryotic species number on earth range from >11,100 (Mora et al. 2011) to $<2 \times 10^6$ in the ocean (Curtis et al. 2002) to 10^9 species of prokaryotes (Dykhuizen 1998, 2005). The highest of these estimates suggest that 99.999% of the existing prokaryotic species thus far have not been taxonomically described.

Fig. 10.3 (continued) sequence similarity). The diagram at the bottom depicts the cumulative number of taxa cultivated by using the infinite number of methods. Fractions A and B represent taxa culturable by standard and non-standard methods, respectively. Fraction C represents rare taxa with medium requirements only met by a few special media. Due to the rareness of the taxa in the environment it is unlikely that representatives of the taxa are present in inoculi used for the few experiments with the suitable media. As a result strains represented by this fraction will remain uncultured. If unculturable taxa really exist, they would be represented by fraction D, but this fraction could not be distinguished from fraction C

10.4 Reasons for Either Lack of or Low Culturability of Prokaryotic Cells

A large number of papers have tried to explain why many prokaryotic cells are refractory to cultivation (Vartoukian et al. 2010; Stewart 2012; Rappe and Giovannoni 2003; Pham and Kim 2012; Alain and Querellou 2009). Most authors assume that a multitude of reasons are responsible for this phenomenon and that reasons might be different from taxon to taxon or even from cell to cell. The known potential reasons for lack of culturability of a certain cell can in principle be classified to three different categories (Fig. 10.3): (1) cells are nonviable, (2) cells are not culturable by the applied method, and (3) cells are viable and could basically be cultured by the applied method but lack at present the appropriate physiological status for growth under the artificial conditions provided by the method.

10.4.1 Nonviable Cells

The first category covers dead (nonviable) cells; however, the status dead is difficult to be defined in prokaryotes and other microorganisms (Roszak and Colwell 1987). Usually the presence of an intact cell membrane is used to distinguish between viable and nonviable cells (Nocker et al. 2006), because an intact membrane is required to maintain the proton motive force essential for prokaryotic energy metabolism. Several cytological methods are used to diagnose the physiological stage of single cells or populations (Kell et al. 1998). A couple of those methods are based on testing the presence of an intact permeability barrier (e.g. life/dead stain), while other methods are based on the detection of messenger RNA, which rapidly disappears in cells lacking transcription. Due to methodical limitations in detection of nonviable cells in environmental samples, only little is known about the abundance of nonviable cells under natural conditions.

10.4.2 Unsuitability of Cultivation Method Employed

The second category covers cases, where the applied cultivation method is unsuitable for cultivation of a certain taxon. The term method circumscribes here again the entire cultivation procedure (Fig. 10.1), which includes potential pretreatment of sample (e.g. dilution or size fractionation), the kind of medium or media used for the cultivation experiment, as well as the employed cultivation conditions (e.g. temperature, gas phase, illumination, agitation). This global view is necessary because a single detail may hinder growth of taxa otherwise able to grow under the offered conditions.

A crucial component of each cultivation method is the medium (or media) employed. The medium should cover all requirements of the organism to be cultivated. As discussed above, media are always selective, and creation of a universal medium is most likely impossible. The literature is full of examples of successful struggle for development of media suitable for cultivation of certain targeted microorganisms (e.g. *Mycobacterium tuberculosis*, *Legionella pneumophila*). The large number of described media (Atlas 2010) emphasizes the importance of media development in cultivation of microorganisms. Media may be unsuitable for cultivation of certain organisms because they lack essential growth factors. *Legionella pneumophila*, the causative agent of Legionnaires' disease, for instance, requires L-cysteine in the medium (Ewann and Hoffman 2006), and the abundant marine bacterium *Candidatus* Pelagibacter ubique requires reduced sulphur compounds (Tripp et al. 2008). Furthermore, media may lack common goods produced under natural conditions only by a small fraction of a microbial community but utilized by a wider range of microbes. Known common goods are siderophores (D'Onofrio et al. 2010) required by some bacteria for iron uptake under iron-limited conditions or enzymes protecting from oxidative stress. Bogosian and colleagues demonstrated that *Vibrio vulnificus* in the viable but nonculturable state can be cultivated with an up to 1000-fold higher efficiency if the cultivation medium is supplemented with catalase an hydrogen peroxide (H_2O_2)-decomposing enzyme (Bogosian et al. 2000), and Morris et al. demonstrated that heterotrophic bacteria (helper bacteria) protect *Prochlorococcus* strains against oxidative stress and increase by this protection the cultivation efficiency for this important marine phototroph (Morris et al. 2008). Recently, Tanaka and colleagues revealed that autoclaving of agar in the presence of phosphate, which is a usual procedure in media preparation, resulted in the formation of H_2O_2 causing a significant reduction of the number of CFU on plates inoculated with environmental samples. Furthermore a negative influence on the diversity of taxa able to grow on the plates was demonstrated (Tanaka et al. 2014). These findings reemphasize the importance of using compounds like catalase or pyruvate known to eliminate H_2O_2 as supplements in microbial media used for isolation of strains from the environment (Stevenson et al. 2004). On the other hand, it remains unknown if only the formation of H_2O_2 caused inferior cultivation efficiencies of plates solidified with agar compared to other solidifying reagents like gellan gum (Janssen et al. 2002; Tamaki et al. 2005).

The abovementioned phenomenon that prokaryotes can only be cultured in the presence of helper cells was repeatedly observed (Kaeberlein et al. 2002; Morris et al. 2008; Jezbera et al. 2009; Garcia et al. 2014), but in most of the cases it is not known which shared goods or service are provided by the helpers. One example is the growth of the freshwater actinobacterium *Candidatus* Limnoluna rubra enabled by the betaproteobacterial helper bacterium *Polynucleobacter* sp. on an agar plated (Hahn 2009). The *Actinobacteria* can only grow in the immediate proximity of colonies of the helper strain (Fig. 10.4), which indicated dependence on a substrate or service provided by the helper. Interestingly, *Polynucleobacter* strains were also found to be efficient helper strains in cultivation of two other but only distantly related *Actinobacteria*. These *Actinobacteria* represent two different lineages within

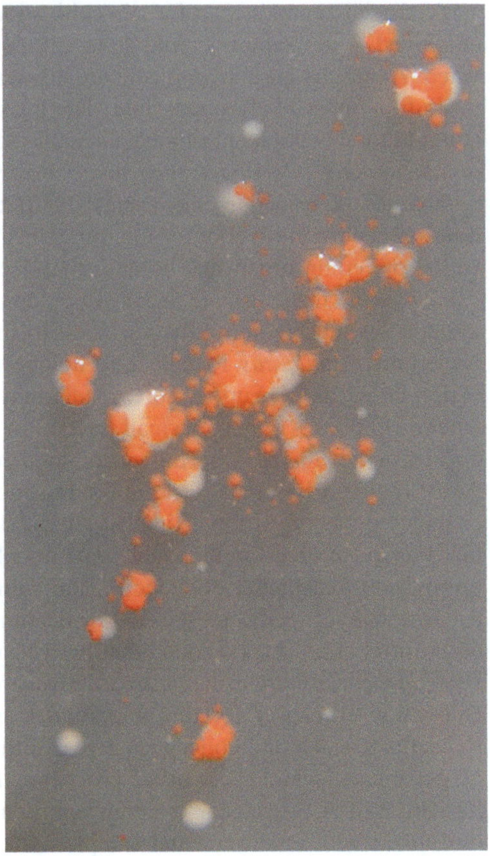

Fig. 10.4 Coculture of the actinobacterium *Candidatus* Limnoluna rubra strain MWH-EgelM2-3 (orange pigmentation) and the betaproteobacterium helper strain *Polynucleobacter* sp. (without pigmentation) on an NSY agar plate. The plate was inoculated with a dense mixed culture of the two strains grown in liquid NSY medium. Despite the sample was evenly spread on the agar plate, colonies of the actinobacterium grew only in the immediate proximity of *Polynucleobacter* colonies. So far, *Candidatus* Limnoluna rubra could only be grown in the presence of the helper strain (Hahn 2009)

the acI *Actinobacteria* clade (Jezbera et al. 2009; Garcia et al. 2014). The Epstein lab systematically exploited the helper concept for cultivation of previously uncultured prokaryotes by using diffusion chamber devices (Kaeberlein et al. 2002; Bollmann et al. 2007). Besides lack of specific growth factors or substrates, other factors may cause a certain medium to be unsuitable for cultivation of a specific prokaryote. Such factors could be the adjusted pH value, a too high substrate concentration and the presence of inhibiting compounds.

10.4.3 Temporary Impairment Due to Physiological Status

The third category includes cases where the organisms are in principle able to grow on the provided medium and at the provided incubation conditions but currently lack the physiological status enabling them to start growth under the provided conditions. Those cells may require an appropriate adjustment of their physiological status (i.e. gene expression pattern) to the provided cultivation conditions. Initiation of the adjustment may be triggered either by a specific stimulus or by an appropriate transition of conditions towards the artificial cultivation conditions. A classic example regarding the influence of the physiological status of particular cells on cultivation success is the viable but nonculturable (VBNC) state of bacteria (Roszak and Colwell 1987). Several pathogenic and non-pathogenic taxa, e.g. *Escherichia coli* and *Vibrio* spp., are known to be able to enter this dormant physiological state if environmental conditions turned unsuitable for their growth (Oliver 2010). As mentioned above cells in this state can be wakened and enabled again for growth on standard media by specific treatments. However, in the genus *Polynucleobacter* representing important freshwater bacteria, the lack of previous cultivation success cannot be explained by hindering due to VBNC states. This ubiquitous and abundant freshwater taxon (Jezberova et al. 2010) was not cultivated before 2003 (Hahn 2003). After successful cultivation of a larger number of strains by a novel method, ten strains were tested for their abilities to grow on plates of ten different standard media (Hahn 2003). Surprisingly, the tested strains could grow by average on 84% of the tested complex media. Interestingly, R2A Medium was found to be the best-performing medium supporting growth of all ten strains with above-average yields which represents a medium very frequently used for isolation of bacteria from environmental samples (see below). Given the abundance and ubiquitous presence of *Polynucleobacter* bacteria in freshwaters and their ability to grow on a very frequently used standard medium, the fact of the previous lack of cultivation of this taxon requires a special explanation. The first isolation of *Polynucleobacter* strains was performed by stepwise acclimatization of bacteria to higher substrate concentrations and exclusion of a large fraction of potential competitors by sample fractionation (filtration). Control experiments with skipped acclimatization and direct plating of fractionated samples on agar plates did not result in growth of *Polynucleobacter* strains (Hahn et al. 2004), leaving only the acclimatization procedure as explanation for successful cultivation. The stepwise acclimatization to higher substrate concentrations may help to avoid substrate-accelerated death known to result from exposure of growth limited bacteria to increased concentrations of the growth-limiting substrate (Postgate and Hunter 1963). A mechanistic but unproven explanation for the effect of the acclimatization procedure could be that the gradual transition to higher substrate concentrations gives metabolically rather inflexible bacteria the opportunity to readjust their metabolism to the strongly changing environmental conditions. It can be assumed that such a gradual transition is especially of importance in bacteria adapted to rather stable environmental conditions. *Polynucleobacter* bacteria are characterized by a rather small genome size of about 2 Mbp resulting in a rather spare metabolic network and only a low number of signal transduction genes (Hahn et al. 2012b). Signal transduction is of crucial importance in sensing of environmental changes; therefore *Polynucleobacter* and other bacteria with limited signal transduction abilities may simply miss the required prompt readjustment of their metabolism to suddenly changed environmental conditions. The demonstrated isolation of *Polynucleobacter* strains by direct plating (without acclimatization) on an improved R2A Medium (Watanabe et al. 2009, 2012) seems to argue against this hypothesis; however, inoculation of an agar plate with 100 μL of a lake water sample [about 10^6 bacteria mL^{-1} with a relative abundance of *Polynucleobacter* bacteria of 1–10% (Jezberova et al. 2010)] results in seeding of 10^3–10^4 *Polynucleobacter* cells on each agar plate but only in growth of a few *Polynucleobacter* colonies (Watanabe et al. 2012). These few colonies may represent a small fraction of *Polynucleobacter* cells possessing by chance a gene expression pattern suitable for surviving a sudden change from in situ conditions to artificial cultivation conditions. Acclimatization enabled cultivation of several previously uncultured bacteria, including, for instance, *Limnohabitans* spp. (Kasalicky et al. 2013) representing important freshwater bacteria (Jezbera et al. 2012), as well as various freshwater *Actinobacteria* characterized by very small genome sizes. Such freshwater *Actinobacteria*, e.g. *Rhodoluna lacicola* MWH-Ta8, possess genomes with only 1.4 Mbp (Hahn et al. 2014), which is only slightly larger than the genome size of *Cand.* Pelagibacter ubique and other SAR11 strains (Giovannoni et al. 2005b). In contrast to SAR11 strains, *R. lacicola* and other freshwater *Actinobacteria* are able to grow on agar plates with high substrate concentrations. Despite the Acclimatization Method attempted for enabling the cultivation of a broad diversity of previously uncultured bacteria, it well may fail in cultivation of obligate oligotrophs like SAR11 bacteria.

Other examples of in principle cultivable but only very rarely cultured bacteria are certain *Verrucomicrobia* and *Acidobacteria* strains. Especially the latter phylum represents bacteria highly abundant in soil systems, but despite their environmental significance, this phylum is extremely underrepresented in culture collections. Some of the few cultivated strains were obtained by sophisticated isolation strategies (Koch et al. 2008), while other strains could be cultivated by using simple direct plating of samples on only slightly modified standard medium (Mannisto et al. 2011). Similar success was reported for novel *Verrucomicrobia* strains (Sangwan et al. 2005). As in the case of the *Polynucleobacter* bacteria, these observations on *Acidobacteria* and *Verrucomicrobia* could be explained by basic culturability of at least some lineages of these phyla but lack of a physiological state (readily culturable) enabling survival of a sudden change to artificial cultivation conditions in the vast majority of cells affiliated with such culturable lineages. Potentially the percentage of readily culturable cells in populations of some taxa can be positively influenced by application of signal compounds. Burns and colleagues demonstrated that the supplementation of media with signal compounds like cAMP and homoserine lactones resulted in an increase of general cultivation efficiency (Bruns et al. 2002, 2003b).

10.5 Does Genome Size Play a Role in Culturability of Prokaryotes?

Genome size of prokaryotes varies from <0.1 Mbp to >10 Mbp. A look on the distribution of genome size among cultivated genome-sequenced bacteria indicates underrepresentation of bacteria with small genome size among the sequenced strains. The average genome size of the 3602 prokaryotic genomes (almost all representing cultivated strains) contained in the Integrated Microbial Genomes (IMG) database (Markowitz et al. 2012) and classified with the sequencing status "finished" is 3.5 Mbs. Only 26% of those genomes possess genome sizes <2 Mbp, and the vast majority of these small genomes represent host-associated (e.g. *Helicobacter*), symbiotic (e.g. *Buchnera*) or pathogenic strains (e.g. *Borrelia*). While small genomes represent about one quarter of all finished genomes, they represent only one-seventh (14%) of all 604 genomes linked to the ecosystem types "marine", "soil" and "freshwater", and remarkably, the average size of genomes associated with strains from these ecosystems is 4.2 Mbp, thus larger than the overall average of all genomes. By contrast, metagenomic investigations (Yooseph et al. 2010; Biers et al. 2009), single-cell genomic analysis (Swan et al. 2013) and genome sequencing of cultivated strains representing abundant taxa (Dufresne et al. 2005) suggested that at least the water column of marine and freshwater systems are dominated by prokaryotes possessing small genome sizes. Cultures of representatives of abundant free-living marine or freshwater taxa characterized by small genome size (e.g. *Cand.* Pelagibacter, *Prochlorococcus*, freshwater *Actinobacteria*, *Polynucleobacter*) were usually not obtained by standard methods, and their cultivation is still considered as being difficult. Obviously, there is a link between small genome size and underrepresentation of respective isolates in culture collections. These free-living prokaryotes share the trait of small genome sizes with bacteria possessing an obligate endosymbiotic or parasitic lifestyle like *Buchnera aphidicola*, an obligate endosymbiont of aphids. In both cases, the free-living pelagic bacteria and the obligately host-associated bacteria, the small genome sizes resulted from reductive genome evolution (Batut et al. 2014; Mira et al. 2001; Giovannoni et al. 2014).

The low culturability of obligate symbionts and parasites is assumed to be a consequence of reductive genome evolution taking place in the evolutionary process transforming free-living or facultatively host-associated taxa to obligately host-associated organisms. Genes required for synthesis of essential compounds essential for free-living stages but provided by host cells to host-associated stages usually lose functionality due to mutations and lack of stabilizing selection. This results in transformation of genes to pseudogenes, which usually undergo further gene erosion and get finally completely lost. Other genes encoding crucial cell functions (e.g. predation defence) not required by host-associated stages may also undergo this process of gene loss. This reductive genome evolution results in streamlined genomes and in physiologies demanding specific compounds and environmental conditions which are provided by their specific host. The details of these demands are usually unknown, and therefore it is difficult to provide under in vitro conditions obligately host-associated strains with all their requirements. A typical example for these difficulties in cultivation is *Buchnera aphidicola* (*Gammaproteobacteria*, *Enterobacteriaceae*) the endosymbiont of various aphid species. These obligate endosymbionts were intensively investigated for decades, they are well represented by complete genome sequences, and detailed knowledge on their physiology and ecology is available but still pure cultures representing this taxon are lacking.

10.6 Cultivation and Description of New Prokaryotic Taxa

Since a few decades, a mandatory requirement for the valid description of new species of prokaryotes is the deposition of cultures of the type strain in at least two public culture collections located in two different countries (Tindall et al. 2006). Journals like the *International Journal of Systematic and Evolutionary Microbiology* (IJSEM), which is the most important journal for publication of descriptions of new prokaryotic taxa (see below), do not accept manuscripts proposing new species for publication before certificates of deposition from two culture collections were presented. Thus, IJSEM is a good source for data suitable for analyses on which new taxa were isolated with what method and medium.

According to the List of Prokaryotic Names with Standing in Nomenclature (LPSN, http://www.bacterio.net) (Parte 2014), the current number of validly described prokaryotic species is about 13,000. About 600 new species and about 100 new genera were described annually during the past 10 years (Fig. 10.5). The vast majority of these new taxa is based on new isolates and only a minor fraction of the new taxa represents revised previously described taxa. Furthermore, most of the descriptions and revisions are published in IJSEM. We analysed all descriptions of bacterial taxa published in volumes 2009 and 2010 in the IJSEM. The analysis of new archaeal taxa was omitted because the number of such descriptions is too small

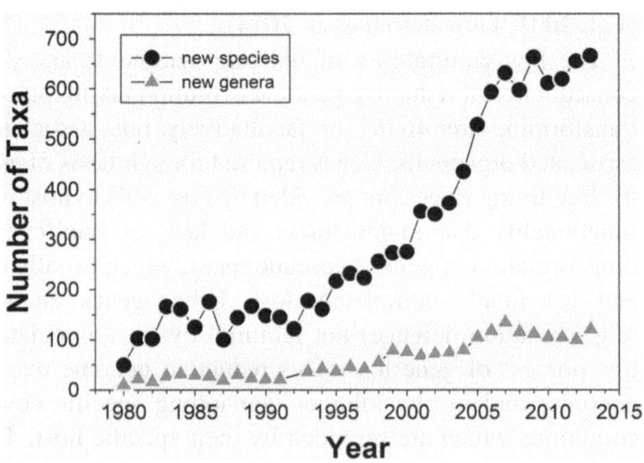

Fig. 10.5 Total number of new prokaryotic species and genera described per year in the period 1980–2013. The graph depicts data obtained from the LPSN webpage (http://www.bacterio.net)

for revealing significant trends. In 2009 and 2010, a total of 1042 new species and 168 new genera were described in this journal (Table 10.1), which represented 82% and 77% of all new species and genera, respectively, described in these 2 years (Fig. 10.5). We analysed the methods and media used for isolation of the type strains representing the new taxa. We distinguished between "standard" and "non-standard methods", whereas the former category includes all isolations performed by direct plating of environmental samples on agar plates, while the latter comprises all cultivation methods deviating from this simple approach, for example, by establishment of an enrichment culture preceding the plating on agar plates or application of advanced cultivation methods.

Examining the information summarized in Table 10.1, it can be seen that 51% of the 1042 newly described species were isolated using standard methods. Nineteen percent of the 1042 newly identified species were isolated by using non-standard methods. The remaining 30% of the newly described species were regarded to have been obtained using isolation methodology reported in the respective papers too imprecise for assignment into either of those two methodological categories (labelled in Table 10.1 as N/A). Comparison of data for standard and non-standard methods revealed that a larger diversity (measured as Shannon indices) is covered by the latter methods. The differences in the calculated Shannon and Shannon's equitability indices (Krebs 1999), respectively, seem to be minor; however, there is a clear trend suggesting that the new species based on isolation by non-standard methods are more equally distributed over genera. With other words, the employed standard methods preferred certain genera and resulted by average in description of 2.1 new species per previously or newly described genus, while the used non-standard methods resulted in a higher evenness regarding the distribution of new species across genera (1.5 new species per genus). Consequently, the used non-standard methods resulted in a more frequent description of new genera (Table 10.1). While by average only each seventh new species obtained by standard methods also represented a new genus, each fourth new species obtained by a non-standard method had to be placed in a new genus. Comparison of sequence similarity of SSU rRNA genes of the type strains of new species with the sequence of their closest related species (Table 10.1) furthermore suggested that non-standard methods obtain by average a bit more distantly related new taxa; however, this difference in sequence similarity may mainly reflect the more frequent isolation of strains representing new genera by non-standard methods. Species descriptions (category "N/A" in Table 10.1) based on type strains isolated by methods neither classified as standard nor as non-standard methods resemble in most of the analysed parameters those descriptions based on standard methods (Table 10.1). The observed differences in trends between standard and non-standard methods (Table 10.1) can at least partially be attributed to a higher diversity of media used by the analysed non-standard compared to the standard approaches (Table 10.2).

More pronounced differences between results of isolations performed with standard and non-standard methods were observed regarding the phylum affiliation of the described new taxa. Almost 97% of the 1042 new species described belonged to the "big four" phyla (Hugenholtz 2002), i.e. *Proteobacteria*, *Actinobacteria*,

Table 10.1 Statistics on methods and media used for isolation of strains described as new species in papers published in the *International Journal of Systematic and Evolutionary Microbiology* (IJSEM) in 2009 and 2010. The two Shannon indices were used to measure the diversity of described new taxa at the genus level

Method or medium	New species		Genera (new and old)	Species/genus	Shannon Index (H)	Shannon's equitability (E_H)	New genera		New genera/new species	Average SSU rRNA similarity (closest species)
	Number	%					Number	%		%
All methods (total)	1042	100.0	520	2.0	5.8	0.93	168	100.0	0.16	96.6
Standard methods	530	50.9	248	2.1	5.4	0.95	80	47.6	0.15	96.7
Non-standard methods	198	19.0	130	1.5	4.9	0.98	47	28.0	0.24	96.2
N/A[a]	314	30.1	142	2.2	4.9	0.95	41	24.4	0.13	96.9
9 Top media, total	450	43.2	250	1.8	5.2	0.95	61	36.3	0.14	97.1
Marine Agar	139	13.3	106	1.3	4.6	0.98	36	21.4	0.26	96.4
R2A Medium	117	11.2	72	1.6	4.1	0.95	10	6.0	0.09	96.8
Trypticase Soy Medium	42	4.0	37	1.1	3.6	0.99	5	3.0	0.12	96.7
Nutrient Agar	42	4.0	31	1.4	3.3	0.96	3	1.8	0.07	98.2
Humic Acid Vitamin Agar	35	3.4	19	1.8	2.7	0.90	2	1.2	0.06	98.6
Luria Broth Agar	35	3.4	25	1.4	3.0	0.95	2	1.2	0.06	97.9
Starch Casein Agar	16	1.5	14	1.1	2.6	0.98	2	1.2	0.13	98.1
MRS Agar	14	1.3	4	3.5	0.8	0.54	0	0.0	0.00	98.3
Plate Count Agar	10	1.0	10	1.0	2.3	1.00	1	0.6	0.10	96.7
Other media, total	592	56.8	334	1.8	5.5	0.94	107	63.7	0.18	97.1

Note that these taxonomic descriptions represent 81.8% of all descriptions of new prokaryotic species in this 2-year period (Fig. 10.5)

[a]N/A, classification in either the category standard or non-standard methods was not be possible due to the lack on specific information on the method

Table 10.2 Comparison of kind of media used for isolation of type strains described as new species either isolated with methods classified as standard or non-standard methods

	Standard methods	Non-standard methods
Total number of described new species	530	198
Total number of different media used	129	119
Shannon index (H)	3.51	4.45
Shannon's equitability (E_H)	0.72	0.93
Most frequently used medium (number of obtained new species)	Marine Agar (110)	Marine Agar (11)
2nd most frequently used medium (number of obtained new species)	R2A Agar (96)	Nutrient Agar (9)

The data refer to descriptions of new species published in the *International Journal of Systematic and Evolutionary Microbiology* (IJSEM) in the volumes 2009 and 2010

Table 10.3 Distribution of the 1042 new species described in 2009 and 2010 in IJSEM across phyla

	All methods	Standard methods	Non-standard methods
	(% of total)	(% of total)	(% of total)
Proteobacteria	38.0	35.9	43.4
Actinobacteria	27.1	28.7	15.7
Firmicutes	17.2	16.8	24.2
Bacteroidetes	14.1	17.7	8.1
Other phyla	3.6	0.9	8.6

Firmicutes and *Bacteroidetes*; however, standard and non-standard methods preferred different phyla. As compared to standard methods, the use of non-standard methods resulted in isolation of a smaller proportion of strains representing new species affiliated with the phyla *Actinobacteria* and *Bacteroidetes* but a higher proportion of *Proteobacteria*, *Firmicutes*, and representatives of other phyla (Table 10.3). Standard methods resulted only in the description of five species (0.9% of new species obtained by standard methods) not belonging to the "big four" phyla, while non-standard methods resulted in the description of 17 new species (8.6% of new species obtained by non-standard methods. The 5 type strains obtained by standard methods represent exclusively *Deinococcus* species (*Deinococcus-Thermus* phylum), but the 17 type strains obtained by non-standard methods represent 9 different phyla (*Aquificae*, *Chloroflexi*, *Deferribacteres*, *Deinococcus-Thermus*, *Ignavibacteriae*, *Planctomycetes*, *Synergistetes*, *Thermodesulfobacteria*, *Thermotogae*) and 1 candidate of a novel phylum (Mori et al. 2009). Four more phyla (*Acidobacteria*, *Caldiserica*, *Fusobacteria* and *Spirochaetes*) are represented by strains obtained with unclassified methods (data not shown).

Analysis of the phylum affiliation of newly described genera even suggests higher selectivity by standard compared to non-standard methods (Fig. 10.6). While non-standard methods enabled description of new genera affiliated with nine

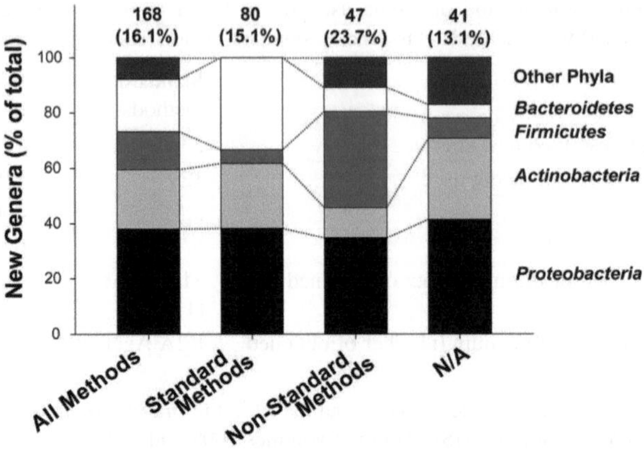

Fig. 10.6 New genera described in the IJSEM journal in 2009 and 2010. Methods used for isolation of the type strains were classified as standard or non-standard methods. For 41 (24.4%) of the new genera, the published information on employed isolation methods were not sufficient for assignment to either of these two categories. These new genera descriptions were assigned to a separate category (N/A, not available). Numbers at the top of the columns represent the total numbers of the described new genera and the percentage of newly described species also representing newly described genera, i.e. the frequency of new genera among the described new taxa (species)

different phyla, standard methods enabled proposals of new genera representing only four different phyla. Furthermore, non-standard methods resulted over-proportionally and under-proportionally in description of new genera affiliated with the phyla *Firmicutes* and *Actinobacteria*, respectively.

A detailed analysis of the media used for isolation of the 1042 type strains representing the newly described species revealed that almost 25% of type strains were obtained by use of only two different media, i.e. Marine Agar (Zobell 1941) and R2A Agar (Reasoner and Geldreich 1985), and almost 43% of type strains were obtained by only nine different media (Fig. 10.7, Table 10.1). Note that the same medium could be employed by standard and non-standard methods (Table 10.2). Among the nine top media, Marine Agar is outstanding regarding the ratio of the number of new genera per newly described species (Table 10.1). Thirty-six strains (25.9% of type strains) of the 139 type strains obtained by Marine Agar represented new genera, while the success of the other eight media for identifying new genera ranged only from 0% to 13% (average 8.0%). Among the nine top media, R2A Agar and Trypticase Soy Medium (TS) resulted in isolation of only four new species not affiliated with one of the “big four” phyla, respectively.

New species obtained by the three top media, i.e. Marine Agar, R2A Agar and Trypticase Soy Medium, showed only small taxonomic overlaps with regard to their genus affiliations (Fig. 10.8). Only three genera received new species from all three media. These genera were *Bacillus* (seven new species), *Pseudomonas* (five species) and *Paracoccus* (three new species). The majority of genera (78%) received one or more new species from only one of the three media. Obviously, these three media

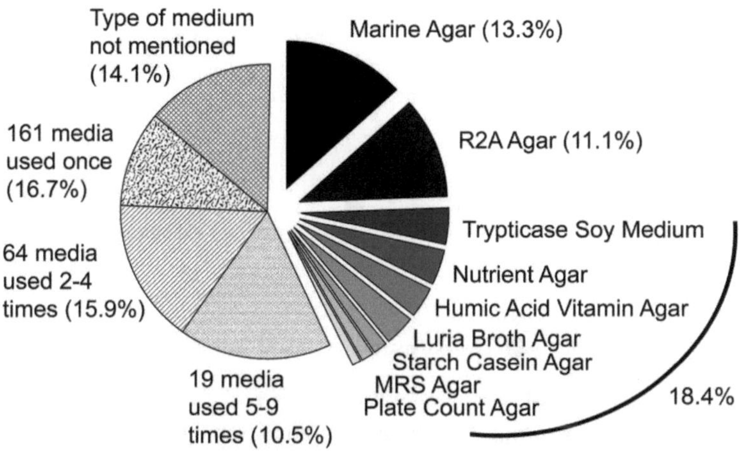

Fig. 10.7 Media used for isolation of the 1442 type strains described as new species in papers presented in the *International Journal of Systematic and Evolutionary Microbiology* (IJSEM) in 2009 and 2010. Media used for description of 5–9 species, 2–4 species or only a single species were lumped together in categories, respectively. Media used for isolation of the type species were not mentioned for 14.1% of the described new species

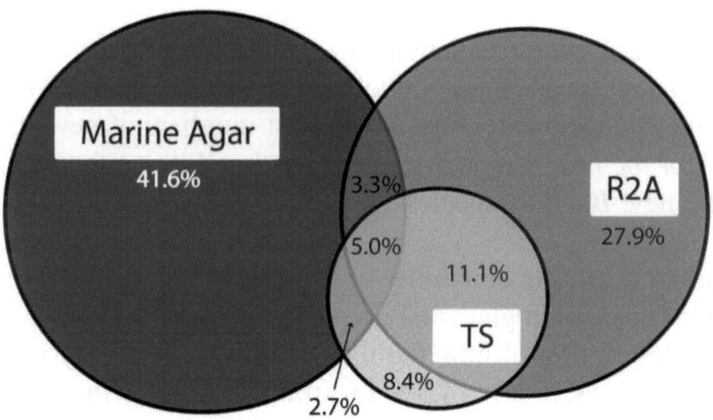

Fig. 10.8 Venn diagram depicting taxonomic overlap of newly described species isolated with the three most frequently used media [Marine Agar, R2A Agar and Trypticase Soy Medium (TS)] regarding their genus affiliation. For instance, 5% of the 298 type strains isolated with one of these three media were affiliated with genera covered by all three media, but 41.6% of strains are affiliated with genera only covered by isolation experiments using Marine Agar. The presented data refer to descriptions of new species published in 2009 and 2010 in IJSEM

strongly differ in their taxonomic selectivity, which could, at least partially, be influenced by application of these media for different kinds of environmental samples. Marine Agar is typically used for processing of samples from marine or salt lake environments, while the other media typically are used for isolation experiments on non-marine samples (e.g. freshwater or soil samples).

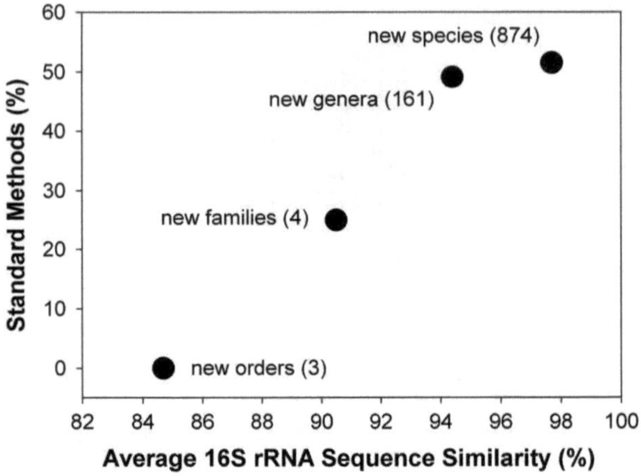

Fig. 10.9 Relationship between percentage of strains obtained by standard isolation methods and the average SSU rRNA (16S rRNA) sequence similarity with the closest related described species in newly established taxa of different rank. Only the highest rank established for a certain type strain was considers. That means that data for type strains representing new orders were not included in the lower categories although these type strains usually also represent new species, genera and families. Numbers in brackets depict the number of taxa established with the respective rank

Across all media used for isolation of type strains of new species, the genera *Streptomyces* (25 new species), *Paenibacillus* (21 new species) and *Bacillus* (20 new species) were the 3 genera receiving the highest numbers of new species. Interestingly, only 8% of the new *Streptomyces* species but 43% of the new *Paenibacillus* and 65% of the new *Bacillus* species were obtained by one of the nine top media (Table 10.1).

A general trend regarding suitability of standard methods for isolation of strains representing new taxa of higher taxonomic rank is obvious (Fig. 10.9). Strains resulting in the description of new families or orders were only rarely isolated by using standard methods. Only 14.3% of strains resulting in description of new families were obtained with a standard method. All other new families and all new orders described during the period under review were obtained either by non-standard methods or methods which could not be classified due to the lack of sufficient information.

The contribution of advanced cultivation methods (see below) to the description of new prokaryotic taxa is still quite small. Some of the previously proposed advanced cultivation methods did not contribute to the isolation of a type strain representing a new species. The high-throughput cultivation (HTC) technique did contribute a substantial number of new species and interesting taxa of higher rank (Table 10.4). High-throughput cultivation combined with dilution-to-extinction and initial cultivation in low nutrient media (frequently accomplished by using simply either autoclaved sea or freshwater as the cultivation medium) as developed in the lab of Stephen J. Giovannoni (Connon and Giovannoni 2002; Stingl et al. 2007) resulted in the description of more than 100 taxa (Table 10.4). Pure cultures of

Table 10.4 Contribution of selected advanced cultivation methods to isolation of type strains described in the period 2000–2014 as new species in IJSEM

Method	Reference for method	Taxa described
HTC dilution-to-extinction	Connon and Giovannoni (2002), Stingl et al. (2007)	>100
Acclimatization Method	Hahn et al. (2004, 2005)	17
MicroDrop technique	Bruns et al. (2003a), Gich et al. (2005)	5
Encapsulation of cells in gel microdroplets	Zengler et al. (2002, 2005)	–
Diffusion growth chamber	Kaeberlein et al. (2002), Bollmann et al. (2007)	–

strains initially grown in low nutrient liquid media typically then were transferred and maintained on high nutrient agar media (usually Marine Agar). Establishment of pure cultures suitable for maintenance on standard solidified media was the crucial requirement for deposition of a strain in public culture collections and subsequent valid description of new taxa. Currently, the research group of Jang-Cheon Cho is probably most active in isolation of new taxa by HTC techniques and their taxonomic description. Some of the highlights among the >50 taxa described by Cho and colleagues are the description of the new class *Opitutae* of the phylum *Verrucomicrobia* (Choo et al. 2007) and the establishment of a new family affiliated with the *Gammaproteobacteria* (Kim et al. 2007). Other taxa of environmental significance, e.g. members of the marine SAR11 cluster (Morris et al. 2002), could be cultivated by HTC techniques (Rappe et al. 2002) as pure cultures in autoclaved sea water, but due to their obligate oligotrophic metabolism, they could not be maintained on media suitable for deposition in public culture collections. Due to this restriction, only the proposal to establish a candidatus species, i.e. *Cand.*, Pelagibacter ubique, was possible so far (Rappe et al. 2002).

It should be noted that the HTC technique is not the only method following a strategy of initial cultivation in sterile seawater, screening of cultures and subsequent seeking transfer of interesting cultures to solidified standard media. A couple of other researchers also described new taxa based on type strains isolated with such strategies; however, these other strategies were not labelled with a brand name like "HTC technique", which makes detailed literature analyses of their contribution to taxonomic advances more difficult. An interesting example for such a cultivation experiment is the isolation of the type strain of *Sphingopyxis alaskensis* (Schut et al. 1993) and its subsequent taxonomic description as a new species (Vancanneyt et al. 2001; Godoy et al. 2003).

The MicroDrop technique (Gich et al. 2005; Bruns et al. 2003a) enabled the description of a new genus affiliated with the phylum *Acidobacteria* (Koch et al. 2008). This phylum contributes in high percentages to soil bacterial communities (Foesel et al. 2014), and cultivation-independent investigations have suggested a huge diversity within this phylum (Eichorst et al. 2007). However, the *Acidobacteria* represent one of the most understudied bacterial phyla currently containing only nine validly described genera with a total of 22 species. Furthermore the type species identifying both a new species and genus representing planktonic freshwater bacteria was isolated by using the MicroDrop technique (Jogler et al. 2013).

The Acclimatization Method (Hahn et al. 2004, 2005) was used for isolation of strains enabling description of nine new species, two new genera and six candidatus species (Table 10.4). This method aims on a gradual transfer of bacteria from their natural low nutrient concentration environments to the artificial high nutrient concentration conditions usually provided by microbiological media. Bacteria present in inoculi from environmental samples are subjected to stepwise increases in nutrient concentrations in liquid media, and strains able to grow at high concentrations are finally transferred to a solidified standard medium (frequently NSY agar plates). This acclimatization procedure is usually combined with an initial selection step, either a filtration or dilution of samples, which shall exclude environmentally rare but under artificial conditions highly competitive taxa able to rapidly overgrow more abundant but slower-growing taxa.

Interestingly, the selection and the acclimatization steps of the Acclimatization Method just represent intermediate steps in the cultivation process, which can be skipped in comparative analyses. Hahn et al. (2004) compared the results of cultivation experiments using lake water as inoculum performed with and without those steps. Thus they compared results of experiments utilizing the Acclimatization Method with results of direct plating of samples on agar plates (standard plating). NSY medium was exclusively used in all experiments performed by using those two methods. The sets of strains (Table 10.5) obtained with these two different methods from the water columns of freshwater lakes and ponds differed substantially regarding their phylogenetic compositions (Fig. 10.10). The majority (68.8%) of strains isolated with the Acclimatization Method belonged to the class *Betaproteobacteria* (phylum *Proteobacteria*) and the phylum *Actinobacteria* (Fig. 10.10), which are known to usually contribute together to the majority of bacteria dwelling as

Table 10.5 Comparison of strains and taxa isolated by the Filtration-Acclimatization Method (FAM) and direct plating on NSY agar

	FAM	Direct plating
Number of strains	73	27
Number of taxa (>97% SSU rRNA simil.)	16	20
Singleton taxa (% all taxa)	62.5	80.0
Taxa closer to uncultured bacteria (% all taxa)	56.3	0
Similarity (%) with closest species (average of all taxa)	94.5	97.6
Similarity (%) range (min. to max.)	86.9–99.5	92.2–100
Taxa with >97% similarity to closest species	12.5	55.6

Both methods employed the same medium (liquid or solidified NSY medium) and resulted in pure cultures maintained on NSY agar plates and subsequently characterized by sequencing of their SSU (16S) rRNA genes. For comparison strains sharing >97% SSU rRNA sequence, similarities were assigned to the same taxon. The relationship to other uncultured and cultured organisms was estimated by performing BLAST analysis with SSU rRNA sequences against public databases. The presented data were adopted from Hahn et al. (2004)

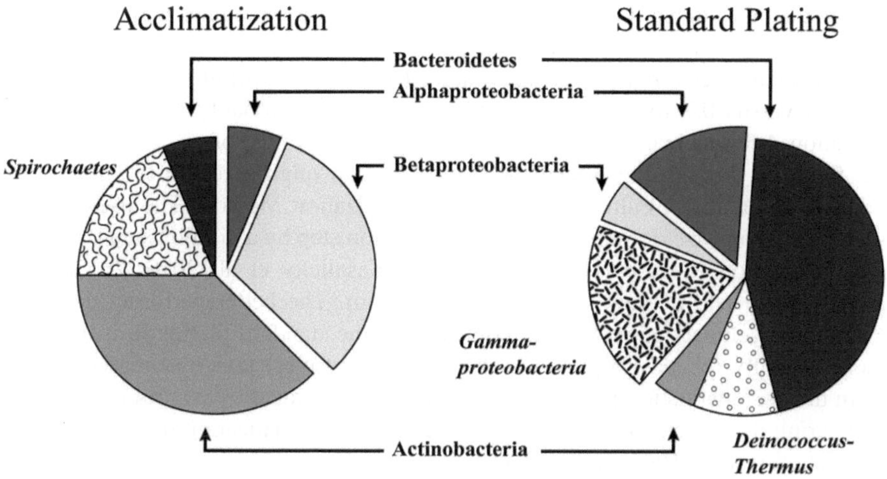

Fig. 10.10 Comparison of the phylogenetic affiliation of bacterial strains isolated with the Acclimatization Method (Hahn et al. 2004) and standard plating. Both methods used the same medium (NSY medium) and resulted in strains able to grow on agar plates. The graph is based on data published by Hahn and co-authors previously (Hahn et al. 2004) and depicts the relative contribution (%) of the respective phylogenetic groups to the total number of strains isolated by the two methods (Table 10.5)

bacterioplankton in freshwater systems (Glockner et al. 2000; Newton et al. 2011). These two taxa were also present among the strains obtained by standard plating (Fig. 10.10) but contributed numerically only 10%; thus they represented only a minor fraction. Other taxa more typically known to usually contribute only minor factions to freshwater bacterioplankton (e.g. phylum *Deinococcus-Thermus* and *Gammaproteobacteria*) were represented as unanticipatedly larger proportions (Fig. 10.10). The presence of a large fraction of spirochetes among the strains obtained by the Acclimatization Method was a result of the filtration step (Hahn 2004; Canale-Parola et al. 1966) but not basically a result of the acclimatization procedure [as suggested by control experiments (Hahn et al. 2004)]. The direct plating resulted in isolation of a higher number of taxa (defined by >97% SSU rRNA similarity) albeit a smaller total number of strains was obtained by this method in comparison to the Acclimatization Method (Table 10.5). However, the latter method resulted by average in isolation of strains and taxa more distantly related to previously characterized isolates and species. Furthermore, while all taxa obtained by the standard plating methods were more closely related to previously described species, about half of the taxa retrieved by the Acclimatization Method were more closely related to so far uncultured bacteria only characterized by cultivation-independent methods (Table 10.5; Hahn et al. 2004). Importantly, this method resulted in isolation of several previously uncultured taxa known to be abundant freshwater bacteria (e.g. *Polynucleobacter* spp., *Limnohabitans* spp., planktonic freshwater *Actinobacteria*), while the standard plating did not obtain such taxa. In general, the Acclimatization Method obtained more candidates for new species and genera than the standard plating method. Some of these candidates were meanwhile described as new taxa (e.g. Hahn et al. 2010, 2012a; Jezbera et al. 2009). The lower ratio of taxa discovered per isolated strain (0.2 for the Filtration-Acclimatization Method versus 0.7 for the standard plating methods) probably resulted from the Filtration-Acclimatization Method entailing too strong of a selection step, i.e. filtration through 0.2 mm filters. This filtration might exclude too many taxa otherwise potentially cultivable by the Acclimatization Method. The use of filters with larger pore size or replacement of the filtration step by a dilution step [Dilution-Acclimatization Method (Hahn et al. 2005; Kasalicky et al. 2010, 2013)] may increase the yield of new taxa per isolated strain. The better performance of the Filtration-Acclimatization Method compared to the standard plating in isolation of more distantly related (to previous characterized species) taxa essentially resulted from the acclimatization procedure. This was demonstrated by control experiments comparing the effects of the filtration step and the acclimatization procedure on the cultivation results (Hahn et al. 2004).

Other advanced cultivation methods, for instance, either the cultivation by using diffusion growth chambers simulating a natural environment and enabling cocultivation (Kaeberlein et al. 2002; Bollmann et al. 2007) or the encapsulation method (Zengler et al. 2002, 2005) have not contributed to the description of new prokaryotic species so far. A potential reason for this lack of contribution might be that these methods, as well as the HTC method (see above), do not primarily aim on isolation of strains able to grow on typical (high nutrient) bacteriological media. The ability of strains to grow on high nutrient media facilitates sustainable maintenance of cultures as well as deposition of strains in public culture collections, but the lack of this ability does not fundamentally prevent the description of new taxa as, for instance, demonstrated by the description of *Mariprofundus ferrooxydans* (Emerson et al. 2007) and the subsequent description of the new class *Zetaproteobacteria* (phylum Proteobacteria), which harbours the genus *Mariprofundus* (Makita et al. 2017).

In summary, the analysis of species descriptions published in IJSEM and the study comparing the Acclimatization Method with standard plating suggested that it is quite easy to isolate and cultivate bacterial strains representing undescribed species. For instance, 20% of the analysed 1042 descriptions of new species resulted from direct plating of environmental samples on either Marine Agar or R2A Agar. About 50% of the new species and numerous of the additional strains identified were successfully isolated by standard methods without requiring either sophisticated pretreatments, enrichment procedures or other laborious non-standard cultivation methods (Table 10.1). Similar trends were observed in the abovementioned direct plating of freshwater samples on NSY agar plates (Hahn et al. 2004). Isolation of 27 strains resulted in nine strains, which reliably represented new candidate species as suggested by the fact of their SSU rRNA sequence values having less than 97% similarity to type strains of the closest related described species (Stackebrandt and Goebel 1994, 2006). An impressive study by Joseph and colleagues also demonstrated that simple cultivation on agar plates (including non-standard media) can retrieve numerous candidates for new species from soil samples, including possible genera and families affiliated with so far understudied phyla like *Acidobacteria*

(Joseph et al. 2003). Obviously, progress in taxonomic description of new bacterial species is not limited by cultivation and deposition of type strains in public culture collections but probably rather by efforts, costs and taxonomic expertise required for successful taxonomic characterization and description of new species (Tindall et al. 2010). New methods and media are currently not required for adding new species and genera to previously established families; however, application of standard cultivation methods results only rarely in description of new taxa of higher rank (Fig. 10.9). Thus, deeper taxonomic exploration of the so far poorly characterized bacterial diversity strongly requires more sophisticated cultivation experiments.

10.7 Strategies for Cultivation

The selection of an appropriate strategy for cultivation experiments largely depends on the project goals. Taxonomists interested in obtaining new strains suitable for description of novel species and genera may use standard methods and media (see above). The use of non-standard methods may increase the yield of taxonomically novel groups; however not all obtained cultures may be suitable for deposition in a public culture collection, which is currently one of several mandatory requirements for description of new prokaryotic taxa (Tindall et al. 2010). Acclimatization procedures in combination with various standard media can be expected to even gain higher yields of novel taxa. Thus far, acclimatization experiments were almost exclusively performed by using only a single medium (Hahn et al. 2004) leaving the potential of this strategy largely unexploited.

Microbial ecologists interested in the study of specific environmentally significant but so far uncultured taxa may first require an appropriate method for high-throughput screening of cultures (Rappe et al. 2002). Fluorescent in situ hybridization (FISH) or PCR with probes or primers, respectively, specific for the targeted organisms represent efficient screening methods. Cultivation experiments should be started with methods and media best mimicking the natural conditions under which the targeted organisms thrive. Initial use of low substrate concentration media, for instance, sterilized sea or lake water, is recommended. If cultivation on such low substrate concentration media is successful, then transfer to standard media can be tried. Acclimatization procedures may help if a single step transfer to substrate-rich media is not successful.

If genomic data on so far uncultured target organisms were obtained by either metagenomics or single-cell genomic methods, then analysis of the available data may help to better understand the physiology and ecology of the targeted organisms. Such knowledge could be used to refine the cultivation strategy and the composition of the employed media. Cocultivation with other organisms thriving in the same habitat may be considered if initial trials on cultivation in pure culture are found to fail. Once a stable mixed culture is established, experiments on supplementation of the used media with substances potentially required by the target organisms could be performed. Sequencing and subsequent analysis of genomes of the organisms contained in an established mixed culture also could be performed. Genomic insights could help to improve the medium so as to release the targeted organisms from their dependence on the helper organisms.

Compliance with Ethical Standards

Funding This study was supported by the Austrian Science Fund (FWF) project 27160-B22 (Mikroevolution eines planktischen Süßwasserbakteriums).

Conflict of Interest Martin W. Hahn declares that he has no conflict of interest. Ulrike Koll declares that she has no conflict of interest. Johanna Schmidt declares that she has no conflict of interest.

Ethical Approval This article does not contain any studies with human participants or animals performed by any of the authors.

References

Alain K, Querellou J (2009) Cultivating the uncultured: limits, advances and future challenges. Extremophiles 13(4):583–594

Alam M, Sadique A, Nur AH, Bhuiyan NA, Nair GB, Siddique AK, Sack DA, Ahsan S, Huq A, Sack RB, Colwell RR (2006) Effect of transport at ambient temperature on detection and isolation of *Vibrio cholerae* from environmental samples. Appl Environ Microbiol 72 (3):2185–2190

Altschul SF, Gish W, Miller W, Myers EW, Lipman DJ (1990) Basic local alignment search tool. J Mol Biol 215(3):403–410

Amann RI, Ludwig W, Schleifer KH (1995) Phylogenetic identification and in situ detection of individual microbial cells without cultivation. Microbiol Rev 59(1):143–169

Atlas RM (2010) Handbook of microbiological media. CRC Press, Boca Raton

Barer MR, Harwood CR (1999) Bacterial viability and culturability. Adv Microb Physiol 41:93–137

Batut B, Knibbe C, Marais G, Daubin V (2014) Reductive genome evolution at both ends of the bacterial population size spectrum. Nat Rev Microbiol 12(12):841–850

Becker S, Vogt J, Knopp S, Panning M, Warhurst D, Polman K, Marti H, von Muller L, Yansouni C, Jacobs J, Bottieau E, Sacko M, Rijal S, Meyanti F, Miles M, Boelaert M, Lutumba P, van Lieshout L, N'Goran E, Chappuis F, Utzinger J (2013) Persistent digestive disorders in the tropics: causative infectious pathogens and reference diagnostic tests. BMC Infect Dis 13(1):37

Biers EJ, Sun S, Howard EC (2009) Prokaryotic genomes and diversity in surface ocean waters: interrogating the global ocean sampling metagenome. Appl Environ Microbiol 75 (7):2221–2229

Blevins SM, Bronze MS (2010) Robert Koch and the 'golden age' of bacteriology. Int J Infect Dis 14(9):e744–e751

Bogosian G, Aardema ND, Bourneuf EV, Morris PJL, O'Neil JP (2000) Recovery of hydrogen peroxide-sensitive culturable cells of Vibrio vulnificus gives the appearance of resuscitation from a viable but nonculturable state. J Bacteriol 182(18):5070–5075

Bollmann A, Lewis K, Epstein SS (2007) Incubation of environmental samples in a diffusion chamber increases the diversity of recovered isolates. Appl Environ Microbiol 73 (20):6386–6390

Braine T (2011) Race against time to develop new antibiotics. Bull World Health Organ 89:88–89

Britschgi TB, Giovannoni SJ (1991) Phylogenetic analysis of a natural marine bacterioplankton population by rRNA gene cloning and sequencing. Appl Environ Microbiol 57(6):1707–1713
Brock TD (1999) Robert Koch: a life in medicine and bacteriology. ASM Press, Washington, DC
Bruns A, Cypionka H, Overmann J (2002) Cyclic AMP and acyl homoserine lactones increase the cultivation efficiency of heterotrophic bacteria from the central Baltic Sea. Appl Environ Microbiol 68(8):3978–3987
Bruns A, Hoffelner H, Overmann J (2003a) A novel approach for high throughput cultivation assays and the isolation of planktonic bacteria. FEMS Microbiol Ecol 45(2):161–171
Bruns A, Nubel U, Cypionka H, Overmann J (2003b) Effect of signal compounds and incubation conditions on the culturability of freshwater bacterioplankton. Appl Environ Microbiol 69 (4):1980–1989
Button DK, Robertson BR (2001) Determination of DNA content of aquatic bacteria by flow cytometry. Appl Environ Microbiol 67(4):1636–1645
Button DK, Schut F, Quang P, Martin R, Robertson BR (1993) Viability and isolation of marine bacteria by dilution culture: theory, procedures, and initial results. Appl Environ Microbiol 59 (3):881–891
Canale-Parola E, Rosenthal SL, Kupfer DG (1966) Morphological and physiological characteristics of *Spirillum gracile* sp. n. Antonie Van Leeuwenhoek 32(2):113–124
Carini P, Steindler L, Beszteri S, Giovannoni SJ (2013) Nutrient requirements for growth of the extreme oligotroph '*Candidatus* Pelagibacter ubique' HTCC1062 on a defined medium. ISME J 7(3):592–602
Cho JC, Giovannoni SJ (2004) Cultivation and growth characteristics of a diverse group of oligotrophic marine *Gammaproteobacteria*. Appl Environ Microbiol 70(1):432–440
Choo Y-J, Lee K, Song J, Cho J-C (2007) *Puniceicoccus vermicola* gen. nov., sp. nov., a novel marine bacterium, and description of *Puniceicoccaceae* fam. nov., *Puniceicoccales* ord. nov., *Opitutaceae* fam. nov., *Opitutales* ord. nov. and *Opitutae* classis nov. in the phylum 'Verrucomicrobia'. Int J Syst Evol Microbiol 57(3):532–537
Connon SA, Giovannoni SJ (2002) High-throughput methods for culturing microorganisms in very-low-nutrient media yield diverse new marine isolates. Appl Environ Microbiol 68 (8):3878–3885
Connon SA, Lester ED, Shafaat HS, Obenhuber DC, Ponce A (2007) Bacterial diversity in hyperarid Atacama Desert soils. J Geophys Res Biogeosci 112(G4)
Crosbie ND, Pockl M, Weisse T (2003) Dispersal and phylogenetic diversity of nonmarine picocyanobacteria, inferred from 16S rRNA gene and cpcBA-intergenic spacer sequence analyses. Appl Environ Microbiol 69(9):5716–5721
Curtis TP, Head IM, Lunn M, Woodcock S, Schloss PD, Sloan WT (2006) What is the extent of prokaryotic diversity? Philos Trans R Soc B Biol Sci 361(1475):2023–2037
Curtis TP, Sloan WT, Scannell JW (2002) Estimating prokaryotic diversity and its limits. Proc Natl Acad Sci U S A 99(16):10494–10499
D'Onofrio A, Crawford JM, Stewart EJ, Witt K, Gavrish E, Epstein S, Clardy J, Lewis K (2010) Siderophores from neighboring organisms promote the growth of uncultured bacteria. Chem Biol 17(3):254–264
Davis KE, Joseph SJ, Janssen PH (2005) Effects of growth medium, inoculum size, and incubation time on culturability and isolation of soil bacteria. Appl Environ Microbiol 71(2):826–834
Davis KE, Sangwan P, Janssen PH (2011) Acidobacteria, Rubrobacteridae and Chloroflexi are abundant among very slow-growing and mini-colonyforming soil bacteria. Environ Microbiol 13:798–805
Delmée M, Van Broeck J, Simon A, Janssens M, Avesani V (2005) Laboratory diagnosis of *Clostridium difficile*-associated diarrhoea: a plea for culture. J Med Microbiol 54(2):187–191
Dufresne A, Garczarek L, Partensky F (2005) Accelerated evolution associated with genome reduction in a free-living prokaryote. Genome Biol 6(2):R14
Dykhuizen D (2005) Species numbers in bacteria. Proc Calif Acad Sci 56(6 Suppl 1):62–71

Dykhuizen DE (1998) Santa Rosalia revisited: why are there so many species of bacteria? Anton Leeuw Int J Gen Mol Microbiol 73(1):25–33

Eichorst SA, Breznak JA, Schmidt TM (2007) Isolation and characterization of soil bacteria that define *Terriglobus* gen. nov., in the phylum *Acidobacteria*. Appl Environ Microbiol 73 (8):2708–2717

Emerson D, Rentz JA, Lilburn TG, Davis RE, Aldrich H, Chan C, Moyer CL (2007) A novel lineage of proteobacteria involved in formation of marine Fe-oxidizing microbial mat communities. PLoS One 2(7):e667

Ewann F, Hoffman PS (2006) Cysteine metabolism in *Legionella pneumophila*: characterization of an L-cystine-utilizing mutant. Appl Environ Microbiol 72(6):3993–4000

Ferrari BC, Binnerup SJ, Gillings M (2005) Microcolony cultivation on a soil substrate membrane system selects for previously uncultured soil bacteria. Appl Environ Microbiol 71 (12):8714–8720

Ferrari BC, Winsley T, Gillings M, Binnerup S (2008) Cultivating previously uncultured soil bacteria using a soil substrate membrane system. Nat Protoc 3(8):1261–1269

Foesel BU, Nägele V, Naether A, Wüst PK, Weinert J, Bonkowski M, Lohaus G, Polle A, Alt F, Oelmann Y, Fischer M, Friedrich MW, Overmann J (2014) Determinants of Acidobacteria activity inferred from the relative abundances of 16S rRNA transcripts in German grassland and forest soils. Environ Microbiol 16(3):658–675

Garcia SL, McMahon KD, Grossart HP, Warnecke F (2014) Successful enrichment of the ubiquitous freshwater acI *Actinobacteria*. Environ Microbiol Rep 6(1):21–27

Gich F, Schubert K, Bruns A, Hoffelner H, Overmann J (2005) Specific detection, isolation, and characterization of selected, previously uncultured members of the freshwater bacterioplankton community. Appl Environ Microbiol 71(10):5908–5919

Giovannoni SJ, Bibbs L, Cho JC, Stapels MD, Desiderio R, Vergin KL, Rappe MS, Laney S, Wilhelm LJ, Tripp HJ, Mathur EJ, Barofsky DF (2005a) Proteorhodopsin in the ubiquitous marine bacterium SAR11. Nature 438(7064):82–85

Giovannoni SJ, Cameron Thrash J, Temperton B (2014) Implications of streamlining theory for microbial ecology. ISME J 8(8):1553–1565

Giovannoni SJ, Tripp HJ, Givan S, Podar M, Vergin KL, Baptista D, Bibbs L, Eads J, Richardson TH, Noordewier M, Rappe MS, Short JM, Carrington JC, Mathur EJ (2005b) Genome streamlining in a cosmopolitan oceanic bacterium. Science 309(5738):1242–1245

Glockner FO, Zaichikov E, Belkova N, Denissova L, Pernthaler J, Pernthaler A, Amann R (2000) Comparative 16S rRNA analysis of lake bacterioplankton reveals globally distributed phylogenetic clusters including an abundant group of *Actinobacteria*. Appl Environ Microbiol 66 (11):5053–5065

Godoy F, Vancanneyt M, Martínez M, Steinbüchel A, Swings J, Rehm BHA (2003) *Sphingopyxis chilensis* sp. nov., a chlorophenol-degrading bacterium that accumulates polyhydroxyalkanoate, and transfer of *Sphingomonas alaskensis* to *Sphingopyxis alaskensis* comb. nov. Int J Syst Evol Microbiol 53(2):473–477

Gould LH, Bopp C, Strockbine N, Atkinson R, Baselski V, Body B, Carey R, Crandall C, Hurd S, Kaplan R, Neill M, Shea S, Somsel P, Tobin-D'Angelo M, Griffin PM, Gerner-Smidt P (2009) Recommendations for diagnosis of shiga toxin–producing *Escherichia coli* infections by clinical laboratories. MMWR Recomm Rep 58(Rr-12):1–14

Hahn MW (2003) Isolation of strains belonging to the cosmopolitan *Polynucleobacter necessarius* cluster from freshwater habitats located in three climatic zones. Appl Environ Microbiol 69 (9):5248–5254

Hahn MW (2004) Broad diversity of viable bacteria in 'sterile' (0.2 μm) filtered water. Res Microbiol 155(8):688–691

Hahn MW (2009) Description of seven candidate species affiliated with the phylum *Actinobacteria*, representing planktonic freshwater bacteria. Int J Syst Evol Microbiol 59:112–117

Hahn MW, Kasalicky V, Jezbera J, Brandt U, Jezberova J, Simek K (2010) *Limnohabitans curvus* gen. nov., sp nov., a planktonic bacterium isolated from a freshwater lake. Int J Syst Evol Microbiol 60:1358–1365

Hahn MW, Lang E, Brandt U, Wu QL, Scheuerl T (2009) Emended description of the genus *Polynucleobacter* and the species *Polynucleobacter necessarius* and proposal of two subspecies, *P. necessarius* subsp. *necessarius* subsp nov and *P. necessarius* subsp *asymbioticus* subsp nov. Int J Syst Evol Microbiol 59:2002–2009

Hahn MW, Minasyan A, Lang E, Koll U, Sproeer C (2012a) *Polynucleobacter difficilis* sp nov., a planktonic freshwater bacterium affiliated with subcluster B1 of the genus *Polynucleobacter*. Int J Syst Evol Microbiol 62:376–383

Hahn MW, Pockl M, Wu QL (2005) Low intraspecific diversity in a *Polynucleobacter* subcluster population numerically dominating bacterioplankton of a freshwater pond. Appl Environ Microbiol 71(8):4539–4547

Hahn MW, Scheuerl T, Jezberova J, Koll U, Jezbera J, Simek K, Vannini C, Petroni G, Wu QLL (2012b) The passive yet successful way of planktonic life: genomic and experimental analysis of the ecology of a free-living *Polynucleobacter* population. PLoS One 7(3):e32772

Hahn MW, Schmidt J, Taipale SJ, Doolittle WF, Koll U (2014) *Rhodoluna lacicola* gen. nov., sp. nov., a planktonic freshwater bacterium with stream-lined genome. Int J Syst Evol Microbiol 64(Pt 9):3254–3263

Hahn MW, Stadler P, Wu QL, Pockl M (2004) The filtration-acclimatization method for isolation of an important fraction of the not readily cultivable bacteria. J Microbiol Methods 57(3):379–390

Handelsman J (2004) Metagenomics: application of genomics to uncultured microorganisms. Microbiol Mol Biol Rev 68(4):669–685

Hazan R, Que Y-A, Maura D, Rahme L (2012) A method for high throughput determination of viable bacteria cell counts in 96-well plates. BMC Microbiol 12(1):259

Hesse W (1992) Walther and Angelina Hesse-Early contributors to bacteriology. ASM News 58:425–428

Huber R, Burggraf S, Mayer T, Barns SM, Rossnagel P, Stetter KO (1995) Isolation of a hyperthermophilic archaeum predicted by in situ RNA analysis. Nature 376(6535):57–58

Huber R, Huber H, Stetter KO (2000) Towards the ecology of hyperthermophiles: biotopes, new isolation strategies and novel metabolic properties. FEMS Microbiol Rev 24(5):615–623

Hugenholtz P (2002) Exploring prokaryotic diversity in the genomic era. Genome Biol 3(2): reviews0003.0001–reviews0003.0008

Hugenholtz P, Goebel BM, Pace NR (1998) Impact of culture-independent studies on the emerging phylogenetic view of bacterial diversity. J Bacteriol 180(18):4765–4774

Huse SM, Welch DM, Morrison HG, Sogin ML (2010) Ironing out the wrinkles in the rare biosphere through improved OTU clustering. Environ Microbiol 12:1889–1898

Jannasch HW, Jones GE (1959) Bacterial populations in sea water as determined by different methods of enumeration. Limnol Oceanogr 4(2):128–139

Janssen PH (2006) Identifying the dominant soil bacterial taxa in libraries of 16S rRNA and 16S rRNA genes. Appl Environ Microbiol 72(3):1719–1728

Janssen PH, Schuhmann A, Morschel E, Rainey FA (1997) Novel anaerobic ultramicrobacteria belonging to the *Verrucomicrobiales* lineage of bacterial descent isolated by dilution culture from anoxic rice paddy soil. Appl Environ Microbiol 63(4):1382–1388

Janssen PH, Yates PS, Grinton BE, Taylor PM, Sait M (2002) Improved culturability of soil bacteria and isolation in pure culture of novel members of the divisions *Acidobacteria*, *Actinobacteria*, *Proteobacteria*, and *Verrucomicrobia*. Appl Environ Microbiol 68 (5):2391–2396

Jaspers E, Overmann J (2004) Ecological significance of microdiversity: identical 16S rRNA gene sequences can be found in bacteria with highly divergent genomes and ecophysiologies. Appl Environ Microbiol 70(8):4831–4839

Jezbera J, Jezberova J, Koll U, Hornak K, Simek K, Hahn MW (2012) Contrasting trends in distribution of four major planktonic betaproteobacterial groups along a pH gradient of epilimnia of 72 freshwater habitats. FEMS Microbiol Ecol 81(2):467–479

Jezbera J, Sharma AK, Brandt U, Doolittle WF, Hahn MW (2009) '*Candidatus* Planktophila limnetica', an actinobacterium representing one of the most numerically important taxa in freshwater bacterioplankton. Int J Syst Evol Microbiol 59:2864–2869

Jezberova J, Jezbera J, Brandt U, Lindstrom ES, Langenheder S, Hahn MW (2010) Ubiquity of *Polynucleobacter necessarius* ssp. *asymbioticus* in lentic freshwater habitats of a heterogenous 2000 km^2 area. Environ Microbiol 12(3):658–669

Jogler M, Chen H, Simon J, Rohde M, Busse H-J, Klenk H-P, Tindall BJ, Overmann J (2013) Description of *Sphingorhabdus planktonica* gen. nov., sp. nov. and reclassification of three related members of the genus *Sphingopyxis* in the genus *Sphingorhabdus* gen. nov. Int J Syst Evol Microbiol 63(Pt 4):1342–1349

Johnson RC (1977) The Spirochetes. Annu Rev Microbiol 31(1):89–106

Joseph SJ, Hugenholtz P, Sangwan P, Osborne CA, Janssen PH (2003) Laboratory cultivation of widespread and previously uncultured soil bacteria. Appl Environ Microbiol 69(12):7210–7215

Kaeberlein T, Lewis K, Epstein SS (2002) Isolating "uncultivable" microorganisms in pure culture in a simulated natural environment. Science 296(5570):1127–1129

Kasalicky V, Jezbera J, Hahn MW, Simek K (2013) The diversity of the *Limnohabitans* genus, an important group of freshwater bacterioplankton, by characterization of 35 isolated strains. PLoS One 8(3):e58209

Kasalicky V, Jezbera J, Simek K, Hahn MW (2010) *Limnohabitans planktonicus* sp nov and *Limnohabitans parvus* sp nov., planktonic betaproteobacteria isolated from a freshwater reservoir, and emended description of the genus *Limnohabitans*. Int J Syst Evol Microbiol 60:2710–2714

Kell D, Kaprelyants A, Weichart D, Harwood C, Barer M (1998) Viability and activity in readily culturable bacteria: a review and discussion of the practical issues. Antonie Van Leeuwenhoek 73(2):169–187

Kenters N, Henderson G, Jeyanathan J, Kittelmann S, Janssen PH (2011) Isolation of previously uncultured rumen bacteria by dilution to extinction using a new liquid culture medium. J Microbiol Methods 84(1):52–60

Kim H, Choo Y-J, Cho J-C (2007) *Litoricolaceae* fam. nov., to include *Litoricola lipolytica* gen. nov., sp. nov., a marine bacterium belonging to the order *Oceanospirillales*. Int J Syst Evol Microbiol 57(8):1793–1798

Koch IH, Gich F, Dunfield PF, Overmann J (2008) *Edaphobacter modestus* gen. nov., sp. nov., and *Edaphobacter aggregans* sp. nov., acidobacteria isolated from alpine and forest soils. Int J Syst Evol Microbiol 58(5):1114–1122

Köpke B, Wilms R, Engelen B, Cypionka H, Sass H (2005) Microbial diversity in coastal subsurface sediments: a cultivation approach using various electron acceptors and substrate gradients. Appl Environ Microbiol 71(12):7819–7830

Krebs CJ (1999) Ecological methodology, 2nd edn. Addison Wesley Longman, Menlo Park, CA

Kunin V, Engelbrektson A, Ochman H, Hugenholtz P (2010) Wrinkles in the rare biosphere: pyrosequencing errors can lead to artificial inflation of diversity estimates. Environ Microbiol 12 (1):118–123

Lesmana M, Rockhill RC, Sutanti D, Sutomo A (1985) An evaluation of alkaline peptone water for enrichment of Vibrio cholerae in feces. Southeast Asian J Trop Med Public Health 16 (2):265–267

Makita H, Tanaka E, Mitsunobu S, Miyazaki M, Nunoura T, Uematsu K, Takaki Y, Nishi S, Shimamura S, Takai K (2017) *Mariprofundus micogutta* sp. nov., a novel iron-oxidizing zetaproteobacterium isolated from a deep-sea hydrothermal field at the Bayonnaise knoll of the Izu-Ogasawara arc, and a description of Mariprofundales ord. nov. and Zetaproteobacteria classis nov. Arch Microbiol 199:335–346

Mannisto MK, Rawat S, Starovoytov V, Haggblom MM (2011) *Terriglobus saanensis* sp. nov., an acidobacterium isolated from tundra soil. Int J Syst Evol Microbiol 61(Pt 8):1823–1828

Markowitz VM, Chen IMA, Palaniappan K, Chu K, Szeto E, Grechkin Y, Ratner A, Jacob B, Huang JH, Williams P, Huntemann M, Anderson I, Mavromatis K, Ivanova NN, Kyrpides NC (2012) IMG: the integrated microbial genomes database and comparative analysis system. Nucleic Acids Res 40(D1):D115–D122

Mira A, Ochman H, Moran NA (2001) Deletional bias and the evolution of bacterial genomes. Trends Genet 17(10):589–596

Moore LR, Rocap G, Chisholm SW (1998) Physiology and molecular phylogeny of coexisting *Prochlorococcus* ecotypes. Nature 393:464–467

Mora C, Tittensor DP, Adl S, Simpson AGB, Worm B (2011) How many species are there on earth and in the ocean? PLoS Biol 9(8):e1001127

Mori K, Yamaguchi K, Sakiyama Y, Urabe T, Suzuki K-I (2009) *Caldisericum exile* gen. nov., sp. nov., an anaerobic, thermophilic, filamentous bacterium of a novel bacterial phylum, *Caldiserica* phyl. nov., originally called the candidate phylum OP5, and description of *Caldisericaceae* fam. nov., *Caldisericales* ord. nov. and *Caldisericia* classis nov. Int J Syst Evol Microbiol 59(11):2894–2898

Morris JJ, Kirkegaard R, Szul MJ, Johnson ZI, Zinser ER (2008) Facilitation of robust growth of *Prochlorococcus* colonies and dilute liquid cultures by "helper" heterotrophic bacteria. Appl Environ Microbiol 74(14):4530–4534

Morris RM, Rappe MS, Connon SA, Vergin KL, Siebold WA, Carlson CA, Giovannoni SJ (2002) SAR11 clade dominates ocean surface bacterioplankton communities. Nature 420 (6917):806–810

Mullins TD, Britschgi TB, Krest RL, Giovannoni SJ (1995) Genetic comparisons reveal the same unknown bacterial lineages in Atlantic and Pacific bacterioplankton communities. Limnol Oceanogr 40(1):148–158

Muniesa M, Blanch AR, Lucena F, Jofre J (2005) Bacteriophages may bias outcome of bacterial enrichment cultures. Appl Environ Microbiol 71(8):4269–4275

Newton RJ, Jones SE, Eiler A, McMahon KD, Bertilsson S (2011) A guide to the natural history of freshwater lake bacteria. Microbiol Mol Biol Rev 75(1):14–49

Nocker A, Cheung CY, Camper AK (2006) Comparison of propidium monoazide with ethidium monoazide for differentiation of live vs. dead bacteria by selective removal of DNA from dead cells. J Microbiol Methods 67(2):310–320

Oliver JD (2010) Recent findings on the viable but nonculturable state in pathogenic bacteria. FEMS Microbiol Rev 34(4):415–425

Pace N, Stahl D, Lane D, Olsen G (1986) The analysis of natural microbial populations by ribosomal RNA sequences. In: Marshall KC (ed) Advances in microbial ecology, vol 9. Springer US, pp 1–55

Page KA, Connon SA, Giovannoni SJ (2004) Representative freshwater bacterioplankton isolated from Crater Lake, Oregon. Appl Environ Microbiol 70(11):6542–6550

Parte AC (2014) LPSN–list of prokaryotic names with standing in nomenclature. Nucleic Acids Res 42(D1):D613–D616

Petri RJ (1887) Eine kleine Modification des Koch'schen Plattenverfahrens. *Cntralbl Bakteriol Pasitenkunde* 1:279–280

Pham VH, Kim J (2012) Cultivation of unculturable soil bacteria. Trends Biotechnol 30 (9):475–484

Postgate JR, Hunter JR (1963) Acceleration of bacterial death by growth substrates. Nature 198 (4877):273–273

Puspita ID, Kamagata Y, Tanaka M, Asano K, Nakatsu CH (2012) Are uncultivated bacteria really uncultivable? Microbes Environ 27(4):356–366

Rappe M, Giovannoni S (2003) The uncultured microbial majority. Annu Rev Microbiol 57:369–394

Rappe MS, Connon SA, Vergin KL, Giovannoni SJ (2002) Cultivation of the ubiquitous SAR11 marine bacterioplankton clade. Nature 418(6898):630–633

Reasoner DJ, Geldreich EE (1985) A new medium for the enumeration and subculture of bacteria from potable water. Appl Environ Microbiol 49(1):1–7

Rippka R (1988) Isolation and purification of cyanobacteria. Methods Enzymol 167:3–27

Roszak DB, Colwell RR (1987) Survival strategies of bacteria in the natural environment. Microbiol Rev 51(3):365–379
Sait M, Davis KE, Janssen PH (2006) Effect of pH on isolation and distribution of members of subdivision 1 of the phylum *Acidobacteria* occurring in soil. Appl Environ Microbiol 72 (3):1852–1857
Sait M, Hugenholtz P, Janssen PH (2002) Cultivation of globally distributed soil bacteria from phylogenetic lineages previously only detected in cultivation-independent surveys. Environ Microbiol 4(11):654–666
Sangwan P, Kovac S, Davis KE, Sait M, Janssen PH (2005) Detection and cultivation of soil *Verrucomicrobia*. Appl Environ Microbiol 71(12):8402–8410
Schleifer KH (2004) Microbial diversity: facts, problems and prospects. Syst Appl Microbiol 27 (1):3–9
Schloss PD, Handelsman J (2005) Introducing DOTUR, a computer program for defining operational taxonomic units and estimating species richness. Appl Environ Microbiol 71 (3):1501–1506
Schloss PD, Westcott SL (2011) Assessing and improving methods used in operational taxonomic unit-based approaches for 16S rRNA gene sequence analysis. Appl Environ Microbiol 77 (10):3219–3226
Schut F, Devries EJ, Gottschal JC, Robertson BR, Harder W, Prins RA, Button DK (1993) Isolation of typical marine bacteria by dilution culture: growth, maintenance, and characteristics of isolates under laboratory conditions. Appl Environ Microbiol 59(7):2150–2160
Sogin ML, Morrison HG, Huber JA, Mark Welch D, Huse SM, Neal PR, Arrieta JM, Herndl GJ (2006) Microbial diversity in the deep sea and the underexplored "rare biosphere". Proc Natl Acad Sci U S A 103(32):12115–12120
Stackebrandt E, Ebers J (2006) Taxonomic parameters revisited: tarnished gold standards. Microbiol Today 33:152–155
Stackebrandt E, Goebel BM (1994) A place for DNA–DNA reassociation and 16S ribosomal-RNA sequence-analysis in the present species definition in bacteriology. Int J Syst Bacteriol 44 (4):846–849
Staley JT, Konopka A (1985) Measurement of in situ activities of nonphotosynthetic microorganisms in aquatic and terrestrial habitats. Annu Rev Microbiol 39:321–346
Stevenson BS, Eichorst SA, Wertz JT, Schmidt TM, Breznak JA (2004) New strategies for cultivation and detection of previously uncultured microbes. Appl Environ Microbiol 70 (8):4748–4755
Stewart EJ (2012) Growing unculturable bacteria. J Bacteriol 194(16):4151–4160
Stingl U, Tripp HJ, Giovannoni SJ (2007) Improvements of high-throughput culturing yielded novel SAR11 strains and other abundant marine bacteria from the Oregon coast and the Bermuda Atlantic Time Series study site. ISME J 1(4):361–371
Swan BK, Tupper B, Sczyrba A, Lauro FM, Martinez-Garcia M, Gonzalez JM, Luo HW, Wright JJ, Landry ZC, Hanson NW, Thompson BP, Poulton NJ, Schwientek P, Acinas SG, Giovannoni SJ, Moran MA, Hallam SJ, Cavicchioli R, Woyke T, Stepanauskas R (2013) Prevalent genome streamlining and latitudinal divergence of planktonic bacteria in the surface ocean. Proc Natl Acad Sci U S A 110(28):11463–11468
Tamaki H, Sekiguchi Y, Hanada S, Nakamura K, Nomura N, Matsumura M, Kamagata Y (2005) Comparative analysis of bacterial diversity in freshwater sediment of a shallow eutrophic lake by molecular and improved cultivation-based techniques. Appl Environ Microbiol 71 (4):2162–2169
Tamaki H, Tanaka Y, Matsuzawa H, Muramatsu M, Meng X-Y, Hanada S, Mori K, Kamagata Y (2011) *Armatimonas rosea* gen. nov., sp. nov., of a novel bacterial phylum, *Armatimonadetes* phyl. nov., formally called the candidate phylum OP10. Int J Syst Evol Microbiol 61 (6):1442–1447

Tanaka T, Kawasaki K, Daimon S, Kitagawa W, Yamamoto K, Tamaki H, Tanaka M, Nakatsu CH, Kamagata Y (2014) A hidden pitfall in the preparation of agar media undermines microorganism cultivability. Appl Environ Microbiol 80(24):7659–7666

Thompson J, Pacocha S, Pharino C, Klepac-Ceraj V, Hunt D, Benoit J, Sarma-Rupavtarm R, Distel D, Polz M (2005) Genotypic diversity within a natural coastal bacterioplankton population. Science 307:1311–1313

Tindall BJ, Kaempfer P, Euzeby JP, Oren A (2006) Valid publication of names of prokaryotes according to the rules of nomenclature: past history and current practice. Int J Syst Evol Microbiol 56:2715–2720

Tindall BJ, Rosselló-Móra R, Busse H-J, Ludwig W, Kämpfer P (2010) Notes on the characterization of prokaryote strains for taxonomic purposes. Int J Syst Evol Microbiol 60(1):249–266

Torsvik V, Goksøyr J, Daae FL (1990) High diversity in DNA of soil bacteria. Appl Environ Microbiol 56(3):782–787

Tripp HJ, Kitner JB, Schwalbach MS, Dacey JWH, Wilhelm LJ, Giovannoni SJ (2008) SAR11 marine bacteria require exogenous reduced sulphur for growth. Nature 452(7188):741–744

Vancanneyt M, Schut F, Snauwaert C, Goris J, Swings J, Gottschal JC (2001) *Sphingomonas alaskensis* sp. nov., a dominant bacterium from a marine oligotrophic environment. Int J Syst Evol Microbiol 51(1):73–79

Vartoukian SR, Palmer RM, Wade WG (2010) Strategies for culture of 'unculturable' bacteria. FEMS Microbiol Lett 309(1):1–7

Vergin KL, Tripp HJ, Wilhelm LJ, Denver DR, Rappe MS, Giovannoni SJ (2007) High intraspecific recombination rate in a native population of *Candidatus* Pelagibacter ubique (SAR11). Environ Microbiol 9(10):2430–2440

Wang Q, Garrity GM, Tiedje JM, Cole JR (2007) Naive Bayesian classifier for rapid assignment of rRNA sequences into the new bacterial taxonomy. Appl Environ Microbiol 73(16):5261–5267

Watanabe K, Komatsu N, Ishii Y, Negishi M (2009) Effective isolation of bacterioplankton genus *Polynucleobacter* from freshwater environments grown on photochemically degraded dissolved organic matter. FEMS Microbiol Ecol 67(1):57–68

Watanabe K, Komatsu N, Kitamura T, Ishii Y, Park H-D, Miyata R, Noda N, Sekiguchi Y, Satou T, Watanabe M, Yamamura S, Imai A, Hayashi S (2012) Ecological niche separation in the *Polynucleobacter* subclusters linked to quality of dissolved organic matter: a demonstration using a high sensitivity cultivation-based approach. Environ Microbiol 14(9):2511–2525

Wilson KH, Blitchington RB (1996) Human colonic biota studied by ribosomal DNA sequence analysis. Appl Environ Microbiol 62(7):2273–2278

Yooseph S, Nealson KH, Rusch DB, McCrow JP, Dupont CL, Kim M, Johnson J, Montgomery R, Ferriera S, Beeson K, Williamson SJ, Tovchigrechko A, Allen AE, Zeigler LA, Sutton G, Eisenstadt E, Rogers Y-H, Friedman R, Frazier M, Venter JC (2010) Genomic and functional adaptation in surface ocean planktonic prokaryotes. Nature 468(7320):60–66

Zengler K, Toledo G, Rappé M, Elkins J, Mathur EJ, Short JM, Keller M (2002) Cultivating the uncultured. Proc Natl Acad Sci U S A 99(24):15681–15686

Zengler K, Walcher M, Clark G, Haller I, Toledo G, Holland T, Mathur EJ, Woodnutt G, Short JM, Keller M (2005) High-throughput cultivation of microorganisms using microcapsules. Methods Enzymol 397:124–130

Zhang H, Sekiguchi Y, Hanada S, Hugenholtz P, Kim H, Kamagata Y, Nakamura K (2003) *Gemmatimonas aurantiaca* gen. nov., sp. nov., a Gram-negative, aerobic, polyphosphate-accumulating micro-organism, the first cultured representative of the new bacterial phylum *Gemmatimonadetes* phyl. nov. Int J Syst Evol Microbiol 53(4):1155–1163

Zobell CE (1941) Studies on marine bacteria. I The cultural requirements of heterotrophic aerobes. J Mar Res 4:42–75

Zwart G, Crump BC, Agterveld M, Hagen F, Han SK (2002) Typical freshwater bacteria: an analysis of available 16S rRNA gene sequences from plankton of lakes and rivers. Aquat Microb Ecol 28(2):141–155

MIX
Papier aus verantwortungsvollen Quellen
Paper from responsible sources
FSC® C105338

If you have any concerns about our products,
you can contact us on
ProductSafety@springernature.com

In case Publisher is established outside the EU,
the EU authorized representative is:
Springer Nature Customer Service Center GmbH
Europaplatz 3, 69115 Heidelberg, Germany

Printed by Libri Plureos GmbH
in Hamburg, Germany